PENGUIN BOOKS

THE SECRET LIFE OF TREES

'The best nature writing' *The Times*

'Page after page of astonishing tree-facts . . . makes us look anew at the familiar, to understand a little more of the hidden and constantly enacted miracles taking place in the woods all around us' *Sunday Times*

'Tudge's delight in the world of trees is infectious' *Herald*

'Set to become a classic reference in the mould of Oliver Rackham's *History of the Countryside*' *BBC Wildlife*

'Tudge's gift as an author lies in being able to explain complex scientific mechanisms in language that the rest of us can understand'
Anna Pavord, *Independent*, Books of the Year

'Magnificent, a minor classic . . . even the most knowledgeable connoisseur of nature will feel themselves in the hands of a witty and erudite guide . . . probably the best general purpose book on the subject published in the last decade' *Oldie*

'Inspiring, a reawakening' *Scotsman*

'Reminds us just what we spend our lives not knowing, and all of it is not only wondrous and important but entirely free' *Guardian*

'A love-letter to trees, written with passion and scientific rigour . . . a pleasure to read. Tudge writes with warmth and wit' *Financial Times*

Colin Tudge started his first tree nursery in his garden aged eleven, becoming an accomplished Cacti grower by the age of eighteen and marking his life-long interest in trees. Always interested in plants *and* animals, he studied zoology at Cambridge and then began writing about science, first as features editor at the *New Scientist* and then as a documentary maker for the BBC. Now a full-time writer, he appears regularly as a public speaker, particularly for the British Council and is a Fellow of the Linnean Society of London and visiting Research Fellow at the Centre of Philosophy at the London School of Economics. His books include *The Variety of Life: A Survey and Celebration of All the Creatures that have Ever Lived* and *So Shall we Reap*. *The Secret Life of Trees* brings together Colin Tudge's knowledge of trees and his fascination with them, built up from trips to the rainforest in Costa Rica, Panama and Brazil, to his time in India, Australia, New Zealand, China, the United States . . . and his own back garden. He is unable to choose a favourite tree, believing that variety's the thing.

COLIN TUDGE

The Secret Life of Trees
How They Live and Why They Matter

PENGUIN BOOKS

To my grandchildren

PENGUIN BOOKS

Published by the Penguin Group
Penguin Books Ltd, 80 Strand, London WC2R ORL, England
Penguin Group (USA) Inc., 375 Hudson Street, New York, New York 10014, USA
Penguin Group (Canada), 90 Eglinton Avenue East, Suite 700, Toronto, Ontario, Canada M4P 2Y3
(a division of Pearson Penguin Canada Inc.)
Penguin Ireland, 25 St Stephen's Green, Dublin 2, Ireland
(a division of Penguin Books Ltd)
Penguin Group (Australia), 250 Camberwell Road,
Camberwell, Victoria 3124, Australia (a division of Pearson Australia Group Pty Ltd)
Penguin Books India Pvt Ltd, 11 Community Centre,
Panchsheel Park, New Delhi – 110 017, India
Penguin Group (NZ), cnr Airborne and Rosedale Roads, Albany,
Auckland 1310, New Zealand (a division of Pearson New Zealand Ltd)
Penguin Books (South Africa) (Pty) Ltd, 24 Sturdee Avenue,
Rosebank, Johannesburg 2196, South Africa

Penguin Books Ltd, Registered Offices: 80 Strand, London WC2R ORL, England

www.penguin.com

First published by Allen Lane 2005
Published in Penguin Books 2006
1

Copyright © Colin Tudge, 2005
Illustrations copyright © Dawn Burford
All rights reserved

The moral right of the author has been asserted

Typeset by Rowland Phototypesetting Ltd, Bury St Edmunds, Suffolk
Printed in England by Clays Ltd, St Ives plc

ISBN-13: 978-0-141-01293-3
ISBN-10: 0-141-01293-5

Contents

I

What Is a Tree?

II

All the Trees in the World

III

The Life of Trees

IV

Trees and Us

Illustrations and Figures

Original drawings by Dawn Burford

Acknowledgements

Over the past half century I have had many illuminating conversations with a lot of people who know a great deal about trees, in at least a score of countries in every habitable continent, and it would be too exhausting to mention everyone who has helped me with this book. Over the years, however, I have been particularly informed by Professor E. R. ('Bob') Orskov, now at the Macauley Research Station in Aberdeen, on Third World agriculture in general and agroforestry in particular. From China, I have particular cause to thank Professor Hao Xiaojiang, director of the Kunming Institute of Botany in Yunnan, China, who introduced me to the extraordinary collection (including 100 species of magnolia) in the Botanic Garden of Kunming; and Dr Ian Hunter, director of the International Network for Bamboo and Rattan, Beijing. In Australia, I spent several excellent days with scientists from CSIRO both in the bush of Western Australia and in the tropical and subtropical forests of Queensland and New South Wales. In New Zealand, Keith Stewart, novelist and columnist, took me to see the kauri forests in the North Island, and introduced me to Tane Mahuta.

For the particular writing of this particular book, I am especially aware of my debt to Professor Jeff Burley, formerly head of forestry at Oxford University, who indeed inspired this whole enterprise (just as he inspired generations of foresters worldwide). Also at Oxford, Dr Stephen Harris read several of the chapters for me; Dr Nick Brown helped me on my way with comments about mahogany; Professor Martin Speight provided fine fresh insights into pests; Professor Andrew Smith instructed me in tree physiology; and Dr Yadvinder Malhi had excellent, original things to say about tropical forests and climate change. At the Royal Botanic Gardens, Kew, I was privileged to be introduced to the latest thinking on conifers by Dr Aljos Farjon. At EMBRAPA in Belém, I was treated royally and introduced to the realities of Brazilian forestry by Ian Thompson; and had particularly illuminating discussions with Dr Mike Hopkins and Dr Milton Kanashiro. In Belém, too, from Johan C. Zweede I learned at least a few of the ins and outs of tropical forest commerce. I was introduced to the Cerrado by Professor Carolyn Proença of the University of Brasilia, and by Dr Manual Cláudio da Silva Júnior of Brazil's forestry department, while José Felipe Ribeiro took my wife and me out into the field to show us ways in which local people can make a much better living from the Cerrado than by growing yet more soya. Hugely instructive, too, as well as enjoyable, was our stay with Robin and Binka Le Breton, at Iracambi, who are seeking among other things to restore at least some of the sadly depleted Atlantic rainforest. At the Smithsonian Tropical Research Institute in Panama I was prodigiously instructed by Drs Anthony G. Coates (continental drift), Neal G. Smith (mainly birds), Stanley Heckadon-Moreno (mangroves), Egbert G. Leigh (particularly on the tropical forest on Barro Colorado Island) and Allen Herre (whose extraordinary researches on figs are a key theme in Chapter 13). Boundless gratitude to all of them, and to Beth King for fixing up the entire trip. At the Tropical Agricultural Research and Higher Education Center (CATIE) in Costa Rica Dr Bryan Finegan was wonderfully generous with his time and showed, as we stood in the rain, how physically hard as well as magical it can be to carry out research in tropical forest; while Dr Muhammad Ibrahim introduced me to CATIE's excellent researches in agroforestry and Dr Wilberth Philips showed me the local trees. Then to India, and in particular to the vast and extraordinary campus and arboretum of the Forestry Research Institute at Dehra Dun, where I have particular cause to be grateful to FRI's then director, Dr Palab Parkash Bhojavid, for excellent discussions and hospitality (including Christmas dinner with his family), and in particular among his colleagues to Dr Sas Biswas. In Latvia, I was introduced to the forest (and its beavers) by Ieva Muizniece of the British Council, Anita Upite, editor of *Hunting, Angling and Nature*, and Monvids Strautins, forester. I am grateful, too, to the British Council in general and to Dr Gavin Alexander in particular, who arranged some of my most illuminating trips.

Overall, I am aware of my debt to my agent, Felicity Bryan; to my editor at Penguin, Helen Conford; and to Jane Birdsell, an outstanding tidier of prose and picker-up of solecisms who has made this book much better than it would otherwise have been. Finally, the book has been much enhanced by Dawn Burford's excellent drawings, almost all taken directly from life. Many thanks to her and to the Birmingham Society of Botanical Artists for introducing me to her. Most of all, I thank my wife Ruth, who introduced me to Oxford and organized and managed most of our travels. Without her heroic efforts I would almost certainly have petered out at Heathrow.

AUTHOR'S NOTE

The following abbreviations have been used throughout the text:

'Judd' refers to Walter S. Judd, Christopher S. Campbell, Elizabeth A. Kellogg, Peter F. Stevens and Michael J. Donoghue (eds), *Plant Systematics* (Sinauer Associates Inc., Sunderland, Massachussetts; 2nd edn, 2002).

'Heywood' refers to V. H. Heywood (ed.), *Flowering Plants of the World* (Oxford University Press, Oxford, 1978).

Preface

Trees inspire: the Buddha received enlightenment under a peepul tree

At Boscobel in Shropshire in the English Midlands stands the Royal Oak, where the provisional King Charles II is alleged to have hidden from Cromwell's men after the Battle of Worcester, which ended his premature attempt to restore the monarchy. Why not? All this happened only about three and a half centuries ago (1651) and oaks may live for two or three times as long as that. Robin Hood and his Merry Men are said to have feasted beneath the Major Oak in

Sherwood Forest in Nottinghamshire – and so they might have, for if they existed at all it was in the time of Richard I, in the late twelfth century, and the Major Oak was alive and well at that time. A yew I met in a churchyard in Scotland has a label suggesting that the young Pontius Pilate may once have sat in its shade – 'and wondered what the future held'. It's an audacious claim. But the tree was there, even if Pilate wasn't – already some centuries old at the time of Christ.

There's a kauri tree in New Zealand called Tane Muhuta (the oldest and biggest kauris are given personal names), with a trunk like a lighthouse, that was 400 years old when the Maoris first arrived from Polynesia. For the first 900 years or so of Tane Mahuta's life the moas, related to ostriches but some of them half as tall again, would have strutted their stuff around its buttressed base, threatened only by the commensurately huge but short-winged eagles that threaded their way through the canopy to prey upon them. Now the moas and their attendant eagles are long gone but Tane Muhuta lives on. Many a redwood still standing tall in California was ancient by the time Columbus first made Europe aware that the Americas existed. Yet the redwoods are striplings compared to some of California's pines, which germinated at about the time that human beings invented writing and so are as old as all of written history. These trees out on their parched hills were already impressively old when Moses led the Israelites out of Egypt, or indeed when Abraham was born. So it is that some living trees have seen the rise and fall of entire civilizations.

Some redwoods, Douglas firs and eucalypts are as tall as a perfectly respectable skyscraper, and there's an extraordinary banyan in Calcutta that would cover a football field. Many are host to so many other creatures that each is a city: as cosmopolitan as Delhi or New York and far more populous than either. Creatures of all kinds may feed on trees, or maraud among their branches. At least, I know of no arboreal octopuses – but there could be, out in the mangroves. There's many a tree-happy crab in the mangroves, as I have seen for myself, and the robbers of the Pacific islands, giant hermit crabs, come on land (as many crabs do) to feed on coconuts. When the Amazon is in flood – deep enough to submerge well-grown trees entirely, over an area not far short of England – the fish feed on fruit and river dolphins race through the upper branches of what should be the canopy, while monkeys hop and swim from crown to crown like ducks. In New Zealand little blue penguins nest in the forest at night with ground parrots (or at least they do on the sanctuary of Maud Island). In the 1970s in the crown of one fairly modest tree in Panama a scientist from the Smithsonian Institution counted 1,100 different *species* of beetle – yet he didn't bother with the weevils, although they are beetles too, or look closely at the host of creatures that are not beetles, or those that were living in the roots. I once found myself in an old kapok tree in Costa Rica in which biologists had thus far listed more than 4,000 different species of creatures.

Yet a tree cannot afford simply to serve as someone else's monument and feeding ground. From the moment the seed falls on to the forest floor (or the sand of the savannah, or a fissure in some mountain crag, or a glacier's edge, or a lakeside, or a tropical seashore) to the moment of its final demise, perhaps a thousand years later, the tree must compete through every second – for water, nutrients, light and space; and to fend off cold, heat, drought, flood, toxicity, and the host of parasites and predators of all conceivable kinds (from a tree's point of view, squirrels or giraffes are 'predators'). A village or a civilization may choose to make a tree their symbol. The entire nation of Brazil is named after a tree – for brazil wood was known to Europeans before the country was. But however we may choose to ennoble it, the tree must fight its corner, a creature like all the rest. If it did not fight it would be dead. Even when it sheds its leaves to ride out frost or drought its cells are still busy beneath its armoured bark. Were it not so the leaves could not burst out as they so spectacularly do when the temperate spring or the tropical rains return – or sometimes in advance of the rains, to the delight of camels and goats, which thus may find green fodder in the depths of drought. In many trees, too, tropical and temperate, the flowers emerge before the leaves – which keeps the path clear for pollinating winds, bees or bats. Since there are no leaves to provide nourishment, the flowers must be fed from the tree's reserves in its trunk and roots. The living timber is multipurpose: a prop, a conduit, a larder.

Flowers, of course – and the cones of conifers – meet life's other demand: not simply to survive and grow, but to reproduce. Here, the trees' immobility is a particular drawback. Many trees reproduce without sex, commonly though not exclusively by root suckers, but all trees (to my knowledge) practise sex as well. For sex, gamete must meet gamete: sperm and egg in the case of animals and primitive plants; pollen and ovule in the case of conifers and flowering plants. Since many flowers of many trees are hermaphrodite (male stamens and female carpels on the same flower), and many trees (like oaks and many conifers) are monoecious (the individual flowers are exclusively male or female, but both kinds occur on the same tree), it may seem easy enough for trees to pollinate themselves. But on the whole they don't. One of the botanical surprises of recent decades (finally proven by genetic studies) is the length to which most trees go to avoid self-fertilization. 'Out-crossing' is the norm: pollination of, and by, other individuals who of course are of the same species but preferably are not too similar genetically. To achieve out-crossing, trees must elicit the help of the wind – or bribe or otherwise coerce a variety of animals, from flies and beetles and bees to birds and bats – to carry their pollen for them. Some temperate trees (like apples and horse chestnuts) are pollinated by animals but most (like oaks and birches and beeches) are content to use the wind. But in tropical forests, where most kinds of trees live, animal pollination is the norm; and because life is competitive, the mechanisms that have evolved for this have become more and more elaborate. Thus for every one of the 750 different species of fig there is a corresponding species of specialist wasp to pollinate it; and each wasp knows its own fig (although, as recent studies have shown, the relationship between figs and their wasps is not quite so cosy as had been supposed). When the ovules are fertilized and become seeds, encased in fruits (or some other kind of fruiting body) they must then be dispersed – sometimes again by wind but often by another, entirely separate, suite of animal accomplices – birds and fruit bats and rodents and orang-utans – whose help must again be actively co-opted.

Thus life is perforce competitive: hordes of creatures of thousands of different kinds are all after the same things, and most live directly at the expense of others. But it is also, just as inescapably, cooperative. Trees are good competitors. But they are also among the world's most exemplary cooperators, forming a host of mutualistic relationships for one purpose or another with an enormous variety of different creatures, from the bacteria and fungi that help them to feed to the many, many different kinds of animal that help them with different stages of their reproduction. Trees do not seem to be aware, as dogs and monkeys are aware. They do not have brains. But they are sentient in their way – they gauge what's going on as much as they need to, and they conduct their affairs as adroitly as any military strategist. Why be 'aware' when you can simulate all that awareness brings? They surely don't think, as animals do. But they orchestrate their fellow creatures nonetheless. A forest is a forest because it has trees in it, not because it may have sloths and toucans or squirrels or chimpanzees. The trees are the prime players and the animals are the dependants.

The human debt to trees is absolute. Modern evolutionary theory has it that we owe our brains – our art, our inventiveness, and presumably much of our deviousness – to our sexuality. We dance and paint and joke and tell stories to impress potential mates – or such at least was the crude beginning of our wits, on which we have built. But pigs and squirrels and elephants are clever too. They also must attract mates. So why have pigs produced no concert pianists, or professors of jurisprudence? Another ingredient is needed – one suggested a long time ago by more conservative biologists. Our brains and our dexterity evolved together: they are an exercise in co-evolution. Pigs are clever, but their hands are hoofs: nothing there with which to express their dreams and insights. We, by contrast, can translate our thoughts into action: our artefacts (as Robert Pirsig put the matter in *Zen and the Art of Motorcycle Maintenance*) are ideas in space. Brains are expensive organs (they require a huge amount of energy) and unless they produced some immediate pay-off, natural selection would select against them. But because we have hands (at the end of long, strong, extremely mobile arms), brains do provide pay-offs, manifest not least in a thousand kinds of tools with which to effect further manipulations. Hands provided the encouragement, the selective pressure, to make our brains even brainier; and the growing brains in turn encouraged more dexterity. But the only reason we have such dextrous hands and whirling arms is that our ancestors had spent 80 million years or so (so some zoologists calculate) in the trees. Arboreal life requires dexterity and hand–eye coordination. Squirrels almost became intellectuals, but not quite. Monkeys and apes came closer – but they stayed up in the trees, where they are obliged to squander their fabulous skills just on getting around. Our ancestors, somewhere in Africa, came to the ground when the climate dried up and the trees retreated. They learned to walk on two legs (which no other primate or any other mammal of any kind has learned to do convincingly) and freed their versatile hands and arms for other purposes. Were it not for that pedigree we would remain as intellectually frustrated as elephants and dolphins sometimes seem to be.

Archaeologists speak of the Stone Ages, and the Bronze Age and the Iron Age and the Steam Age, and now we have the age of the internal combustion engine and nuclear power and space and IT. But every age has been a Wood Age – ours at least as much as any in the past; and perhaps, in the decades to come, even more so. Ice-Age Russians made houses from the bones of mammoths, the Inuit use ice, and the people of the Bronze-Age Orkneys built remarkable villages, with restaurants and mausoleums, from slabs of rock. But great architecture demands wood. The ruins that survive from classical times are all of stone but that's only because wood rots. Architecture in stone and bricks evolved from timber architecture, and needs wooden-handled tools and wooden scaffold for its construction – and timber roofs and rafters. Wood, in this energy-conscious age, may well begin to replace steel, or much of it, in the grandest buildings.

Wood was the first serious fuel, too – and human beings clearly learnt the use of fire at least 500,000 years ago, long before we were as big-brained as we are now. No fuel: no smelting – so no Bronze Age or Iron Age or modern machines. No wood: no ships. No ships: no ocean travel – no human beings in Australia, New Zealand or any other island that could not be reached simply by hitching a lift on floating vegetation (as many a beast is thought to have done, from rats to monkeys and tortoises). No ocean travel: no empires: no modern politics. A woodless world would have had advantages. But we could also say no wood: no civilization.

Yet timber is not the end of it. Trees are the source of drugs, unguents, incense, and poisons for tipping arrows, stunning fish and killing pests; of resins, varnishes, and industrial oils, glues and dyes and paints; of gums of many kinds including chewing gum; of a host of fibres for the rigging and hawsers of great ships (whether made of wood or not) and for the stuffing of cushions – and of course, perhaps above all these days, for paper. All that, plus a thousand (at least) kinds of fruits and nuts and – in traditional agrarian societies – a surprising amount of fodder for animals, including cattle and sheep, which most of us assume live primarily on grass. As a final bonus, the wooden husks of many a tree fruit make instant household pots and drums and ornaments.

In short, without trees our species would not have come into being at all; and if trees had disappeared after we had hit the ground we would still be scrabbling like baboons (assuming the baboons allowed us to live at all).

Perhaps this is why we feel so drawn to trees. Groves of redwoods and beeches are often compared to the naves of great cathedrals: the silence; the green, filtered, numinous light. A single banyan, each with its multitude of trunks, is like a temple or a mosque – a living colonnade. But the metaphor should be the other way around. The cathedrals and mosques emulate the trees. The trees are innately holy. Christians with their one omnipotent God may take exception to such pagan musing: but the totaras and the kauris were sacred to the Maoris, and the banyan and the bodhi and the star-flowered temple trees (and many, many others) to Hindus and Buddhists, and the roots of this reverence, one feels, run back not simply to the enlightenment of Buddha as he sat beneath a bo tree (in 528 BC, tradition has it), but to the birth of humanity itself.

But Christianity did give rise to modern science. The roots of science run far back in time and from all directions – from the Babylonians, the Greeks, many great Arab scholars in what Europeans call the Middle Ages, the Indians, the Chinese, the Jews, and the much under-appreciated natural history of all hunter-gatherers and subsistence farmers everywhere. But it was the Christians from the thirteenth century onwards, with an obvious climax in the seventeenth, who gave us science in a recognizably modern form. The birth of modern science is often portrayed by secular philosophers as the 'triumph' of 'rationality' over religious 'superstition'. But it was much more subtle and interesting than that. The great founders of modern thinking –

Galileo, Newton, Leibniz, Descartes, Robert Boyle, the naturalist John Ray – were all devout. For them (as Newton put the matter) science was the proper use of the God-given intellect, the better to appreciate the works of God. Pythagoras, five centuries before Christ, saw science (as he then construed it) as a divine pursuit. Galileo, Newton, Ray and the rest saw their researches as a form of reverence.

This book is written in that same spirit. Of course, I don't claim to walk on the same plane as Pythagoras and Galileo, but I don't think it's too pretentious to aspire at least to drink at the same spring. This book is mainly about the science of trees – what modern research is telling us about them. The last chapter is about the uses we make of them, and what they do for us, and why for reasons that are purely material they must be conserved: our survival depends on them. But most of this book is not about their usefulness, but about what they *are*: how they came into being; what kinds there are and where they live and why; how they live, competing and cooperating. The revelations build by the week: how they may live and grow huge on what seems like nothing at all; how they draw prodigious quantities of water from the ground, send it up into the atmosphere, and then (so some have claimed) may call it in again, by releasing organic compounds that seed fresh clouds; how they speak to each other, warning others downwind that elephants or giraffes are on the prowl; how they mimic the pheromones of predatory insects, to summon them to feed upon the insects that are eating their own leaves. Every week the insights grow more fantastical – trees seem less and less like monuments and more and more like the world's appointed governors, ultimately controlling all life on land (and in the oceans too, vicariously), but also the key to its survival.

So this book presents science not as it is often presented, as a tribute to human cleverness and power, but truly in a spirit of reverence. I like the idea (I have found that some people don't, but I do) that each of us might aspire to be a connoisseur of nature, and connoisseurship implies a combination of knowledge on the one hand and love on the other, each enhancing the other. Conservation – of all living creatures, including trees – has little chance of long-term success without understanding, which depends in large measure on excellent science. But conservation cannot even get on to the agenda unless people care.

Caring is an emotional response, to which science has often been presented as the antithesis. In truth, science cannot be done properly without a cool head. But when the science is done its primary role (to reverse an adage of Marx's) is not to change the world but to enhance appreciation. That is the purpose of this book. Science in the service of appreciation, and appreciation in the service of reverence which, in the face of wonders that are not of our making, is our only proper response.

I

What Is a Tree?

I

Trees in Mind: Simple Questions with Complicated Answers

Round-leaved and altogether beautiful: the Judas tree

'I never stopped thinking like a child,' said Einstein. Neither should any of us. It's the way to get to the heart of things. Children ask ridiculously simple questions like 'Who made God?' that have kept theologians busy for many a century. In such a vein we might innocently inquire, 'And what, pray, are trees, that anyone should presume to write a book about them?' And, '*Why* do plants grow into trees?' And, 'How many kinds are there?' Childish stuff: but it will serve to mark out the ground.

WHAT IS A TREE?

A tree is a big plant with a stick up the middle.

Everybody knows that. But that statement as it stands requires what modern philosophers would call a little 'deconstruction'.

What, for a start, is meant by 'big'? It's a relative term of course, although if we choose we can put a figure on it – say a minimum height of five or six metres. There is a case for doing this: if you are a forester, or are running a nursery, you need some guidelines. But guidelines are not definitions. They are ways of helping practical people to do practical things. They do not – and are not intended to – capture what Aristotle would have called the essence of nature.

For many trees grow big when conditions are favourable, and stay small when they are not. An oak is a noble tree in a forest or a park but an acorn that falls in a fissure in some Scottish crag may spend a couple of centuries in bonsai'd mode, never more than a twisted stick. Yet it may turn out acorns which, if they should be carried to some fertile field, could again produce magnificence. Is the twisted stick less of an oak because it fell on stony ground? And if it remains an oak, is it not still a tree? Or then again – a different kind of case – the world's many kinds of birch form the genus *Betula*. None are as huge as an oak may often be, but most are perfectly respectable trees. Yet there is one, *Betula nana*, that is adapted to the tundra of the north of Scotland and mainland Europe and is very small indeed. Do we say that all birches are trees except for the tough little *Betula nana*? Or do we say it's a dwarf tree?

What of the stick that runs up the middle, the 'trunk', that holds the 'crown' of the tree aloft? Should there be just one, a solitary pillar, or are several allowed? Many a gardener and forester has insisted that plants with a lot of supporting sticks should be called shrubs. Again, for practical purposes such distinctions can be useful. If Alice's Queen of Hearts had instructed her long-suffering gardeners to plant her an arboretum and they'd come up with a shrubbery, their heads would surely have come off. But wild nature is not so easily pinned down. In the Cerrado of Brazil – the vast dry forest, about the size of France, in the middle of the country to the south and east of Amazonia's rainforest – there are trees that form bona fide, big, one-trunked trees when they grow along the banks of the occasional rivers, but become multi-stemmed, short shrubs where it's drier. The shrub is not merely stunted, like the oak in the rock. It is a discrete life form. Many organisms exhibit what biologists call 'polymorphism', meaning 'many forms'. Many kinds of fish, for example, have dwarf forms and full-size forms; some butterflies and snails are highly variable. Here we see a polymorphic tree – one form for the forest, another for the open ground.

Then again many big trees including some cedars, many a mulberry, or the beautiful blue-flowered jacaranda, may grow from ground level with several solid trunks of equal magnitude. Each may be as big as a respectable oak. Are they trees, or big shrubs? The family of the heathers, Ericaceae, also includes the rhododendrons from the Himalayas, and the beautiful flaky, yellow/pink/grey-trunked madrone trees of the United States (which add yet more colour to the already wondrous hills of California). Rhododendrons tend to have many stems while madrones are commonly content with one. But the rhododendrons can be just as big and solidly wooden as the madrones. In nature, in short, trees and shrubs are not distinct. Why should they be? Nature was not designed to make life easy for biologists.

Must the central stick be of wood? That, after all, is what we generally mean by 'stick'. How, then, should we categorize banana plants? In general shape they resemble palm trees, with a thick central stem and a whorl of huge leaves at the top. But the stem of the banana plant is not of wood. Its stem is formed largely from the stalks of the leaves, and its strength comes from fibres which are not bound together as in pines or oaks or eucalypts to form true timber; its hardness is reinforced, as in a cabbage stalk, by the pressure of water in the stem. So botanically the banana plant is a giant herb. But it looks like a tree and competes with trees on their own terms, as a big plant seeking the light (although like the trees of cocoa and tea and coffee, the banana prefers a little shade).

In fact there are many lineages of trees – quite separate evolutionary lines that have nothing to do with each other except that they are all plants. Many plants, in many of those lineages, have independently essayed the form of the tree. Each achieves freedom in its own way.

'Tree' is not a distinct category, like 'dog' or 'horse'. It is just a way of being a plant. The different kinds have much in common and it is good and necessary to have some feel for what is essential. But the essences of nature will not be pinned down easily. In the end, *all* definitions of nature are simply for convenience, helping us to focus on the particular aspect that we happen to be thinking about at the time. There is no phenomenon in all of nature – whether it's as simple as 'leg' or 'stomach' or 'leaf' or more obviously conceptual like 'gene' or 'species' – that does not take a variety of forms, and which cannot be looked at from an infinite number of angles; and each angle gives rise to its own definition. A horse cannot be encapsulated as Charles Dickens' Thomas Gradgrind insisted in *Hard Times* as 'A graminivorous quadruped'. There is more to horses than that. The way we define natural things influences the way we treat them – whether we speak of wild flowers or of weeds, of Mrs Tittlemouse or of vermin. But in the end nature is as nature is, and we must just try with different degrees of feebleness, and for our own purposes, to make what sense of it we can.

For the purposes of this book, the child's definition of 'tree' will serve – albeit with slight elaboration: 'A tree is a big plant with a stick up the middle – or could be, if it grew in the right circumstances; or is very closely related to other plants that are big and have a stick up the middle; or resembles a big plant with a stick up the middle.' It is clumsy, but it will have to do. So to the next childish question.

WHY BE A TREE?

A non-living thing is passive. The atoms of which a stone is composed sit there for as long as it endures – until it is melted in some volcano, or dissolved by acid rain. But living things are restless, through and through. As soon as some living cell has constructed some protein, as part of its own fabric, it starts to dismantle it again. This constant self-renewal, powered by an endless intake of energy, is called 'metabolism'.

Metabolism – the basic business of staying alive – is half of what living things do. The other half is to reproduce. It is not vital to reproduce in order to stay alive – indeed, reproduction involves sacrifice; reproduction, as we will see later in this book, is often the last fling: many a tree dies after one bout of it. But it is essential nonetheless. At least, all creatures that do not reproduce die out. However successfully an organism may metabolize, sooner or later time and chance will finish it off. Everything dies. Only those that reproduce endure – or at least, their offspring do. All individuals are part of lineages – offspring after offspring after offspring.

But then, too, each creature finds itself in the company of other creatures, of its own kind and of different kinds. To some extent they are its rivals, to some extent it needs them – for food, shelter, mates, or whatever. Each successful creature, then – each one that survives at all, that is – must come to terms with the others around it.

All of life's requirements – metabolism, reproduction and the business of getting along with others – are difficult. Each creature must solve life's problems in its own way. There is no perfect, universal life strategy. Each has its own advantages and drawbacks.

So it can pay a creature to be very small; or it can pay to be big. Each mode has its pros and cons. A plant that is big like a tree can stretch further up into the sky, and so capture more of the sun's energy; and reach further down into the earth, for water and minerals. This is the upside. But it takes a long time to achieve large size, and whether you are an oak tree or an elephant or a human being, the longer you take to develop the more likely you are to be killed before you reproduce.

Being big is difficult, too. To hold a ton of leaves aloft in the sun and air requires enormous strength: specialist material like wood, and clever architecture. All trees have wood, by definition (apart from those granted honorary status, like bananas); but as we will see, wood is subtle stuff, requiring much chemistry and micro-geometry. Trees between them have essayed many architectural forms. Ginkgoes and conifers are built from repeats of a single simple module: a straight trunk up the middle with circles or spirals of branches at intervals. In others, like the elm, the lead shoot bends over and the next shoot in line takes over the lead until it too bends away and the one below that takes over. In others (particularly some tropical trees) the branches that grow upwards from the horizontal branches repeat the form of the whole tree – it's as if a new, miniature forest grew aloft, from the horizontal branches of the giants below. Others, like oaks or chestnuts, are more free-flowing. There are many basic designs. The point is, though, that such design is *necessary*. Being big requires a lot of engineering as well as a lot of chemistry, and it takes a long time to put in place. But the bigger trees grow, the more they are vulnerable to wind – and tropical storms regularly cut swathes as big as an English county through the world's rainforests.

For the purposes of reproduction, creatures in general pursue one of two main strategies. Some, known as 'K-strategists', produce just a few offspring at a time which in general are large at the time of their birth to give them a good chance in life; after they are born, typically, the parents take good care of them. K-strategists in general are long-lived and reproduce several times in their life, often at long intervals. Orang-utans, elephants, eagles and indeed human beings are classic K-strategists. Other creatures, known as 'r-strategists', produce an enormous number of offspring. Inevitably, each individual offspring is small, and so has little chance of survival. But there is safety in numbers. Codfish are noted r-strategists. They produce up to 2 million eggs at a time. The newly hatched fish live for a time as plankton, floating fairly helplessly – and most perish: they just get eaten. But so long as each pair of codfish manage to produce just two surviving offspring in the course of their lives, then the lineage of cod will carry on. Despite the enormous prodigality of their reproductive strategy, its fantastic wastefulness, codfish are immensely successful – or at least were until the North Sea fishermen became too technically proficient, and too 'competitive', and disastrously reduced their numbers. Cod live a long time. But many r-strategists, like flies, run through their entire life cycle in a few weeks – birth, growth, reproduction, death. Thus populations of flies may rise and fall from near zero to plague proportions in what seems like no time at all.

Trees seem to get the best of both worlds. Many – most – produce huge numbers of seeds and may do so repeatedly. A mature oak or beech may produce many millions of seeds in a good year (good seed years are known as 'mast' years) and although they won't do this every year, they may well have scores or even hundreds of prolific years in the course of their lives. They are r-strategists indeed – in a good year at least as prolific as codfish. Yet many trees – including oaks – produce seeds that are large and which do not need to germinate immediately: each has a very good chance of survival. To this extent they are K strategists too. To combine the advantages of the K- and r-strategy an organism must be truly mighty. Yet there is a downside too: most trees must grow for several years, and many must endure for several decades, before they can reproduce at all; and all the time they are growing, without yet scattering their seed, they are vulnerable.

We don't think of trees as r-strategists, because they are so big and long-lived. Their populations do not boom and bust like those of flies. They cannot, we imagine, leap to take advantage of newly created environments as a fly or a mouse may do. Yet we can see that they can and do do this – once we venture beyond our own puny timescale, and take the long view. Thus when the last ice age ended in the northern hemisphere around 10,000 years ago the forests of birches and alders that had been whiling away the time further south were able virtually to race towards the Pole in the wake of the retreating glaciers; and they will surely resume their advance as global warming reduces the polar ice still further. By the same token the huge tropical rainforest of Queensland in the southern hemisphere has not been there for ever, as it may seem. Like the Great Barrier Reef, which stands just off the Queensland coast and is as long from end to end as Great Britain, the rainforest of Australia grew up only after the last ice age, and is a mere 10,000 years old. Macbeth was shocked to see the Great Wood of Birnam shift a few miles across the moor to the Hill of Dunsinane. But if we could take a time-lapsed view of all the world this past few million or tens of millions of years, as cold has followed warm has followed cold, we would see vast and apparently immovable forests flitting over the surface of the globe like the shadows of clouds.

Thus the advantages of treedom are both manifold and manifest. Big plants can metabolize more effectively because they command so much earth and sky; and they can produce literally tons of seed, to be scattered far and wide. Small wonder that a third of all land is covered in forest. But being big is complicated – all that chemistry and architecture – and it is risky, because all the time a tree is growing, time and chance and other creatures are working on its downfall. So it is that many other plants, such as mosses and liverworts, never acquired the means to be big at all; but still they have made a very good living this past 400 million years, just by sticking to damp and easy places. Then again, trees cannot grow where it's too dry or the soil is too thin, and so they leave scope for many smaller plants that can. So the world's grasslands are vast too, like the savannahs of the dry tropics, or the prairie of temperate North America and the pampas of subtropical South America, and the steppes of Asia. These grasslands at best have scattered trees, though they grade into open woodland – many small trees but with big, mainly grassy spaces in between, as in the dry, tropical Cerrado of Brazil. Furthermore, trees are classic 'keystone species': simply by existing and doing their thing, they create niches where other creatures can live. Hence forests create endless scope for small, quick-growing plants – herbs and ramblers – to occupy the ground in between the trees; and a vast variety of plants of all kinds (mosses, liverworts, ferns, and many kinds of flowering plants including many relatives of the arum lily and of the pineapple, some cacti, and most of the orchids) grow on the trees themselves, as epiphytes. Overall, too, there is more room for small plants than for big ones. Whole, viable populations of small plants may need only a few square metres, while a population of wild trees that is numerous enough to endure will generally need many hectares. So although there are tremendous theoretical advantages in being a tree, the species of trees are outnumbered by non-trees by about five to one. The non-trees live in places where trees cannot – and in the niches created by trees.

So now to the third childish question.

HOW MANY KINDS OF TREE ARE THERE?

A simple question indeed – but of course there are complications. To begin with, as the more irritating kind of philosopher would say, 'It depends what you mean by *kind*.'

In this context, 'kind' most obviously means 'species'. The common oak is a species: *Quercus robur*. So is the Scots pine: *Pinus sylvestris*.

The common birch is *Betula pendula*. And so on. What's the problem?

One problem is, how you tell the different kinds apart. Any one species is liable to be highly variable, and sometimes different species resemble each other very closely. Sometimes there is more variation *within* species than there is between species. Or then again: many creatures can be identified definitively only by their reproductive organs, which in the case of flowering plants (including most trees) means flowers. But many trees are not in flower at the time you come across them – a particular problem in the tropics, where flowering often seems to be erratic (or at least, the tree knows when it is appropriate to flower, but the biologist does not). But some trees with similar flowers have different leaves, and both may be needed to make the identification. Willows, however, tend to produce their flowers before they produce leaves – so you never find flowers and leaves on the same tree at the same time. If you want to know what species a particular willow belongs to, you may have to make two visits.

But biologists do not define species purely in terms of what they look like. Much more fundamental, they very reasonably feel, is who mates with whom. If different individuals breed together, then it is reasonable to declare that they are of the same species. *Betula pendula* will happily breed with other *Betula pendula*, but not with *Quercus robur*. So they are different lineages of creatures, living separate lives. Easy.

Still, there are snags. Many species can and do interbreed with other species, and so form hybrids. The example that everyone knows is the mule: the issue of a male donkey and a mare. But horses and donkeys seem to be very different kinds of animals. If they can breed together, doesn't this mean they are of the same species? No – for although the mule is a powerful animal and 'mean', as cowboys were wont to complain, it is nonetheless sexually sterile. Strong though it is, it is not, as a biologist would say, 'viable'. So we can extend our definition slightly: 'Two or more individuals can be considered to be of the same species if they can mate together to produce *fully viable* offspring.' 'Fully viable' implies sexual potency; and also implies that the offspring should be able to compete successfully in the wild. For there are some hybrids (for example among frogs) that are sexually fertile yet generally fail in the wild, unable to compete with either of their parent species. Again, it is reasonable to rank the parent types as separate species, since the hybrids they produce between them are (relative) failures.

Still, there are problems. For example, two apparently different species, which look different, may fail to interbreed in the wild simply because they live in different places. Bring them together, and they may interbreed perfectly happily. Trees provide scores of examples – among oaks, willows, poplars, and many more. Many hybrids have arisen in gardens, where human beings bring plants from very different areas together, perhaps for the first time in many thousands of years. Among the most striking examples is the London plane, *Platanus × hispanica* (the × indicates its hybrid status). It is tremendously successful in cities, not simply in many a London street and square, but throughout the northern hemisphere. Because it sheds its outer bark (as a eucalyptus or a madrone does), it gets rid of all the soot and other pollutants that can make life so difficult for many other kinds of tree. It is the offspring of the oriental plane from southern Europe and Turkey, *Platanus orientalis*, and the western plane from North America, *Platanus occidentalis*, and arose, so tradition has it, in the Botanic Garden of Oxford University, in the seventeenth century. An offspring of the first-ever London plane now stands in a courtyard in Magdalen College, which is next to the Botanic Garden. That offspring, now several centuries old, is huge. For those who would be connoisseurs, it is well worth a diversion (assuming the porters will let you in).

Then there is the extremely important phenomenon of 'polyploidy'. Genes, as everyone knows these days, are aligned along chromosomes. Every kind of organism has its characteristic number and arrangement of chromosomes. Eggs and sperm (or the appropriate cells in ovules and pollen) contain only one set of chromosomes, and are said to be 'haploid'. When they fuse in the act of fertilization, the resulting embryo has two sets of chromosomes and then is said to be 'diploid'. Most organisms (at least of the most familiar kinds) are diploid: for example, human beings have 46 chromosomes – 23 acquired via the egg of the mother and 23 from the sperm of the father. Chimpanzees have 48 chromosomes – 24 from each parent.

Sometimes, however, apparently spontaneously, the chromosome number will double (the chromosomes divide in the normal way they do in preparation for cell division – but then the cell fails to divide). Then the diploid cell becomes tetraploid, with four sets of chromosomes. This does not apparently happen much in animals (or not, at least, in mammals) but it is extremely common in plants. The newly-formed tetraploid organism can breed successfully with other tetraploids of its own kind but it cannot usually breed successfully with either of its parents. So it forms an instant new species. Many plants in nature turn out to be tetraploid, and many more tetraploids have been produced in cultivation. The common potatoes grown in Europe are tetraploid derivatives of diploid potatoes that grow wild (and are cultivated) in the Andes. Many trees, wild and cultivated, are tetraploid. Sometimes the chromosomes of the tetraploid plant double again to produce octoploids. The octoploids again form new, discrete species – generally unable to interbreed with the tetraploid parents who gave rise to them. 'Polyploid' is the general term that describes any organism with more than two sets of chromosomes. Sometimes the complications become too much even for the plants and they finish up with an odd number of chromosomes (some having been lost among all the cell divisions and matings). Plants with anomalous numbers of chromosomes are said to be 'aneuploid'. Aneuploidy in animals generally leads to various degrees of disorder – aneuploid animals usually die and if they live they tend to be compromised at least to some extent. But many plants put up with aneuploidy. Sugar cane is aneuploid; but that doesn't stop it being an extremely vigorous, major crop.

There is one further complication. As we have noted, diploid organisms that are of different species sometimes mate to produce fully viable offspring (as the eastern and the western plane trees evidently did). But usually such crosses fail, and often this is because the chromosomes of the two parents are incompatible. The two different sets of chromosomes might be able to support body cells that work well enough (as in the mule). But even if cells with two different kinds of chromosomes succeed this far, they will not necessarily produce sound gametes (eggs and sperm, or ovules and pollen) because this requires close cooperation between the chromosomes.

But if a hybrid organism doubles its chromosomes, it often *can* produce viable gametes. So we find diploid parents of different species mating to produce diploid, hybrid offspring that are sterile; but the hybrids then double their chromosomes and become tetraploid – and the hybrid tetraploids are then fertile. This happens a lot among plants, and has produced many, many new plant species, both in the wild and in cultivation. Indeed the complications seem endless. For instance, a tetraploid plant might mate with a closely related diploid plant to produce a triploid offspring – two sets of chromosomes from the tetraploid parent, and one set from the diploid parent. Triploids are sterile – they cannot produce gametes at all – but they may still form viable plants. Thus the cultivated banana is triploid. Because it is sterile, its fruits contain no seeds (as wild banana fruits do). So the domestic banana has to be reproduced vegetatively, by planting cuttings. In other cases, though, triploid hybrids double their chromosomes to become hexaploid (with six sets of chromosomes). The most famous and important hexaploid organism of all is bread wheat (as opposed to pasta wheat, which is tetraploid).

If you have been brought up with animals, and are innocent of botany, you may find all this fantastical. But among trees, hundreds of examples of polyploids are now known: the more that botanists look, the more polyploids they find. Some of the polyploids simply represent a doubling (or redoubling) of chromosomes within one species. Others are polyploid hybrids. For good measure, breeders have produced many hundreds of polyploids by artificial means. (Some chemicals induce polyploidy almost to order.)

Willows, genus *Salix*, provide many fine examples of polyploid trees. There seem to be around 400 species – although there must be many more that are yet unknown, including an entire phalanx in western China, yet to be properly studied. Some willow species have a haploid number of 19 chromosomes, so that the diploids have 38 (2 × 19). But another group of willows has a haploid number of 11 (diploid 22) and the third group has 12 (diploid 24). There doesn't seem to be much hybridization between willows with different haploid numbers, but there is a great deal of hybridization between different species with the same haploid number, and this has produced a whole array of polyploids with up to 224 chromosomes. Most of those polyploid hybrids are fertile, and some willows have been bred artificially from combinations of up to fourteen different species. For good measure, many of the hybrids are all of one sex and reproduce by suckers, so that all the members of such 'species' in fact form a clone (of which more later). Thus, the hybrid known as *Salix* × *calodendron* is all female. Many willows, too, both wild and in cultivation, are aneuploid. All in all, identification of the multifarious willows – the diploid types and all their polyploid hybrids – is a nightmare (even when they are not tucked away on some remote Chinese hillside).

Acacias show a similar picture. Acacias are those lovely, lonely, sprawling trees of tropical grasslands worldwide that provide such essential shade and fodder for giraffes, camels, gazelles, and the domestic cattle and goats of nomadic pastoralists. *Acacia* is a huge and messy genus with 1,300 species – which should probably be further subdivided, perhaps into five or more different genera. Be that as it may, the basic haploid number of the whole group is 13, so the default diploid number is 26, but there are polyploids with up to 208 chromosomes – sixteen times the haploid number. In some of these, it's clear that the ancestor simply doubled (and then sometimes redoubled) its chromosomes. Others clearly arose as polyploid hybrids.

In birches, the haploid number is 14, so the diploid number is 28 – but some species have up to 112 chromosomes, which means they are octoploid: and there are some aneuploid hybrids in cultivation. In northern Europe, the silver birch, *Betula pendula*, and the downy birch, *B. pubescens*, can look very similar, and some have suggested they are the same species. But silver birch is a diploid with 28 chromosomes and downy birch is a tetraploid with 56. Downy birch presumably arose from silver birch, but now, following polyploidy, it is very clearly a separate species. Alders, too, show much the same kind of thing. Clearly the variety depends in part on past hybridization of what had been separate species.

How many more species of trees will turn out to be hybrids or polyploids, or fertile polyploid hybrids? Another century or so of serious study will throw a great deal more light. Science takes time.

There is one final set of complications. If different populations of trees become isolated one from another, then eventually they may evolve into separate species. But in the shorter term, the separated populations may remain similar enough to breed easily together – that is, they are still the same species – and yet become genetically distinct to some extent, and may look different. Then biologists say that the two populations are different 'races' or 'varieties' of the same species; and if the variety is really distinct, they may call it a 'subspecies'. Varieties of plants that arise through informal selection on traditional farms are called 'landraces'. Domestic varieties of plants that have been produced through formal breeding programmes are called 'cultivars' (and domestic races of animals are called 'breeds').

Sometimes, both in the wild and in domestication, 'variety' simply means a subset of the species. Among domestic crops, the different varieties of runner beans are of this kind: subsets of the runner bean species as a whole, but breeding sexually (by seed), and genetically still diverse. But many plants also reproduce vegetatively as well, by means of bulbs or tubers – or, as with many trees, by suckers from the stem or roots. A tree produced vegetatively in the wild may remain attached to its parent so that parent and offspring together form an entire copse (as in English elms, or groves of giant redwoods). Indeed, the parent tree and the offspring that grow from its suckers may cover many hectares, as in the aspens of Canada. Growers and foresters often reproduce their favoured trees by cuttings, which of course they separate from the parent. Whether they are separated from the parent or remain attached, all the offspring that are produced vegetatively are genetically identical with each other, and with their parent (who of course is a single parent). All the offspring are then said to be 'clones' of each other, and of the parent: and the whole genetically identical group is collectively called a 'clone'.

Thus among apples, all the Cox's orange pippins there have ever been are a clone: cuttings of cuttings of cuttings that were taken from the first ever Cox's orange pippin tree that was produced (from a tree grown from a pip) in the nineteenth century. Cox's orange pippin is only one of many hundreds of apple varieties, each with its own special character – Egremont Russet; Bramley; Beauty of Bath; Worcester; Discovery; and so on and so on. Each of those varieties is simply a clone. All belong to the single species, *Malus domestica*.

So how do we answer the very simple question, 'How many kinds of tree are there?' Well, in the wild (as in cultivation) you may find that what you construe to be different 'kinds' are indeed different species; or they may be different varieties of the same species; or they may be hybrids of other pairs of species – hybrids that in the fullness of time may be perfectly capable of hybridizing again with some other, apparently quite separate, species. But then again, you may find two patches of aspen (or elm, or willow) that look quite different – and then find that each patch is simply a clone; and that the two clones are really from the same species and might even have arisen from seeds produced by the same parents. And if you ask a grower or a forester how many kinds of tree there are, they may well suggest that the number is virtually infinite – since they regard each of their cultivars as distinct, and know that there could be as many different kinds as breeders care to produce.

So let us be more specific and ask with what surely is irreducible simplicity, 'How many *species* of trees are there?' At this point the biologists must surely stop prevaricating and provide a clear answer. But the only honest answer is: 'Nobody knows.'

STILL COUNTING

In truth we can never know for sure how many species of tree there are. As J. S. Mill pointed out in the nineteenth century, it is impossible to know, in science, whether you know everything there is to know. However much you know, you can never be sure that nothing has escaped you. With trees, there are many good reasons to think that a great deal *has* escaped us. Every so often some highly conspicuous tree turns up which either has never been seen before, or is known only from fossils, and has long been presumed extinct. Two classic examples are discussed in Chapter 4: *Metasequoia*, the dawn redwood and *Wollemia*, regrettably dubbed the Wollemi pine.

But there is also a practical reason for ignorance. Most kinds of tree, like at least 90 per cent of organisms of all kinds, live in tropical forests, and tropical forests are very difficult to study – largely because there are so many trees in the way. It requires hundreds of person-years, and heroic years at that, to list the species even in relatively small areas of tropical forest; and despite the best efforts of legal and illegal loggers, the tropical forest that remains to us is still mercifully vast – so that all of Switzerland, for example, could easily be lost in Amazonia. (Amazonia is the forest that surrounds the Amazon River: it occupies the western half of Brazil, and extends into Peru, Colombia, Bolivia and Venezuela. With a total area of more than 4 million square kilometres, it is about a hundred times bigger than Switzerland, which is a mere 41,000 square km. Amazonia is also about sixteen times bigger than the United Kingdom, which is around 240,000 square km.)

So it is that from the sixteenth century onwards a succession of naturalists-cum-conquistadors, administrators, soldiers, traders and priests, became obsessed with the flora and fauna of tropical America, and set out to identify, describe and collect what was there. Dedicated research expeditions were mounted from the eighteenth century, driven by scholarship and supported by empire and commerce – not least in search of new and valuable crops, of which rubber became the jewel. Greatest of all the explorers, so many believe, was the German Alexander von Humboldt who, together with the French physician and amateur botanist Aimé Bonpland, travelled 10,000 kilometres in South America between 1799 and 1804, on foot and by canoe. They collected 12,000 specimens of plants, including 3,000 new species, and hence doubled the number known from the western hemisphere. On their return they published the thirty volumes of *Voyage aux régions équinoxiales* at von Humboldt's expense (it cost him his entire fortune), of which von Humboldt wrote twenty-nine volumes and Bonpland contributed just one, although von Humboldt insisted that they share the authorship of the whole. The book was first published in English between 1814 and 1829 in five volumes as *Narrative of Travels to the Equinoctial Regions of the New Continent during the years 1799–1804*. The great revolutionary Venezuelan general Simon Bolivar (1783–1830) commented that 'Baron Humboldt did more for the Americas than all the conquistadores'.

The young Charles Darwin loved von Humboldt's writings, and carried the *Narrative* with him on his journey on the *Beagle* in the 1830s which changed his own life and went on to change the world. The *Narrative* also lured Alfred Russel Wallace to the Amazon, to which he set sail in 1848 with Henry Walter Bates, an inspired amateur collector of beetles. Wallace stayed for four years before malaria and gut trouble forced him to return to England – although he set off to the Malay archipelago a couple of years later, in 1854, and stayed for eight years. Bates stayed in the Amazon for eleven years and among other things described a form of mimicry in which innocuous and tasty butterflies are protected by their wondrous resemblance to other butterflies that are noxious and toxic. He also collected an estimated 14,712 *species* from Amazonia, including 14,000 insects, and 8,000 of his creatures were new to science.

The Yorkshireman Richard Spruce (with whom Wallace corresponded from Malaysia) stayed in South America even longer than Bates – for fifteen years – and gathered more than 30,000 specimens from 7,000 species. Spruce, Wallace, Bates, von Humboldt, Bonpland and many more were iron men, obsessively collecting, bottling, pickling, pinning, pressing and drying for year after year, always recruiting the help of the local people who were and are naturalists par excellence because their lives depend on knowing the creatures around them. Yet I believe that Spruce spoke for all of them when one day on the Amazon, aboard the steamer *Monarca*, he wrote: 'There goes a new *Dipteryx*, there goes a new *Qualea* – there goes a new "Lord knows what."' All that effort over many years provided but a glimpse of what was out there.

Now, of course, the solo naturalists, the upper-middle-class (though far from rich) von Humboldt, the upper-middle-class (and significantly rich) Darwin, and the self-made artisan collector-naturalists like Wallace, Bates and Spruce, have been replaced by teams of scientists from the world's great universities and government institutions, relentlessly quartering the Amazon and everywhere else and systematically recording all there is. Now, we might suppose, all is more or less sewn up. In truth, a century and a half after Spruce, his lamentation seems almost as cogent as ever. We have very little idea indeed what's out there. Estimates even of the total number of species in the world as a whole differ by an order of magnitude, from a possible 4 or 5 million to 30 million or more (though neither figure includes bacteria). Most biologists opt for a compromise of around 5 to 8 million. After several hundred years of conscientious natural history and a century of formal science the task even of listing all there is seems hardly to have begun. Nature is very big, and very various indeed.

Thus it is impossible to count all the different species of tree – or to be sure that they have all been counted. But biologists can at least guess.[1] Extrapolating from what is known, they estimate that there are around 350,000 species of land plants in general. At least 300,000 of them are flowering plants. Around one fifth of these are trees. There are also some non-flowering trees, among which the conifers are by far the most important; but there are only about 600 different species of conifer, so they don't much affect the overall statistics. So there are probably around 60,000 species of trees in the world, plus quite a few thousand hybrids; although any of the species or hybrids might be further subdivided into an indefinite number of wild races or cultivars. Sixty thousand seems a good working number.

Most of those species are in the tropics. Britain may seem to have hundreds of different species of tree, but most of them have been imported by human beings. Only thirty-nine are believed to be true natives (and one of them, the common juniper, may in fact have been brought in by ancient people). The vast boreal forests of northern Canada are dominated by only nine tree species – the quaking aspen, and a handful of conifers. The total of trees that are native to the US and Canada just about exceeds 600. Yet the New World tropics (the 'neotropics'), stretching south from the Mexico border as far as Chile and Argentina, contain tens of thousands of species, sometimes with hundreds of different species per hectare. Why the tropics are so much more various is discussed in Chapter 11.

Meanwhile, two more immediate questions present themselves. First, how on earth can anyone – the most astute of hunters and gatherers, or the most learned of professors – keep tabs on 350,000 or so species of plant, including around 60,000 trees? How can we begin to comprehend so many? Secondly, how did the enormous complexity that is entailed in being a tree come about? These matters are addressed in the next two chapters.

2

Keeping Track

How can we make sense of so much diversity?

We share this world with millions of other species, and engage directly with many thousands of them – for food, shelter, medicines, aesthetic pleasure, and sometimes just because we need to stay out of their way. At least at a few stages removed, *all* of them affect us so some extent, and we in turn affect them. For the purposes of both exploitation and conservation, we need to know who's who. So first we must try to identify and describe what species are out there – and so far biologists have listed nearly 2 million; perhaps one in four of the total. Then we must ascribe names to each, partly as an aide-mémoire, but mainly so as to communicate our findings with others. Thirdly, we must classify: place the creatures we have identified into groups, and then nest those groups in larger groups, and so on. Without classification, naming becomes ad hoc, and we could not hope to keep track of more than a few hundred different kinds, and probably a lot fewer.

The reasons for all this endeavour are not purely practical. Science is an aesthetic and spiritual pursuit. The more that is revealed, the more wondrous nature becomes. The more we know about living creatures, the more deeply we can engage with them. This is the appetite, as Hamlet said, that grows from what it feeds on.

But the problems of identification, naming and classification are many and diverse. This after all was the first task that God gave to Adam (Genesis 2: 19), and although a lot of Adam's descendants have been hard at it ever since, there's still an awfully long way to go.

WHO'S WHO?

Identification is the beginning of all natural history. Nature appears to us as the grandest conceivable theatre, endlessly unfolding. There can be no understanding at all until we have at least some inkling of the cast. We must be able, again to quote Hamlet, to tell a hawk from a handsaw.

But identification can be difficult for all kinds of reasons – even of trees, which are so big and conspicuous, and which do not run away. We have already seen the practical problem posed by some willows: that both leaves and flowers may be needed for identification, but the two may not be present at the same time. Yet whatever problems may confront us in temperate climes, we can be sure that the tropics will pose far worse. In tropical forest, flowers, which are the principal guide to botanical identification, are usually absent. In seasonal rainforest (with a distinct wet and dry season), many trees gear their flowering to the rains, so flowering is to some extent predictable. But much rainforest (as in much of Amazonia) is non-seasonal, and trees may flower at any time. To be sure, different trees of the same species generally flower simultaneously, for if they did not, they could not pollinate each other. So they must be responding to signals from the environment at large, or else (or in addition) they must be communicating with each other. But what those signals are is unknown, at least to us. To the human observer, the flowering seems random. In any case, in tropical forest (at least in 'secondary forest', which is forest that is regrowing after previous harvesting or clearance) the trees grow close as a football crowd, and most are remarkably thin, like poles, and grow straight up and disappear into the gloom, twenty metres overhead. Even if there are flowers, you won't necessarily see them.

The leaves may not be too helpful either, at least when viewed from the ground. Rainforest trees all face the same kinds of conditions, and have adapted in the same general kinds of way. Rainforests are wet by definition. But in some there is a dry season, and even when there isn't it doesn't rain all the time. The forest floor may be moist right enough, but the topmost leaves of the canopy are far above it, and are exposed to the fiercest sun. I have spent time on towers in quite a few rainforests and remember in particular in Queensland, near Cairns, how lush and green it all was on top – but also how uncompromisingly desert-like it felt. So the uppermost leaves must resist desiccation. Yet from time to time, and in due season every day, they must also endure tremendous downpours. Leaves that can cope with such contrasts tend to be thick and leathery (to resist drought), oval in shape, and have a projection at the end like a gargoyle, known as a 'drip tip', to shoot off surplus rain. Many hundreds of trees from dozens of only distantly related families have leaves of this general type. But even if you can see the leaves, it is hard to be certain if they belong to the tree you are interested in or to the one next to it, or to some epiphyte or liana slung over its branches. Often, in short, you have nothing to look at but bark. The trunks of tropical trees are sometimes highly characteristic, deeply furrowed or twisted like macramé, but in most species the bark is simply smooth and grey, dappled with lichen and moss.

In temperate forest you can be fairly sure that any one tree is the same species as the one next to it – or at least, it will be one of a cast list that is unlikely to exceed more than half a dozen (oak with ash in much of Britain; lodgepole pine with aspen in much of Canada; alder,

Scots pine and spruce in the Baltic; and so on). But in Amazonia in particular you can be fairly sure that any one tree is *not* the same species as the one next to it. Often there is half a kilometre between any two trees of the same species and there can be up to 300 different species of tree in any one hectare. So the task, often, is to identify an individual tree that may be not much thicker than your arm from the appearance of its bark, out of a total cast of several hundred (or thousand) possibilities – which may well include some that haven't previously been described, so that there is nothing to refer back to.

In practice, there are three main routes to identification, whether of trees or of any living creature, and in practice botanists and foresters use them all together. The first is to make use of botanical 'keys'. These are lists of characteristic features with a series of decision points, which you work through like a flow chart, as in: 'Do the flowers of this particular plant have four petals, or five?' If four, you move to question X, and go on from there; if five, you go on to Y, and so on. Such keys first became popular in the eighteenth century. The great French biologist Jean-Baptiste Lamarck, best remembered for his pre-Darwinian theory of evolution, was particularly good at devising them. They are very much in the spirit of the Enlightenment.

The second, ultra-modern way to identify a tree – or any living creature – is to sample its DNA. Modern scientists working in tropical forest commonly take a bore of cells (a biopsy) from the cambium under the bark, which is the most reliably accessible living tissue in a tree. They then send the sample back to the laboratory (although field tests of DNA are becoming available).

Ultimately, both of these standard methods – the diagnostic keys, and the DNA printouts – rely on some kind of central reference point of information. The principal kind of reference point in botany is the herbarium, a central collection of plants kept in dried and mounted form (though some bits might be pickled, and there might be micro-scopic slides, and so on). There are many herbaria worldwide, each specializing to various degrees. Some, like that of the Royal Botanic Gardens, Kew, are extremely wide-ranging. Others focus on the par-ticular plants of a particular region. Scattered among the various herbaria are the 'type specimens': the first examples of the species ever to be formally described.

But there is a third way to identify plants – and this is simply to *know*. This is how we recognize friends and family. Recognition is unconscious; and the unconscious does not need to check out the shortlist of diagnostic features that the key-maker or the forensic image-makers must refer to, but takes into account a dozen (or a hundred?) secondary cues – quirks of speech, expression, and so on. Many people grew up in tropical forests and some of them recognize trees from the feel of their leaves, their scent, or indeed the texture of their bark, as readily as any of us recognize our cousins and our aunts. In Brazil these indigenous experts are called *mateiros*. Mike Hopkins of EMBRAPA (Brazil's research centre for forestry and agriculture) told me that he has sometimes known *mateiros* disagree with specialist botanists. Typically, the botanist says that two similar trees are the same, while the *mateiro* says they are different. Resolution becomes possible when DNA from the cambium is analysed, or the trees bear flowers and fruit. In such disputes, says Dr Hopkins, 'I have never known the *mateiros* to be wrong.'

The *mateiros* are wonderful but they have their limitations. A *mateiro* who is supremely at home in one place, may be less at ease in another, which may well have a different complement of trees. Then again, the subtleties of the subconscious are not easily conveyed. It's important that everyone who visits the forest with a serious purpose – scientist, environmentalist, forester – should be able to identify what they see. To convey the information formally and reliably, we have to go back to basic diagnostic features that can easily be described: whether the leaves are alternate or opposite, the texture of the bark, and so on. In short, identification needs formal field guides and/or keys, and herbaria in scholarly centres where difficult material can be assessed definitively, and preferably a lab for molecular biology – *and* the *mateiros*. Sometimes all are available. Sometimes not.

To see how hard life can be for botanists in tropical forest we have only to look at the wondrous Reserva Florestal Adolfo Ducke (which rhymes roughly with 'booker'): a hundred square kilometres (roughly 10 × 10 km) of pristine forest in northern Brazil, about a thousand miles up the Amazon river, just outside Manaus. The Ducke Reserve has been studied intensively for decades. Sir Ghillean Prance, formerly director of the Royal Botanic Gardens, Kew, has been working there on and off since 1965. When he first arrived the total species list of ferns and flowering plants that had by then been identified stood at a little over 1,000, of which about 60 per cent were trees. He and the other resident and visiting botanists steadily added to the species list and by 1993 it ran to 1,200. A high proportion of their additions were orchids.

The botanists aimed from the outset to produce a key to the native plants but, says Professor Prance, 'all of us got diverted', and they never got around to it. But at last, in the early 1990s, helped by the National Research Institute of the Amazon (INPA) and a grant from the British government, botanists were recruited specifically to provide the much-needed flora and guide. Mike Hopkins was appointed as one of the two coordinators. In truth he was not a botanist but an insect specialist (an entomologist) from Wales (via Oxford). But he pointed out to the appointments board that the guide was intended to be read and used by all who were interested – and if it was prepared by botanists, then, he argued, only botanists would be able to follow it. 'But if I can understand it,' he said, 'then anybody can.' Ghillean Prance supported his case and Hopkins got the job. A succession of Brazilian scientists have also acted as coordinators.

It took five years of focused study to produce the guide – and that five years more than doubled the species list that first confronted Ghillean Prance in the mid 1960s. The total inventory now stands at 2,200, of which about 1,300 are trees. Not all of the additions are new to science, of course. But some of them are (including more orchids). From all this, three points are abundantly clear. First, identification really can be difficult – for it had taken decades to produce the species list of 1965 that included only half the species in the Ducke Reserve. Secondly, in the tropics at least, the more that botanists look, the more they find. Thirdly, the tropics really are extraordinarily diverse. The Ducke Reserve is about two thousand times smaller than Britain (which has 2.3 million square kilometres), but it has forty times as many native trees.

Anyway, in 1999 the team that Dr Hopkins coordinated finally produced the *Flora da Reserva Ducke*, a magnificent work that drips with colour photographs and diagrams of everything pertinent, and allows identification not simply from the flowers and fruits that are commonly taken to be definitive but are usually absent, but also from the leaves, twigs and all-important bark. The guide would be among my favourite reading, if only it wasn't in Portuguese.

Practical, hard-headed forestry of the kind that supports economies and ultimately supports the whole world depends on such fine-tuned botany. If foresters don't know what's what, they can finish up making horrible mistakes. It is still necessary to harvest at least some trees from the wild, and possibly always will be. In temperate countries where the trees in any one forest tend to be of less than half a dozen species (and sometimes only one) it is easy at least conceptually to take a proportion without doing terminal damage. But when the forest contains hundreds of species, and no two next to each other are alike, problems abound. Some tropical trees provide timber of immense value – worth several thousand dollars a cubic metre. Some, which may look very similar from the ground, may be good for nothing but firewood – and yet may be immensely important to the other creatures of the forest. Make a mistake, and you waste time and effort, and do damage for no reason. When valuable trees are harvested, it is important not to take too many and in particular to leave 'mother trees' that will seed the next generation. The sad loss of mahogany from the West Indies over the past two centuries is just one example of many of what happens when foresters are careless. Often, though, we find that the target species has relatives that are very similar (and may indeed provide timber that is just as good) but are much rarer. If the forester is careless about identification, he (tropical foresters are usually 'he') may harvest the rare, related species alongside the main target species – and so may wipe out the rare one altogether, again with immense and unnecessary ecological damage. Of course, much of the logging in the tropics is still carried out illegally (even in Brazil, where the forest is as well managed as in most places, an estimated 60 per cent of all logging is carried out illegally), and illegal loggers generally don't give a damn. But the trend to sustainable harvesting is increasing – and depends absolutely on good identification.

But the Amazonian tree known as the angelim, much valued for its fine, strong timber, shows that good identification and careful harvesting are not yet the norm. The angelim is a legume: one of the vast family formerly known as Leguminosae and now called Fabaceae, which includes the acacias and laburnums among trees, and gorse, peas, beans and clover among non-trees. But what exactly is an 'angelim'? At EMBRAPA, Mike Hopkins has found that foresters apply this hallowed name to well over a dozen different species from at least seven genera. Admittedly, all the commonest 'angelims' come from the right family (Fabaceae), but still they come from more than one subfamily from within the Fabaceae.[1] Similarly, the excellent timber that is marketed under the general name of 'taurai' commonly includes at least five species (and probably many more) from the Brazil-nut family, Lecythidaceae. Perhaps the most notorious confusion of all – much of it deliberate obfuscation – surrounds mahogany. The term should refer to one or at best several species of the genus *Swietenia*, in the family Meliaceae. In reality an enormous variety of brownish timbers are marketed as 'mahogany'.

Whatever is identified must also be named. Names are an aide-mémoire but more than that they are vital for precise communication. But after God had expelled Adam from the Garden of Eden, so Genesis tells us, he created the Tower of Babel. Perhaps that was where the trouble began: in any case, naming has always been, and remains, a huge bugbear.

WHAT'S IN A NAME?

Different people speak different languages of course; in aboriginal societies there is commonly one language per tribe, and the world as a whole has many thousands. Since particular species of trees may be widespread, many finish up with a hundred or more different local names – to which travellers, notably Europeans, have been wont to add a few more of their own. For those who simply want to know about trees, and enjoy them – aspiring connoisseurs – this can make life very difficult; although for those who enjoy words *qua* words, the variety is also most intriguing.

Local names to those who understand their roots are of course instructive: they reflect what the tree means to the people who coined them. Some gave rise to their common English equivalents. The toon tree is *tun* in Hindi and Bengali. The sacred fig known as the peepul is *pipala* in Sanskrit – and is also known in English as the bo or bodhi, apparently from the Burmese *nyaung bawdi*. 'Tamarind' comes from the Arabic *tamr-hindi*, meaning 'date of India'. Neem is from the Bengali *nim* or the Hindi *nim balnimb*. Teak is *tek* in Tamil. Some local names have simply been adopted straight into English – including, in recent years, the Maori names of native conifers; totara, rimu, miro, matai, kahikatea (which is the tallest tree in New Zealand), and kauri (the most massive New Zealander).

But although local names often mean a great deal to the people who coined and use them they can be of much less use to outsiders. It may be, for instance, that the Maoris recognize some deep similarity, practical or spiritual or whatever, between, say, the kahikatea and the kotukututu, yet see no worthwhile parallel between the kahikatea and the rimu. Certainly, at least to the foreigner, 'kahikatea' sounds similar to 'kotukututu', while 'rimu' is altogether different. You could not guess, as an outsider, that the kotukututu is the tree fuschia, *Fuschia excorticata* – the only fuschia that still has the form of a (very lovely) tree; and that the kahikatea and rimu are both tall conifers in the same botanical family, the Podocarpaceae. It may be that by not speaking Maori, outsiders miss a great deal that is instructive. But it is at least possible that many local names in Maori and a thousand other languages are not meant to express particular relationships at all. After all, traditional societies – or at least the specialists within them – typically know their local flora and fauna as well as the rest of us know our friends and family. When you know everyone individually, you do not need to name them in ways that express particular relationships. Bill is Bill and Sarah is Sarah and Romesh is Romesh. Why should their names express more than who they are?

Some societies, however, including many Europeans this past few thousand years, have tended to travel the world and have actively sought to find relationships between whatever they came across. I really do not know when such a way of thinking first arose, but Aristotle's pupil and colleague Theophrastus refers to different kinds of oaks, growing in different places. Even Britain with its paltry inventory of native trees has two distinct native oaks: the common oak, *Quercus robur*, and the sessile oak *Quercus petraea*. ('Sessile' means 'sitting', and refers to the way the cups of the acorn sit directly

on the twig – they don't have a stalk of their own as in common oak.)
As European botanists began to travel the world, from the sixteenth
century onwards but particularly from the eighteenth, they found
more and more oaks throughout Europe, Asia and north America –
and up to now have listed a somewhat astonishing 450 different
species of oak, including both the deciduous kinds (as the two British
natives are) and a host of evergreen types (for example, the cork oak,
Quercus suber and the holm oak, *Quercus ilex*). Although, as we will
shortly see, many non-oaks are also casually called 'oak', in general
the term expresses a true biological relationship in a way that is not
always evident in local languages. It's not that some languages are
superior to others. It's just that different languages serve different
purposes. They have different agendas. They express what different
people feel is important.

But English and other such global languages raise problems of their
own. English names have been conferred by many different groups
of people with different traditions and for different purposes: by
local people everywhere; and by gardeners, nurserymen, naturalists,
foresters, traders, carpenters, makers of pulpits and pianos, and even,
in recent years, marketing people. Thus the same tree and its timber
may have several different English names (which are often different
too in English and American); and also – which can be especially
damaging – many different trees may finish up with the same name.
It's as if the word 'dog' was also applied as the mood took to horses
or ants or goldfish; and goldfish were sometimes called 'butterfly' or
'baboon'.

(Please skip the next four paragraphs if you have no wish to be
confused: they are written to illustrate the prevailing confusion.) Thus,
the various trees that are peremptorily called 'tulip trees' include
Liriodendron tulipifera from the magnolia family, Magnoliaceae; and
Spathodea campanulata of Africa which is from the catalpa family,
Bignoniaceae. But Brazilian tulipwood is *Dalbergia frutescens* – a
member of the Fabaceae (alias Leguminosae), the family of peas and
acacias. On the other hand *Dalbergia* is best known for various species
of 'rosewood', though of course these have nothing to do with roses,
which are in the family Rosaceae. Almonds and plums do belong to
the Rosaceae. But the 'wild almond' of India is *Sterculia foetida*, a

relative of cocoa in the Sterculiaceae family (now included in the Malvaceae); and the Java plum is *Syzygium cumini*, in the eucalyptus family, Myrtaceae.

Australians – or, rather, the British who first colonized Australia – seem to specialize in confusion. Thus true oaks belong to the genus *Quercus* in the family Fagaceae. But the Tasmanian oak is a eucalypt, *Eucalyptus delegatensis* in the family Myrtaceae; and the silky oak is *Grevillea robusta*, one of the Proteaceae. The genus *Flindersia* belongs to the family of oranges and lemons, Rutaceae. But various species of *Flindersia* are known as Queensland maple (although true maples belong to the Aceraceae); and *Flindersia schottiana* is known as Southern silver ash (although true ashes are related to olives, in the family Oleaceae). On the other hand, Australia's 'mountain ash' – the world's tallest broadleaved tree – is another eucalypt. Britain's 'mountain ash', better known as the rowan, is *Sorbus aucuparia*, a member of the Rosaceae. Hmm.

But the populist namers of trees saved their full powers of obfuscation for the conifers. It's good to acknowledge that members of the Pinaceae family have important characters in common, and to call them all 'pines'. But foresters and timber traders in particular have applied the term 'pine' to just about anything that has needly leaves and is evergreen. Thus, while the Pinaceae predominates among the northern conifers, the two great families of the southern continents are the Podocarpaceae and the Araucariaceae. The Podocarpaceae includes the New Zealand conifers: the kahikatea, the matai, the miro, the rimu. But the pioneer British foresters called the kahikatea the white pine, the matai the black pine, the miro the brown pine, and the rimu the red pine. The Araucariaceae includes three genera: *Agathis*, *Araucaria* and *Wollemia*. *Agathis australis*, the kauri, is still widely known as the kauri pine. The parana pine, much favoured by do-it-yourselfers, is *Araucaria angustifolia* from Argentina and surrounding areas. *Araucaria heterophylla* is commonly called the Norfolk Island pine. (*Araucaria araucana* is the monkey puzzle tree, from Chile and Argentina.) Even *Wollemia*, long thought to be extinct and only recently rediscovered in New South Wales, was immediately called the 'Wollemi pine' although in modern times we should know better. On the other hand the Scots pine, *Pinus sylvestris*, which really is a pine, is sometimes known in the trade as redwood; although the trees more commonly known as 'redwoods' are the Californian giants of the genus *Sequoia*, which is now included in the cypress family, the Cupressaceae. But then again, the tree that Californians more commonly call the 'giant sequoia' is of the related genus *Sequoiadendron* (formerly sometimes known as *Wellingtonia*). Colour is a particular obsession of the timber trade. Trees called 'blackwood' come from at least three families – the ebonies, Ebenaceae; the Fabaceae; and the mahogany family, Meliaceae. I won't even begin to bore you with the number of species and their timbers known as 'whitewood'.

I can't resist one last example. The British tend to feel that cedars are conifers of the genus *Cedrus*, relatives of the pines in the family Pinaceae. But in America a whole range of lovely trees are called 'cedar' including various *Calocedrus*, *Thuja* and *Chamaecyparis* from the Cupressaceae, the family of the cypresses and junipers. West Indian cedar is *Cedrela*, from the Meliaceae, another relative of mahogany. Thus the common names may flit not only from family to family but also leap the enormous gulf between conifers and broadleaves.

THE PROS AND CONS OF LATIN AND GREEK

The formal scientific names of living creatures are often called 'Latin'. There is truth in this. The medieval naturalists who laid the foundations of modern nomenclature were classical scholars first – and we should be grateful that they were: the formal names can seem very long but they have an elegance that would be difficult to achieve in most languages. Of course, though, they are not strictly Latin. They are compounded primarily of Latin plus Greek, but they also in their modern form incorporate bits of scores of other languages from Swahili to Inuit, plus the names of people and places, as in *Taiwania* and *cunninghamii*. Above all, the 'Latin' names are consistent. Each name is first mooted by whoever first describes the creature in question, and then must be approved by committees of specialist taxonomists, who decide what is appropriate.

Even so, there are snags.

Firstly, in recent years the specialist taxonomists who look after plants decided to rename many of the families. In the old days, most plant families ended with the suffix '-aceae', as in Fagaceae (the family of the oaks and beeches) and Betulaceae (the family of the birches). But some plant families, for historical reasons, had different endings, as in Leguminosae (the pea, bean, laburnum and acacia family) and Compositae (the family of daisies and thistles). But a few years ago the powers that be decreed that *all* plant families must end in '-aceae', and so the ones that didn't were renamed. The Leguminosae became Fabaceae. Compositae became Asteraceae. Palmae, the palms, became Arecaceae. Gramineae, the grasses, became Poaceae. Labiatae, the family of mint, basil and some surprising trees (which in the interests of suspense I will discuss later) became Lameaceae. The family of carrots and celery, traditionally known as the Umbelliferae after its umbrella-shaped inflorescences, became the Apiaceae. (Incidentally, '-aceae', should be pronounced with all three syllables: 'ace' (as in 'trace') – ee – ee.)

Secondly, as briefly explained in this chapter and as will become apparent in Part II, taxonomy as a whole (the craft and science of classification) has upped its game of late. New techniques have come on board: notably the discipline of cladistics; direct investigation of DNA; and the use of the computer, which has vastly increased the amount of data that taxonomists are now able to take account of. These new techniques have given rise to a flurry of reclassification in recent years as older ideas, based mostly on the skill and experience of specialist individuals, have succumbed to new rigour. Worse: when DNA studies first became possible in the 1970s, biologists tended to assume that they would provide the royal road to truth. In reality, those studies have caused at least as much controversy as the traditional classifications did.

Perhaps (with luck) the worst of the turmoil is over. The new classification, based on the new techniques, is beginning to settle down. Even so, there is still a lot of shuffling. Many species are still being transferred from family to family. Some families are currently being split into more than one, and others are being fused. All of this to-ing and fro-ing is liable to entail some name changes.

Finally, the Latin names can be rather long, and sometimes too similar for comfort. If you're sitting up late with a 40-watt bulb it's easy to confuse, say, the Myrtaceae, Myricaceae, Myrsinaceae and Myrsticaceae. But then, many languages that people use every day are enormously polysyllabic – like German and some of those of Sri Lanka. Gardeners love to dazzle their employers with polysyllables and small children revel in the names of European football teams and dinosaurs. If you can say Munchengladbach and *Tyrannosaurus rex*, you can say *Sequoiadendron*. The only trouble is that at present you cannot be sure that *Sequoiadendron* will still be called *Sequoiadendron* in ten years' time. But it probably will. The Latin names have their drawbacks, but they are worth it, and we should be grateful to the old-time biologists who first began to put them in place.

Naming, however, is only the first step. Classification requires another order of endeavour.

GETTING SORTED

Classification at its most basic is an exercise in convenience; and if convenience is all we are interested in, then any of us is free to carve up the world as we choose. So it is that fishmongers and chefs recognize the category of 'shellfish', which includes anything that lives in water and is crunchy on the outside and soft on the inside – in practice an astonishingly mixed bag of crustaceans (such as shrimps and crabs) and molluscs (such as whelks and oysters). Timber merchants label all conifers 'softwoods' and all broadleaved trees 'hardwoods'. They do this even though some conifers are a lot harder than many hardwoods, and the softest woods of all are in truth 'hardwoods'.

But there seems to be an innate order in nature; and some at least – perhaps most – of the terms by which most peoples classify living things do seem to reflect that underlying order; something more than mere convenience. Thus in English as surely in most languages we recognize the category of 'insect' and on the whole take 'insect' to be different from 'spider'. 'Birds', 'horses', 'dogs', or indeed 'conifers' and 'flowers' (meaning flowering plants) are vernacular categories but again (unlike 'shellfish') they do seem to reflect some real quality of nature: a true orderliness. In short, deep in the human psyche (and deep in the psyche of animals, too, as can be shown in laboratory trials) is the belief that nature *is* orderly, at least to some extent. Behind the terms 'robin', 'duck', 'eagle' and 'canary' lies, very clearly, the broad general concept of 'bird'.

This simple musing raises a series of deep questions – deep enough to have kept philosophers and biologists occupied for thousands of years. Firstly, is the order that we think we perceive, 'real'? Intuitively it seems obvious that shrimps and oysters are very different, even if they are lumped together as 'shellfish', while ducks and robins really are variations on a single theme that we might as well label 'bird'. But can we trust our intuition? Might it not be that all creatures are entirely independent of each other, and that ducks in reality are no more similar to robins than shrimps are to oysters? Intuition tells us that there is indeed order; but we know that our intuitions can be wrong.

If the order is 'real', not just a trick of our minds, how do we pin it down? Insects for instance are immensely various, so why do we call them all 'insects'? By what criteria do we place butterflies and beetles and grasshoppers in the same grand category, which we take to be different from the grand category of spiders? Are those criteria valid?

Then, thirdly, there's the issue that has especially exercised theologians (and a great many biologists) this past two hundred years: *why* should nature be orderly? Where does the order come from? Is it orderly simply because it is the Creation of God (as Genesis tells us), and God has a tidy mind? Or are there other feasible or indeed necessary explanations?

If the order in nature is 'real'; *if* it reflects some deeper underlying intent or force, then it would (would it not?) be very good to reflect this in the classification. A classification based purely on convenience (shellfish, softwood, hardwood) is just a temporary device, a throwaway thing, that meets the needs of particular trades at particular times. A classification that reflects the true order of nature – a 'natural' classification – provides true insight: insight, so many philosophers have opined, into the mind of God; or insight, so others have insisted, into forces that have brought order into being, whether inspired by God or not. Thus, the idea of providing a truly 'natural' classification has engaged philosophers since – well: at least since the beginnings of philosophy.

Plato and his pupil Aristotle are commonly taken as the twin founders of modern western philosophy and they had different ideas about where 'order' comes from – and both ideas have been reflected in the attempts of later biologists to provide a natural classification. Thus Plato thought that everything on earth is merely a copy, and a flawed one at that, of some 'ideal' counterpart that exists in what might be called 'heaven'. These ideals were in fact more 'real' than the things we see around us. Plato's ideas were absorbed into Christianity and Christianity has been a driving force in western science, so biologists until well into the twentieth century were wont to think, Platonically, that all earthly things and creatures are ideas of God. Thus in the late nineteenth century Louis Agassiz, then an extremely influential professor of biology at Harvard, declared that each separate species is a 'thought of God'.

Aristotle, Plato's pupil, was on the whole more down-to-earth, and rejected Plato's 'ideals'. Instead he spoke of 'essence': there is no ideal insect, of which beetles and butterflies are reflections; what we see is what there is. Nonetheless, all insects share some 'essence' of insecthood. Aristotle, unlike Plato, was a naturalist, who liked to look at nature; and he was the first philosopher that we know about who tried to devise a 'natural' classification that truly reflected the essences of different forms. In doing so, he set out the most basic rules of taxonomy – and identified some of the key problems. Thus, he said, if we really want to see who belongs with whom, then we have to see what features they have in common. More specifically, the taxonomist must pick out particular 'characters' (the biologist's term for 'characteristics') of each of the creatures in question, and then see which and how many of those characters they share with other creatures.

This is fine as far as it goes. Feathers are a very clear character of birds, and all living creatures with feathers may reasonably be classed as birds. But what about, say, number of legs? That is a clear character, too. But birds have two legs, and so do humans. Do birds and people belong together? Everything else about humans seems to suggest that we are closer to dogs, monkeys and other mammals: like them we have hair rather than feathers, and we produce live babies rather than eggs, and women suckle their babies as other female mammals do and birds do not. So what do we make of our two-leggedness? Well, the broad generalization is that in seeking the true order of nature, some characters are more informative, or less deceptive, than others. Feathers are a good guide. Number of legs is a less good guide. Or at least this is true in this instance. When it comes to telling insects from spiders, the number of legs is a very good guide indeed.

From the time of Aristotle, and with many a diversion, the art and craft of taxonomy shuffled along, as naturalists and apothecaries, and anyone else with an interest in nature, tried to classify the creatures they dealt with, and to some extent at least tried to create systems of classification that were 'natural', and reflected the true order of nature. The medieval herbalists made great progress, describing an impressive variety of plants, with Latin (or latinesque) descriptions to match. In the Middle Ages emerged the idea that different species of similar plants can be grouped together into genera (singular, 'genus'); and this thinking is reflected in the names they gave to their plants. They did not have enough data; communications were not good, and they tended to work semi-independently; and they had few robust principles to guide their thinking. But they did a lot of vital groundwork nonetheless.

The seventeenth century saw the birth of recognizably modern science, both in method and philosophy. The method included close, repeatable, quantified observation, and orderly experiment. The philosophy included the final acknowledgement of the idea that the universe is indeed orderly. It was run, so Galileo and Newton and other great seventeenth-century physicists averred, according to natural 'laws', an idea which is still with us, at the heart of science. Naturalists quickly came in on the act. Living creatures are far more various in form and in their behaviour than are the planets, or the mechanical devices that the physicists and engineers played around with. But even so, the naturalists felt, biology should have its 'laws' too. This general feeling reinforced the idea that the apparent orderliness of nature, that is reflected in general terms like 'bird' and 'insect', did indeed have deep origins.

John Ray was outstanding among the seventeenth-century naturalists who sought to broaden the scope of classification; to include many more creatures than the herbalists had, and to devise ground rules for finding the true order of nature that lies behind appearances. Notably, in our present context, he distinguished two great categories of flowering plants – a distinction that still persists. Some flowering plants, he pointed out, have long narrow leaves, like lilies and grasses; and others have broad leaves. More than a century later the French taxonomist Antoine Laurent de Jussieu pinned down the deep difference that lies behind this distinction. The embryos of all flowering plants, still within their seeds, have leaves, known as 'cotyledons'. The embryos of narrow-leaved flowering plants, such as lilies, grasses and palm trees, have only one cotyledon. The embryos of broadleaved plants, like oak trees and daisies, have two cotyledons. Hence the two great groups of flowering plants: monocots and dicots (much more of this in Chapter 5). Jussieu's discovery illustrates another great principle, in line with Aristotle's musing over the number of legs: that the characters that really count, and really show who is related to whom, are often ones that are *not* particularly obvious; are indeed 'cryptic'.

Jussieu was a child of the Enlightenment, in which thinkers of all kinds sought to integrate all the wisdom of the world into one grand 'rational' framework. The Enlightenment was centred in France, and Jussieu was only one of a host of late eighteenth-century French biologists who made an enormous and lasting impact. Best known of them all was Jean Baptiste Lamarck who was a fine botanist and devised keys to aid identification. But the Enlightenment touched all of Europe, and perhaps the most influential Enlightenment biologist of all was a Swede, Carolus Linnaeus or Linneus (whose name is sometimes Germanized for no good reason to Carl von Linné). Linnaeus was primarily a botanist, and led several expeditions deep into Europe, much of which in his day was still very wild and woolly, discovering many hundreds of new species. He was also a marvellous extrovert, and led botanical expeditions from his native Uppsala with the local band out in front and everyone dressed in a uniform of his own design. This demonstrates once more how much easier it is to be a botanist than a zoologist. Animals faced with such a mob would surely have packed themselves off to Russia.

More to the point, between the 1730s and the 1750s Linnaeus built upon the ideas of his contemporaries and predecessors, to create the system of classification that is with us still, and is called 'Linnean'. In truth, since Linnean classification has been significantly modified over the years, it should surely these days be called 'neolinnean'. But so far as I know, I am the only person to use the term 'neolinnean' (and will continue to do so until it catches on).

At the root of Linnaeus's classification is the 'binomial' system of naming living creatures. Each creature has two names, as in *Quercus robur* or *Homo sapiens*. The first name is 'generic', denoting the name of the genus, and the second is the species. In truth, Linneus did not invent the binomial system from scratch – it is evident in the medieval herbalists – but he made it formal. It remains one of the few items of language that is universally acknowledged worldwide. Absolutely unbreakable convention rules that these scientific names are always written in italic; the generic name always begins with a capital letter; and the specific name always begins with lower case, even when it is based on the name of a country (as in *indica* or *africana*) or a person (as in *williamsii* or *cunninghamii*). (Newspapers almost invariably get the convention wrong.) The names are properly called 'scientific' but are often known as 'Latin' even though they commonly include just as much Greek, and also may incorporate the names of people and places, or bits of Swahili or Inuit or what you will.

Linnaeus also proposed that similar genera should be contained within larger groups – orders; and similar orders should be grouped into classes; and classes into kingdoms. He regarded kingdoms as the biggest grouping of all, and recognized only two: Plantae and Animalia. He was not a good microscopist (even though microscopes were very popular in the eighteenth century) and had little to say about the creatures that cannot be seen without them (such as protozoa and bacteria). Somewhat perversely (he should have known better) he rammed fungi in with the plants.

Early in the nineteenth century the English anatomist Richard Owen provided one more, crucial, conceptual advance: one that answered Aristotle's problem of how to distinguish important characters from less important. The important characters, said Owen, are 'homologous' – features which may have different functions in different creatures, but nonetheless clearly have a common origin. Thus the wings of birds, the front legs of horses and the arms of people serve very different functions, yet they all originate as forelimbs. This can be seen from common observation – and can be seen unequivocally when you look at the embryos. The wings of flies serve the same general purpose as the wings of birds, but are clearly very different. They arise as projections from the back, quite independently of the limbs. Bird wings and fly wings are merely 'analogous'. The creatures with homologous, shared features should be grouped together (and birds, horses and people are all classed as 'vertebrates', with flies in the separate category of 'insects').

Cases like this are obvious. But when biologists are looking at unfamiliar structures in unfamiliar plants – and especially at fossils that are reduced to fragments – the crucial distinction between homologies and analogies can be very hard to make. Even in what may seem like clear-cut cases, the distinction may not be easy. Charles Darwin wondered whether flowers are homologous with the cones of conifers. They have a roughly similar structure (at least when compared with primitive flowers, such as magnolias) and they do the same job. The general consensus today is that they are not homologous. Conifers and flowering plants invented their sexual organs separately.

The trek from Aristotle to Linnaeus, with the additional insight of Owen, takes us halfway to modern taxonomy. But even by the time of Linnaeus, a sea-change was in the offing.

THE FINAL ROAD TO MODERNITY

Most European and American biologists until well into the nineteenth century took it for granted that life began on earth in the way described in Genesis. God created everything. He made each creature separately: the enormous diversity reflects the fertility of his mind. He placed each one in the environments to which it was best suited – shaggy bears in the north, smooth-haired bears in the tropics (Malaysia, South America), and so on. Each creature is 'adapted' to its environment – for if it were not, it could not live there, and the general phenomenon of adaptation was explained by God's beneficence. He moulded creatures to thrive in whatever conditions he placed them in. Of course he did. He is benign.

But Genesis also implied that the world was created quickly – on the first day, on the second day, and so on. Furthermore, in the seventeenth century a zealous Irish bishop called Ussher added up the reported ages of all the patriarchs listed in the early books of the Old Testament, and concluded that the earth must have been created in 4004 BC, which made it less than 6,000 years old. Genesis also describes the Flood, in which Noah rescued a male and female of all the creatures. Present-day creatures are all descended from the couples that Noah took on to his ark. Clearly the creatures that lived before the Flood were the same as the ones that live now.

The general rationalism of the eighteenth century, the huge exercises in civil engineering that ate deep into the bedrock, and the new, growing, formal science of geology, nibbled away at the details offered in Genesis. It was clear by the end of the eighteenth century that the earth was much older than 6,000 years (even though the geologists who discovered this, such as Scotland's James Hutton, typically remained as devout as ever). In the early nineteenth century formal collections of fossils, most spectacularly of dinosaurs and other ancient reptiles, showed beyond reasonable doubt that a huge range of creatures existed *before* the Flood, yet did not survive it – and also suggested that many of the creatures that surround us now, like elephants and oak trees, did not exist at the time of the dinosaurs. Clearly there had not been a once-for-all creation of plants and beasts that had remained unchanged ever since. Clearly the ones that were created first are long gone, replaced by others. Either there had been a series of separate creations (not recorded in Genesis) or the initial creatures had changed over time, to give rise to those of the present day. The idea that creatures might change over time was, and is, the idea of evolution.

Many people floated general notions of evolution in the eighteenth century. Even Linnaeus, it seems, though on the whole content with conventional theology, was veering towards it at the end of his life. Several formal descriptions and explanations were published in the late eighteenth and early nineteenth centuries, of which the best known is that of Lamarck. What was lacking, though, was a plausible mechanism: a way of explaining how and why there are so many different creatures on earth and how each one is adapted to its surroundings;

and also how there could be change over time even though all creatures in general give rise to offspring who resemble themselves ('like begets like').

The biologists who finally provided the convincing account, and the plausible mechanism, were two Englishmen: Alfred Russel Wallace and Charles Darwin. Independently, they came up with the idea that Darwin called 'natural selection'. Creatures do give birth to offspring that are like themselves – but the offspring (if sexually produced) are not identical with their parents. There is variation. Some variants, inevitably, will be better adapted to the prevailing conditions than others. But not all can survive, because all creatures are able to produce more offspring than the environment can support. The survivors, therefore, are the ones that are best adapted to the conditions. To the Victorians, the word 'fit' meant 'apt'. So the ones that were best adapted were the 'fittest'. In the 1860s Herbert Spencer, a philosopher who at that time was extremely famous, summarized the idea of natural selection as 'survival of the fittest', a phrase that Darwin later adopted.

In 1858 Darwin and Wallace presented their ideas in a joint paper, which was read on their behalf to the Linnean Society of London. The Linnean is a august society of biologists, still with its headquarters in Piccadilly, that was founded to commemorate Linnaeus. Darwin and Wallace's paper was surely the most momentous ever presented to them – indeed it was one of the most momentous ever presented anywhere. But the Linnean's president, in his annual report for 1858, dourly reported that nothing much of interest had happened that year. In 1859 Darwin (who had been thinking about the ideas for longer, and had a much broader scientific background) expounded the ideas more fully in *On the Origin of Species by Means of Natural Selection*, generally referred to as the *Origin*. The *Origin* changed the course of modern biology, and also changed all philosophy and theology. In it, Darwin speaks of 'descent with modification'. Most other biologists preferred, and prefer, the term 'evolution'.

In truth, Darwin made four outstanding contributions that are central to our theme. First, he established once and for all that evolution is a fact. Secondly, he provided the plausible mechanism – natural selection. Thirdly (a separate issue) he argued that species are not as the Platonists still conceived them to be – once-for-all creations that could not be changed. Over evolutionary time, he said, species could change into other species, and the lineages could branch so that any one species could give rise to many different types that would all then evolve along separate lines.

Finally, he proposed that all the creatures that have ever lived on earth are descended from the same common ancestor that lived millions of years in the past (although Darwin did not know how many millions). We share a common ancestor with robins and mushrooms and oak trees. This at a stroke answers the deepest problem: *why* there is order in nature. To be sure, we can say that God designed butterflies and bees along similar lines simply because he has a tidy mind. But we can also argue that butterflies and bees are similar because, in the deep past, they shared a common ancestor: the first ever insect. Deeper back in time, the first ever insect shared a common ancestor with the first ever shrimp – and insects and shrimps clearly have a lot in common. Even before that, the common ancestor of insects and shrimps shared a common ancestor with spiders. So although insects are clearly very different from spiders, they, too still have quite a bit in common.

Since all creatures are literally related, they can all be represented on one great 'family tree'; although a family tree drawn on such a scale is more properly called 'phylogenetic', from the term 'phylogeny', which refers to the evolutionary relationship between different groups of creatures (it comes from the Greek *phylos*, meaning 'tribe'). This idea chimes beautifully with Linnaeus's classification. Linnaeus's kingdoms represent the great boughs of this all-embracing phylogenetic tree. The classes and orders are the thinner branches. The individual species are the twigs.

Some were offended by Darwin's grand view of phylogeny. Some continue to argue that it is blasphemous, because it seems to contradict Genesis, which states that God created human beings separately from all other creatures, in his own image. Others are affronted by Darwin's particular suggestion that human beings are most closely related to apes. The Creationist movement is still strong worldwide – not just in the United States. Some professional biologists are Creationist fundamentalists. In absolute contrast, many modern biologists and philosophers argue that since evolution by means of natural selection seems to offer a plausible alternative to the account in Genesis, then this means that religion in general is obsolete and God is dead.

In truth, neither of these extreme positions is valid. It makes no sense to reject evolutionary ideas; and it makes no sense to try to use those ideas to justify atheism. Leading churchmen of the late nineteenth century knew this (and Darwin is buried in Westminster Abbey). Many modern biologists who are steeped in evolutionary theory remain devout. Many take the wondrousness and subtlety of evolution as further proof that God is indeed marvellous, and demands reverence. Many indeed continue to argue in the spirit of the seventeenth century that the true purpose of science is to enhance appreciation of God's works. For my part, I feel that Darwin's is a glorious vision. I love the notion that we are literally related to all other creatures: that apes are our sisters, and mushrooms are our cousins, and oak trees and monkey puzzles are our distant aunts. Conservation, on such a view, becomes a family affair.

Conceptually, too, with Darwin's great insight the task of taxonomy became easier. All the taxonomist has to do is identify creatures that share common ancestors. The way to do that is to identify shared characters that are homologous. In fact, Richard Owen remained wedded to the conventional theology of his day and never fully accepted the idea of evolution – and yet, ironically, his idea of homology provides a principal clue to evolutionary relationships. But in practice it can be very difficult to decide which of the characters that different creatures share are truly homologous – and even if the difficulties are overcome there is still one theoretical snag.

The snag is as follows – and again for simplicity I will use an example from animals rather than from plants; but the principle applies universally. Suppose you wanted to work out whether human beings were more closely related to horses, or to lizards. Suppose you decide to count the number of toes – a perfectly good 'character'. Then you would conclude that the human and the lizard are closer – because both have five toes. The horse, with one toe, is the odd one out. Yet everything else about horses, human beings, and lizards, suggests that horses and people belong together (in the class of the mammals) and that lizards are the odd ones. This is the same kind of problem that Aristotle identified. Owen's idea of homology is not all that helpful in this context. After all, the feet of lizards, horses and human beings are all homologous.

One further idea is needed to sort this out, and this was pinned down formally in the 1950s by a German biologist (in fact an entomologist) called Willi Hennig. He distinguished between homologous characters that are 'primitive', and those that are 'derived'. 'Primitive' characters are those that are inherited from the very earliest ancestor of *all* the creatures in question. Thus lizards, horses and human beings are all distant descendants of some ancient amphibian that lived about 350 million years ago – and that ancestor had five toes. For all the descendants of that amphibian ancestor, the default position is also to have five toes. But some of those descendants have lost at least some of the toes – as birds have done and so (quite separately) have horses. Horses have lost four of the five fingers – all except the middle one. So too have asses and zebras. The point is that horses, asses and zebras all inherited their one-toed-ness from the same ancestor, the first ever one-toed equine, who lived somewhat more than 5 million years ago. Although human beings have many 'derived' features – including enormous brains – we happen to have retained the five-toed limbs of the first amphibian ancestor – the 'primitive' feature. So have lizards. But the fact that lizards and people have such characters in common does not show any special relationship. However, our big brains and our forward-looking eyes are 'derived' features, which were not present in that ancient amphibian ancestor, or indeed among the first ancestral mammals. They arose only among primates. They are among the characters that show our special, close relationship to chimpanzees.

By the same token, we can see that oaks, chestnuts and beeches all belong together (in the family Fagaceae) because all enclose their seeds within very similar casings (the cup of the acorn, the shell of the beech and chestnut). This casing is a derived feature, that shows their affinity. All three also, of course, have green leaves. But the leaves are primitive features, also found in magnolias and eucalypts, or indeed in pines and araucarias. The mere presence of leaves tells us nothing about the relationships of oaks, chestnuts and beeches – beyond the fact that all three are plants.

Hennig provided a whole list of rules for deciding whether shared homologous features are primitive or derived, and his general approach is known as 'cladistics', from the word 'clade', meaning all the descendants of a common ancestor. Cladistics has become the taxonomic orthodoxy only in past few decades. Much of the traditional classification in conventional textbooks does not incorporate Hennig's ideas. Traditional taxonomists sometimes (quite often in fact) treated primitive and derived features together, and so created many groupings that look convincing – since all the creatures in the various groups do have plenty of characters in common – but in fact, if you look closely, are no more convincing than a classification would be that placed humans and lizards together, and excluded horses. In the chapters that describe the various groups of trees, you will find many instances of reclassification. This is partly because botanists are now revisiting old territory, and distinguishing more clearly than was often done in the past between the shared, homologous characters that are derived, which denote true, close relationships; and characters that are merely primitive.

Thus taxonomy has advanced conceptually over the past few decades – and it has advanced, too, in technique. From earliest times taxonomists looked at the obvious, 'gross' anatomy of creatures. From the seventeenth century onwards they could refine their observations with the help of microscopes – which also helped to reveal the insights provided by embryos. From the 1930s they could look even closer, with electron-microscopy, and home in on 'microanatomy'. The fossil record has grown wonderfully, too, these past few decades. Some recently discovered fossils, recovered by modern techniques, offer the same microanatomical detail as living tissue. Brilliant. Then, of course, there are DNA studies – exploring and comparing the detailed chemical structure of genes.

But all these approaches have their drawbacks. All are subject to the trap that has beset all taxonomists since Aristotle: divergence and convergence. That is, creatures that are very closely related may adapt rapidly to different circumstances and finish up looking very different; and creatures that are not at all related may adapt to similar circumstances, and finish up looking much the same. Thus it transpires (when you look closely) that the family of oaks, beeches and chestnuts

(Fagaceae) is closely related to that of cucumbers, melons, squashes and marrows (Cucurbitaceae): a fine case of divergence. On the other hand, as we have seen, many tropical rainforest trees have leaves that look very similar even though they may not be closely related, simply because all are adapted to dryness on the one hand and downpours on the other: a striking example of convergence. Fossils can be wonderfully instructive – but although some fossils show fine detail, most are to some extent fragmented and the fossil record as a whole is, as the palaeontologists say, 'spotty'. Only one in many millions of extinct creatures gets to be fossilized and then discovered, and whole vast groups must have gone missing. Thus everything we know suggests that flowering plants and conifers share a common ancestor but it is very hard to find truly convincing links between the two within the fossil record, vast though it has become.

This, too, is why DNA studies do not provide the royal road to truth that was hoped for. Genes, too, may diverge or converge, just as anatomical features do, and so they can deceive. Even more to the point, different genes in the same organism may tell different stories. Thus studies in the 1980s suggested that the genes of red seaweeds are very different indeed from those of green plants, and that the two groups should be placed in different kingdoms that were miles apart on the grand Darwinian phylogenetic tree. But later studies in the 1990s, which looked at a different set of genes within red seaweeds and green plants, suggested that the two are very closely related – so closely that the two groups are, as taxonomists put the matter, 'sisters'. The later studies are probably more accurate than the earlier ones – but it is always hard to be sure. Judgement and experience play as much part in modern taxonomy as they always did in the past, and there will always be disagreements about who is really related to whom. In taxonomy, as in science as a whole, there *are* no royal roads to truth. Some of the continuing debate is reflected in Chapters 4 to 8.

All these new approaches have caused taxonomists to modify Linnaeus's original classification more than somewhat. In particular, modern taxonomists have greatly increased the number of kingdoms. Early twentieth-century biologists decided that all single-celled creatures that are not green are 'protozoa', and put them in with the animals; and all single-celled organisms that are green they called 'single-celled algae', and put them in with the plants. Fungi and similar creatures such as slime moulds were also rammed in with the plants. So were the brown seaweeds (wracks) and red seaweeds. Now it is clear that there is huge variation within the protozoans and the single-celled 'algae', so that they are now divided among about a half a dozen or a dozen different kingdoms (depending on who is doing the dividing); and some of those newly defined kingdoms contain both 'protozoa' and 'algae'. Fungi, the various groups of slime moulds, and red and brown seaweeds now each have their own kingdoms; plants and animals are just two kingdoms among many, albeit by far the most conspicuous. Broadly speaking, all the kingdoms seem to divide into two great blocks, one including the plants (and red and green seaweeds and others) and the other containing the animals and fungi (and a lot of smaller types). In the early twentieth century, too, no one knew quite what to do with bacteria, although a few brave souls put them in a kingdom of their own. Now they are found to be so different that they are given their own 'domain' (though in truth this is divided into two domains). The kingdoms of the plants, animals, fungi, seaweeds and so on together form a third domain.

So now, the ranks that Linnaeus first described (species, genus, order, class, kingdom) have been increased to eight. The rankings now run: species, genus, family, order, class, phylum, kingdom and domain (though botanists commonly substitute the term 'division' for what the zoologists call 'class' and/or 'phylum'). Often these basic eight ranks are further subdivided or bunched together as in 'subfamily' or 'superorder'. (But this can be overdone!)

All this may seem cumbersome, but it is all extremely useful. The family names make it easy to keep track – which is the most basic purpose of classification. Thus the 600 or so living species of conifer are now divided into eight families – and although 600 is too many for non-professionals, eight is straightforward; and if you can place conifers in their families, tell a pine from a swamp cypress, that is a lot better than nothing. The 300,000 or so species of flowering plant divide into 400 or so families – still too many for comfort; but the 400 or so are further grouped within about forty-nine orders, of which about thirty or so contain significant trees. It isn't hard to get your head around this number (especially if you focus for starters on the top dozen or so, and then work outwards). Thus with just a small sense of modern taxonomy the whole bewildering world of trees, all 60,000 or so of them, begins to become tractable.

Then, too, the modern phylogenetic tree that includes all living creatures is, in effect, a graphic summary of their evolutionary history. If you know what group a creature belongs to, then you also know who its ancestors were, and who it is related to. We also know where at least some of the groups originated. Some began in the southern hemisphere – even way down in Antarctica, which once was forested. Some started life in Asia, and then spread west across Europe, and were blown as seeds across the Atlantic, and into America; or spread east across the Pacific, again into America. Some began in South America and tracked through the whole world. Most of this happened millions of years ago – long before human beings came on the scene (and stirred the pot even more). It is a wonderful thing to contemplate a living tree, or a fossil one, or any other creature. It is even more moving when we add the fourth dimension, of time, and see in our mind's eye how the ancestors of the tree that grows in the field next door first saw the light in some remote corner of the globe millions or hundreds of millions of years in the past, and floated on its respective bit of continent as the continent itself circumnavigated the globe, and skirted round the glaciers of the Ice Age, and perhaps sweated it out in some primeval, long-gone swamp, with alligators around its feet and the world's first hawks and kingfishers scouting from its branches.

This is why I have been so keen to root this book in phylogeny – in modern taxonomy. It is an aide-mémoire to be sure, but more than that, it reflects evolution: and evolution reminds us of the glorious past of all living creatures; without it, as the Russian-American geneticist Theodosius Dobzhansky said, biology makes no sense at all. The next chapter looks at a few of the historical details: how, in practice, modern trees are thought to have come into being.

3

How Trees Became

Tree ferns once abounded. Some, like this Dicksonia, *are still with us*

A tree is a big plant with a stick up the middle – and a big plant with a stick up the middle is not an easy thing to be. Darwin spoke of evolution as 'descent with modification' and it took a lot of descending – several billion generations – and a great deal of modification to get from the world's first life, to the world's first plants, to the creatures we recognize as oaks and monkey puzzles and eucalypts.

This chapter runs rapidly through the key events. It isn't meant to be philosophical, but a point of philosophy continues to absorb me nonetheless. For when all Western thought was dominated by theology, change over time (like all aspects of nature) was seen to be part of God's plan. Late-nineteenth-century and some twentieth-century theologians and scientists who were unhappy with Darwin's particular idea that human beings descended from apes consoled themselves with the thought that those early apes were destined to become us; that they were mere prototypes, and prototypes are inevitably crude. In the same way, descriptions of evolution sometimes imply that the first land plants, say, somehow *knew* that their descendants would be vines and roses, redwoods and oaks, and saw themselves as a rehearsal. Again, the notion was that the course of evolution had been prescribed – or as the Moslems say, 'It is written'.

But Darwin argued differently (which was one of the ways in which he irritated the theologians and the clerically-inclined naturalists of his day). He suggested that evolution was merely opportunist; that each generation simply tried to solve its own problems as best it could; that as lineages of creatures unfolded (and evolution literally *means* 'unfolding') they might wander off in any direction. Thus a bear might evolve into a whale. The descendants of some of the apes that lived 10 million years ago in the Miocene did become us – but it would have been impossible to know at the time which ones were going to do so; and with the flip of a coin the creatures that in fact were our ancestors might instead have become more ape-like, or simply gone extinct (which is the fate of most lineages). Many late-twentieth-century biologists, not least the eloquent Harvard professor and writer Stephen Jay Gould, saw evolution as a 'random walk'. Lineages of creatures over time, he argued, go every which way. There is no pattern to it; and there can therefore be no prescription, and nothing resembling destiny.

Hard-nosed biology is at present more fashionable than theology, so notions of random walk now prevail over those of destiny. But fashion is a poor guide to truth. One stunning and undeniable fact of evolution is the phenomenon of convergence: the way in which lineage after lineage of creatures have independently reinvented the same body forms and, often, the same kind of behaviour. There may be no literal prescription for how life should turn out – but any two creatures in the same kind of environment tend to evolve along much the same lines. Sharks, bony fish, ichthyosaurs and whales all independently reinvented the general form of the fish (and so for good measure, did penguins and seals). Water poses its own particular problems to which there is one optimal solution, which they all adopt. Among plants, lineage after lineage has independently reinvented the form of the tree. A tree, after all, is a good thing to be.

So nature may not be literally prescriptive. But it is not random either. Living creatures are in perpetual dialogue with all that surrounds them – with the other creatures that they encounter minute by minute, and with climate and landscape; which means they are in perpetual dialogue with the whole world, which in turn is subject to the influence of the whole universe. Whatever other creatures may do, however the world changes, each individual must take everything else into account. Each of us is engaged in this dialogue with other creatures and with the universe at large from conception to the grave. Furthermore, what applies to individuals also applies to whole lineages of creatures, as they evolve over time: all lineages of living creatures, whether of oaks or dogs or human beings, are engaged in this dialogue from inception to extinction. All creatures might in principle be able to evolve in an infinite number of ways as Darwin suggested, but if they are to survive along the way then each must solve the particular problems of its own environment at all times – and to each problem there is a limited number of solutions. There is something about the universe, at least as it is manifest on earth, that seemed to demand the emergence of fish and of trees (and perhaps – who knows? – of human intelligence). The physicist David Bohm spoke of the 'implicate order' of the universe. Fish, like trees (and human intelligence) reflect this innate, implicit orderliness. They are its manifestation.

What follows is an outline of what's known about the historical (evolutionary) events that led to modern trees. I will discuss it as a series of what I will call 'transformations'.

TRANSFORMATION I: LIFE

The first transformation on the path to treedom was the evolution of life on Earth – probably more than 3.5 billion years ago (the Earth itself seems to have begun about 4.5 billion years ago). So how did life begin?

In modern body cells, whether in people or in trees, the genes, in the form of DNA, sit in the middle, ensconced within the nucleus, like the chief executive in his office. They give out orders which are relayed by RNA (a smaller molecule akin to DNA) to the rest of the cell outside the nucleus (the cytoplasm), where these orders are carried out. Accordingly, DNA and RNA are often taken as the starting point of life itself; as if there is not, and never could have been, anything that could lay claim to life before DNA and RNA came on the scene.

Look closer, however, and we see that the flow of information within the cell is two-way – the genes themselves (the DNA) are turned on and off by signals from the cytoplasm, which in turn relay messages from the world at large. In short, DNA is in dialogue with cytoplasm, with all its intricate chemistry. Even at its most fundamental level, life is innately dialectic.

It follows from all this that life could *not* have begun with DNA. DNA cannot survive by itself; it cannot function at all except in dialogue with cytoplasm and all that goes on in it. Furthermore, the DNA molecule is itself extremely intricate and highly evolved. It could not have been the first on the scene. RNA is simpler and can make a better fist of independent living – but RNA too is a highly evolved molecule. DNA and RNA were not the prime movers, therefore. We might as soon say that by the time these two aristocrats had come on board, the hard work had already been done. At least, the absolute beginnings had been left far behind.

In truth the essence of life is metabolism – the interplay of different molecules to form a series of self-renewing chemical feedback loops that go round and round and round: and they do this simply because, chemistry being what it is, such a modus operandi is chemically possible, and what is possible sometimes happens. The first life, so it is

The Names and Times of Ages Past

era	period	epoch	millions of years since begining
Cenozoic	Quaternary	Holocene	0.01
		Pleistocene	1.8
	Tertiary	Pliocene	5.4
		Miocene	24
		Oligocene	34
		Eocene	55
		Palaeocene	65
Mesozoic	Cretaceous		142
	Jurassic		206
	Triassic		249
Palaeozoic	Permian		290
	Carboniferous		354
	Devonian		417
	Silurian		443
	Ordovician		495
	Cambrian		545
Precambrian			4,600

widely argued, was simply a metabolizing slime that spread over the surface of the earth, which in those early days was a very different place from now. Indeed it was a nightmare world, at least by our standards: hot, steamy, volcanic, with an atmosphere absolutely devoid of oxygen and probably full of gases such as ammonia and hydrogen cyanide that would snuff out almost all life of the kind we know today in a trice. The hot springs of present-day Yellowstone, New Zealand and Iceland, and the perpetually out-gassing vents in the depths of the great oceans, give some idea of what it was that early world was like. An extraordinary variety of creatures live within today's hot springs – most of which would be poisoned by oxygen if they were ever exposed to it. Anthropocentrically, we think of ourselves as 'normal', and call the creatures of the hot springs 'thermophiles' – heat lovers. But historically speaking, they are the normal ones. *We* – human beings and dogs and oak trees – are the highly evolved anomalies: the cold-loving 'aerobes', utterly dependent on the hyper-reactive gas, oxygen, that would have laid our earliest ancestors flat.

TRANSFORMATION 2: ORGANISMS

Life today is not a continuous slime. For at least 3 billion years the substance of life has been divided into discrete (or fairly discrete) units, each known as an 'organism'. Of course we don't know how, in practice, this separation came about – and never can, until someone builds a time machine. But we can speculate.

For natural selection would have been at work within the original slime, just as it is today and always has been. Inevitably, some bits of the slime would have metabolized more efficiently than others. Some of the endlessly cycling chemical feedback loops would have harnessed energy, and processed raw materials, more rapidly than others. The bits that worked best would have been held back by the bits that worked less well. Natural selection would surely have favoured the bits that were not only more efficient, but which also cut themselves free from the rest, surrounding themselves with membranes to monitor and filter all inputs from the world at large.

So the first organisms came about: the first discrete creatures. After a time (probably a long time) these primordial creatures developed the general kind of structure that is still seen in present-day bacteria and archaes (pronounced 'arkeys' – creatures with a similar general form to bacteria which in fact have a quite different chemistry). We tend to say that bacteria are 'simple', not least because they are small. In truth, of course, nature is far more wondrous than anything we could cook up and bacteria in reality are more complex than battle-ships, and a great deal more versatile.

TRANSFORMATION 3:
MODERN-STYLE CELLS

Bacteria compared with us (or indeed with mushrooms or seaweeds or flowering plants) *are* simple. In particular they keep their DNA loosely packaged, hanging around the cell. In our own body cells (and those of mushrooms, seaweeds and flowering plants) the DNA is neatly contained and cosseted within a discrete nucleus, cocooned in its discriminating membrane. Cells of this modern kind are said to be 'eukaryotic' (Greek for 'good kernel'). The nucleus is surrounded by cytoplasm, and within the cytoplasm there is a series of bodies known as 'organelles' that carry out the essential functions of the cell. Among these organelles are 'mitochondria', which contain the enzymes responsible for much of the cell's respiration (the generation of energy). These are found in all eukaryotic cells (apart from a few weird single-celled organisms which live as parasites, but they belong in another book). Plant and other green cells contain a unique kind of organelle known as the 'chloroplast'. This contains the green pigment chlorophyll which mediates the process of photosynthesis.

I am treating all this in some detail because herein lies a tale of immense importance, which is crucial to all ecology, and is discussed again in Chapter 12. For the eukaryotic cell evolved as a coalition of bacteria and archaes. Broadly speaking, the cytoplasm seems to have originated as an archae. Either this ancient archae then engulfed some of the bacteria around it, and/or the bacteria invaded it. Either way, some of those engulfed or invading bacteria became permanent residents – and evolved into the present-day organelles. Mitochondria and chloroplasts both contain DNA of their own. The DNA of mitochondria most closely resembles that of present-day bacteria of the kind known as proteobacteria. The DNA of chloroplasts resembles that of the bacteria that still manifest as cyanobacteria (in the past erroneously called 'blue-green algae'). Cyanobacteria, not plants, were the inventors of photosynthesis.

In his notion of evolution by means of natural selection, Darwin emphasized the role of competition. Soon after Darwin published *Origin of Species* the philosopher and polymath Herbert Spencer summarized natural selection as 'the survival of the fittest', which was taken by post-Darwinians to imply that evolution proceeds by the stronger treading on the weaker. Two decades before Darwin, Lord Tennyson wrote of 'nature red in tooth and claw'; and 'Darwinism', extended backwards to embrace Tennyson and forwards to Spencer, is commonly perceived these days as an exercise in strong bashing weak. But Darwin stressed too that we also see collaboration in nature – he made a particular study of the long-tongued moths that alone are able to pollinate deep-flowered orchids: two entirely different creatures, absolutely dependent on each other.

Yet we see a far more spectacular illustration of nature's collaborativeness within the fabric of the eukaryotic cell itself – the very structures of which we ourselves are compounded. For the eukaryotic cell is a coalition. It was formed initially by a combination of several different bacteria and archaes which hitherto had led separate lives (and others are probably involved, besides the proteobacteria and cyanobacteria). Over the past 2 billion years or so the eukaryotic cell, innately cooperative, has proved to be one of nature's most successful and versatile creations. There could be no clearer demonstration that cooperation is at least as much a part of nature's order as is competition. They are two sides of a coin.

The ancestors of today's plants arose from the ranks of the general mêlée of eukaryotic cells. These first ancestors contained chloroplasts, and were green, and these can properly be called 'green algae'. Many single-celled green algae are still with us (they often turn ponds bright green).

It's a reasonable guess that the first green algae appeared on Earth about a billion years ago. Thus it took about 2.5 billion years to get from the first living things to single-celled green algae; and only another 1 billion to get from single-celled algae to oaks and monkey puzzles. Still we tend to think of algae as 'simple' and primitive. If we took the long view and considered all that life entails, we might rather argue that by the time the first green algae evolved, it was all over bar the shouting. (Though there was still an awful lot of shouting to be done.)

TRANSFORMATION 4: ORGANISMS WITH MANY CELLS

Organisms that have only one cell are doomed to be small. There are many advantages in smallness: there is more room for small organisms than for big ones, and a virtual infinity of niches to exploit. Single-celled organisms are easily the most numerous and always have been – living free wherever there is moisture, in oceans and lakes and soil, as inhabitants of bigger creatures' guts, and as parasites of bigger creatures.

But there are advantages in being big, too. A whole range of ways of life are open to big creatures, whether trees or people, that small ones cannot aspire to. To become large, organisms must become 'multicellular'. Creatures like oak trees and us have trillions of body cells.

Multicellular organisms must originally have arisen from single-celled organisms. At its simplest, a multicellular 'organism' is little more than a collection of cells that have divided, but failed to separate. The real transition comes about when the different cells in the bunch begin to take on specialist functions – some producing gametes, some not; some photosynthesizing, some not; and so on. Then we see real division of labour, and real teamwork. Then you have what the great English biologist John Maynard Smith was wont to call a 'proper' organism, with each cell dependent on all the rest, and groups of cells cooperating to form organs, such as lungs and livers, or leaves and flowers. This degree of collaboration requires enormous self-sacrifice: to be a member of a bona fide organism, each cell must give up some of its own ability to live by itself. Each cell has to 'trust' the others. Any cell in the organism that goes berserk and tries simply to do its own thing destroys the whole, and ultimately destroys itself. Such cells in medical circles are said to be cancerous.

In fact, there is a spectrum of compromise positions between cells that can live perfectly well by themselves (as single-celled organisms), and cells that are utterly dependent on those around them (like human brain cells). Thus many cells from many organisms (including many of ours) can be grown indefinitely in special cultures. Many cells from many plants, once cultured, can then be coaxed to develop into whole new organisms. Indeed, many plants (including many of the most valued trees, such as coconuts and teak) are now cloned by cell culture. On the whole, though, the generalization applies. True multicellularity is possible only because the individual cells give up their autonomy, each relying on the rest for its survival, and for the replication of its genes.

TRANSFORMATION 5: PLANTS COME ON TO LAND

The first plants which can loosely be called 'algae' ventured on to land around 450 million years ago. On land they faced, for the first time, the problems of gravity and desiccation. Some of the earliest algal pioneers evolved into mosses, liverworts and hornworts, known collectively as 'bryophytes'.

None of the bryophytes has ever come properly to terms with the special difficulties posed by life on land. They duck the issue of gravity by staying squat and small, and hence extremely lightweight. They never solved the desiccation problem. They remain confined to damp places – but because there are plenty of damp places, they are extremely successful. Mosses in particular abound on damp walls and rocks just about everywhere. They are a huge presence in forests, as epiphytes. Some, particularly the sphagnum or peat mosses, form vast swards in the wet tundra and tend to prevent other plants from growing there. Mosses in general overcome desiccation not by resisting it, as a leathery-leaved holly tree or a spongy, water-packed baobab will do, but by putting up with it. They can be dried to a virtual crisp, and yet spring back to life.

An aside is called for on the reproduction of mosses – for they illustrate one of the fundamental phenomena of botany, and without some inkling of it, we cannot properly understand the reproduction of the plants that mainly concern us in this book: the conifers and flowering plants. The phenomenon is known as 'alternation of generations'. The moss that is a permanent presence on walls and tree trunks is called the 'gametophyte generation' because it produces eggs and sperm (gametes) which fuse to produce embryos, which grow into the 'sporophyte generation'. (It is odd to think of plants producing eggs and sperm – but that is what the primitive types do). The sporophytes appear among the general background of 'leafy' moss as little upright structures that commonly resemble tiny lamp-posts – the 'lamps' at the top contain spores. Spores are little more than packets of unspecialized cells, encased in some protective coating. They are dispersed by various means (not least by water), and if they land in some comfortably damp spot, they multiply and differentiate to produce new mosses of the gametophyte type. The sporophytes, which produce the spores, cannot live independently. They depend entirely upon the gametophyte.

Thus the gametophyte practises sexual reproduction, while the sporophyte practises asexual reproduction. Both ways of reproducing have their advantages and drawbacks – and plants practise both, in alternate generations. In this they are ahead of us. We (together with most but not all large animals) reproduce only by sex.

Bryophytes could never have given rise to trees. Their overall body structure is too simple. They have no proper roots, merely anchoring themselves by projecting 'rhizoids', which have no special role in absorbing nutrients and water. Most mosses look as if they have leaves, but they are not true leaves: just green scales. Most importantly, bryophytes have no proper, specialist conducting tissue within them, to fast-track water and nutrients from one part of the plant to another (or at best they have very rudimentary conducting tissue). Lacking specialist plumbing, they are bound to remain small.

Evolutionarily speaking, bryophytes may be seen as a dead end. The ancestors of modern trees are not to be found among their ranks.

All Land Plants

```
                          ┌─── liverworts ─┐
              ┌───────────┼─── hornworts   ├── bryophytes
              │           └─── mosses ─────┘
 all land ────┤
 plants       │           ┌─── lycophytes
              └── trachaeophytes ─┤
                                  │           ┌─── ferns and horsetails
                                  │           │
                                  └── euphyllophytes ─┤      ┌─── cycads ─┐
                                                      │      ├─── ginkgo  │
                                                      │      │            ├── gymnosperms
                                                      │      ├─── conifers │
                                                      │      └─── gnetales ┘
                                                      │
                                                      └─── angiosperms
```

This shows the presumed relationship between all the groups of *living* land plants. Some extinct groups (notably the Bennettitales) are left out. In the old days, liverworts, hornworts and mosses were thought to form a single, coherent group that was formally called Bryophyta. But in truth, nobody knows the relationship between the three, or whether they are closely related at all. They are best treated as separate groups, although they can still be referred to informally as 'bryophytes'. The Trachaeophyta – the plants with specialist conducting tissue (internal plumbing) – are a coherent group, a true clade. Note however that the bryophytes are not the ancestors of the trachaeophytes. Bryophtes and trachaeophytes are shown here simply as 'sister groups'

The option of being big was left to other lineages, which did develop plumbing.

TRANSFORMATION 6: PLANTS WITH 'VESSELS' – AND THE FIRST STIRRINGS OF WOOD

Some time around 420 million years ago, in the late Silurian, other groups of land plants emerged that did solve the problems of being big. These were the first 'vascular plants', with columns of cells that act as conducting vessels, providing them with a plumbing system – comparable with the bloodstream of animals – which allows them to grow big, and for different parts of them to become specialized without losing touch with one another.

The early vascular plants also invented lignin. Chemically speaking, lignin is not spectacular. It is a fairly small molecule, but it serves to toughen the cell walls of plants, which are made of cellulose. Pure cellulose is flexible – it is the stuff of cotton – but cellulose spiked with lignin is tough and hard. Lignin, in short, is what turns floppy cellulose into wood. Plants that lack lignin (or have only small amounts) are called 'herbs'. They can grow fairly tall, like the stems of tulips, say. They can stay upright because each of their cells is filled with water under pressure, and this water-pressure ('turgor') gives them resilience, like a well-inflated football. But such plants wilt when their water supply fails. Plants with lignin to help them can outride dry periods, and can grow far bigger than any herb. Many herbs have some lignin that toughens them here and there, yet they remain primarily herby. Bona fide wood requires special architecture – the lignin-toughened cells meticulously stacked and interlaced. With lignin *and* appropriate architecture, then truly we have wood. Although we may admit bananas as honorary trees for the purposes of discussion, in truth it is wood that makes trees. In practice, it is mainly the cells of the conducting vessels that become lignified, and they and their surrounding, supporting cells are the main stuff of timber. Creatures like us have a blood supply to carry water and nutrients around the body, and a separate skeleton to keep us upright. The woody plumbing system of trees serves both purposes.

Full-blown treedom, though, took a long time to achieve. The very earliest vascular plants were little bigger than matchsticks (and not as stiff), as they emerged from swamps. Among the oldest of them are the rhyniophytes, (named after the Scottish village of Rhynie where their fossils were first discovered), which date from around 420 million years ago. They and their various successors have long gone, but shortly after they first appeared, one of their number gave rise to the two great lineages that are still with us today, which between them include *all* the living plants that are larger than moss. Both of these lineages, quite independently, invented the form of the tree – and one of them at least reinvented the tree form several times.

The Two Great Lineages of Big Land Plants

The first of these two great lineages are the lycophytes (Lycophyta). The surviving types are small – club mosses, *Selaginella* (also moss-like) and quillworts (which look like sprouting onions). But in the deep past, spanning the Carboniferous period and lasting well into the Permian (from about 360 million years ago to around 270 million years ago), the lycophytes produced a range of forest trees. Their architecture was primitive: their roots and branches divided simply, each into two equal parts, like a 'Y'. But some of these ancient trees were magnificent. *Lepidodendron* could be up to 40 metres high; as tall as most modern forest giants, and as high as a twelve-storey building. The straight columnar trunks of *Lepidodendron*, patterned all the way up with leaf scars shaped like diamonds, could be two metres across at the base. They formed great swampy forests. Among the strange animals that roamed within them were eurypterids, like giant scorpions – some aquatic, some land-bound, and some more than two metres long; the size of a small rowing boat. The ecology of those lycophyte forests was doubtless as intricate as that of modern forests, and doubtless was played out by much the same rules – and yet the cast list of players was utterly different. Some of those early forest creatures have left descendants but others (including the eurypterids) have not. They have had their hour upon the stage.

So it was among the lycophytes, plants that are now known only to botanists as also-rans, that some of the world's first trees emerged – perhaps the very first – and some of them were magnificent. Yet, like the bryophytes, the lycophytes lack true leaves. In lycophytes, the organs that resemble leaves are really just scales. It was left to the second great group of vascular plants to invent true leaves. These were, and are, the euphyllophytes ('good leaf plants'), which contain all our living trees. The euphyllophytes, like the lycophytes, had a magnificent past. Unlike the lycophytes, they also have a magnificent present.

The earliest euphyllophytes, like the bryophytes and the lycophytes, continued to reproduce sexually by means of eggs and sperm, and asexually by means of spores. But somewhere around 400 million years ago (by now in the Devonian) the euphyllophytes divided again into two great groups. One group, now known as the monilophytes, continued to reproduce in the traditional way – a generation that produces eggs and sperm, then an alternate generation that produces spores. The other group gave rise to the spermatophytes – the group that reproduce by seeds. Both groups independently gave rise to trees – and indeed in both groups the form of the tree arose several times.

The monilophytes include present-day ferns and horsetails. Ferns nowadays are hugely various, and include many tree-like forms: 'tree ferns' form significant forests in much of the tropics and subtropics (which I have been privileged to walk among in New South Wales – a must for all connoisseurs). More accessibly, they also turn up in botanic gardens throughout the world – even in England (as in Cornwall's Lost Gardens of Heligan).

Present-day horsetails are modest plants that are often to be found on waste ground, where they resemble the swagger sticks that indeed are carried with considerable swagger by sergeant-majors, but with rings of needle leaves at intervals along them, like tutus. Their stems have ridges, like Ionic columns, and along the ridges are spicules of silica. In earlier times, when people made use of whatever grew, horsetails made excellent pan scourers. Only about fifteen species are known, all placed in the single genus *Equisetum*, but in Carboniferous times in particular some of the horsetails grew into fine trees. *Calamites* is among the best known. It could be up to 10 metres tall, shaped like a torch, with a straight thick stem and a crown pointed like a flame. *Calamites* grew like irises – or indeed like modern horsetails – from thick, creeping underground stems (known as 'rhizomes').

So the monilophytes invented the form of the tree at least twice – tree ferns and tree horsetails. Only one group of spore-bearing trees, the tree ferns, is still with us, but we should be very grateful to the extinct types – *Calamites* and *Lepidodendron* and their relatives. In fossil form, the horsetail and lycophyte trees formed much of the coal that gave rise to the Industrial Revolution. Indeed, it was mining that made them known to the world. Worldwide, in the deep past, those spore-bearers were very significant players.

But now we will put them, and the tree ferns, to one side. It was left to the seed-bearers to produce the world's grandest trees in the greatest variety, and they must dominate the rest of this book.

TRANSFORMATION 7: PLANTS WITH SEEDS

A little more than 360 million years ago, in the late Devonian, there appeared the first plants that reproduced, not by spores, but by seeds. Seeds were, and are, a marvellous innovation. Spores obviously do a good job. The plants that make use of them include many that were and are hugely successful. But although it has become politically correct to argue that there is no progress in evolution, there very clearly is, of a technological kind; and seeds, beyond doubt, are a technological improvement. Spores are little more than groups of relatively undifferentiated cells wrapped in a protective coating, light enough to be carried away by wind or water. Unless they land somewhere very favourable indeed (and in particular very damp), they perish. Spores are like children setting out on a wild adventure with nothing but high spirits and a bag of toffees. Seeds, by contrast, contain embryos that have already developed significantly while still attached to the parent plant, and are equipped with a food store of carbohydrate, protein and fat. The embryo and its attendant hamper is encased within a coat (a 'testa') that is custom-built for the circumstances that are liable to be met, and commonly contains (chemical) instructions on when to germinate (sometimes, both in trees and herbs, including devices to delay germination for several years, for not every season is favourable). To continue the metaphor, seeds are like commandos, beautifully equipped with iron rations – in some cases able to grow for weeks after germinating, before receiving any fresh nutrient from outside – and with a well-worked-out survival strategy to boot (the strategy being encoded within their DNA).

There is one final subtlety: alternation of generations. This occurs not only in mosses, but in *all* plants. In ferns and horsetails, the plant you see all the time is the sporophyte, the generation that produces the spores. The spores then germinate to produce a small gametophyte (which typically resembles a liverwort), where sexual exchange takes place, producing a new sporophyte generation (a new fern or horsetail).

In seed plants too the main plant is the sporophyte, but instead of spores it produces small collections of cells which represent the entire gametophyte generation. In the male flower (or the male part of a hermaphrodite flower) this collection of cells is contained within a protected package, the pollen. The pollen is then carried to the female flower by wind, animals or water. The female gametophyte remains within the ovary and manifests as the ovule. I like the whimsical notion that since pollen contains the entire male gametophyte it is, botanically speaking, flying moss.

So that's it. By the time we have seed plants, all the transformations required to take us from inchoate clouds of noxious gases to plants that can manifest as oaks and redwoods have taken place. There were many refinements still to come, including the evolution of flowers. But the basics were in place at least 150 million years before the time of the first dinosaurs. Such antiquity is hard to comprehend; yet, botanically speaking, it was the beginning of modernity.

Many lineages of seed plants have appeared during that long, long time. Most are long extinct. But five are still with us. Two of them – the conifers and the flowering plants – dominate the terrestrial ecosystems, and account for at least 99 per cent of all trees. These two occupy all of the rest of this book. But the other three remaining lineages also contain trees, including some highly attractive and sometimes magnificent ones. They deserve passing mention.

CYCADS, THE GINKGO AND THE MYSTERIOUS GNETALES: THREE NOBLE ALSO-RANS

Of the five remaining lineages of seed-bearing plants the most ancient is that of the cycads – the Cycadales. Beyond doubt you must have seen them on your travels – although you may have mistaken them for something else. Some have thick wooden trunks like giant woody pineapples, with a mop of spiky dark green leaves at the top. Others have somewhat more cylindrical trunks, and superficially resemble palms. They are widely cultivated for their exotic beauty in warm countries.

Another way to be a tree: cycads look like palms but are quite different

The cycads first came into being around 270 million years ago in the early Permian, the age just before the dinosaurs appeared. But they became most various and abundant in dinosaur times, and were doubtless staple dinosaur fare. About 130 species are left to us.[1] They have many unusual features. For one thing they have spherical seeds – often large, and with a fleshy, coloured coat. Individual cycads are either male or female (known as 'dioecious'). Their reproductive apparatus is neither a cone like a conifer's, nor a flower. It is a 'strobilus'. The female strobili bear the seeds, and the males bear pollen. Male or female, the strobilus is often very large, like the head of a drum major's mace, and sometimes brightly coloured. Strobili function as flowers, but they are not homologous with them: they are a separate invention. Like flowering plants, present-day cycads employ insects to effect pollination and various animals to help scatter their seeds. Indeed, the first ever symbiosis between plants and insect pollin-ators was probably between cycads and beetles; and the flowering plants, which evolved later and independently, would have cashed in on the beetles that had already evolved to service cycads. In nature, one thing leads to another. Evolution is opportunistic and everything builds on what was there before.

The pollen of cycads is odd. It invades the would-be seed by sending out a multitude of 'roots' like a fungus; and then at the end of this invasion, and quite unlike a conifer or a flowering plant, the pollen produces a giant sperm – a sperm with many tails. Both these features may be primitive – they possibly represent the way that very early seed plants in general conducted their affairs.

You cannot miss cycads, as you stride along the avenues and prom-enades of Florida or California or Spain – or at least you could, if you did mistake them for palms. It is worth looking closely. It would be a shame to miss out on life's subtleties.

Second most ancient of the surviving seed-plants are the Ginkgoales, which first appeared in the fossil record around 260 million years ago again in the Permian. In the past there have been many species, which were highly various. But only one is left to us: the ginkgo or maiden-hair tree (*Ginkgo biloba*) with its curious and absolutely character-istic, half-moon-shaped leaves. The ginkgo too may be extinct in the wild, but human beings cosset and cultivate it for its physical beauty and curiousness – around temples in China, and in gardens, parks and avenues in all the temperate world. The outer layer of the skin that surrounds its seeds is fleshy and smelly, and the Chinese gather the

Ginkgoes too were once diverse. Now only one species remains

seeds for cooking (as indeed they were doing in New York's Central Park, the last time I was there).

It is lucky that ginkgoes are so quaint: humanity has driven many other, less striking trees to extinction. Peter Raven, director of the Missouri Botanical Gardens in St Louis, and one of botany's most original thinkers, says that if you want to save a plant from extinction, you should put it into the horticultural trade: and the ginkgo is a case in point. This option does not work quite so well for animals. No one could give a satisfactory home to a blue whale.

The third of the five remaining groups of seed-plants is, or are, the Gnetales. They are, taken all in all, seriously weird. The whole group contains only about seventy living species in three genera – which look nothing like each other to the untrained eye but seem, nonetheless, to be truly related. One is *Welwitschia*, which grows in the extremely dry coastal desert of Angola, Namibia and South Africa. Most of the plant stays buried in the sand. The bit that shows is a massive, woody, concave disc bearing two enormous leaves that are never shed and never stop growing, are ragged at the ends with wear and tear, and look permanently dead. *Welwitschia* is a wan creature: botany's answer to A. A. Milne's Eeyore. But, like Eeyore, it endures, and indeed does quite well. The remaining Gnetales belong to the genus *Ephedra* (mostly shrubs, highly branched, with small inconspicuous scaly leaves superficially like horsetails) or to *Gnetum* (some of which are vines – but others of which are trees, with big leathery leaves).

The Gnetales taken all in all are now a minor group (if very curious) and so far as can be seen, they always have been. While other kinds of plants took over vast stretches of the globe, the Gnetales just jogged along.

This leaves just two main groups of seed plants, which between them contain well over 99 per cent of all trees. Before we survey them all (Part II) and look at the ways they live (Part III), we should look more closely at the wondrous material that they have in common and which enables them to grow so big, and live so long, and has enabled them to occupy a third of all the world's land: wood.

4

Wood

A young yew – with perhaps another 2,000 years to live

If all the greatest aesthetes and engineers that ever lived were assembled in some heavenly workshop and commissioned to devise a material with the strength, versatility and beauty of wood I suggest they would fall far short. Wood is one of the wonders of the universe. Of course, human architects create structures that are bigger than any tree and sometimes, like the great cathedrals and mosques, are of great beauty. But a cathedral or a mosque is built: it does not grow.

Until it is complete it is useless, and probably unstable. It must be held up by scaffold. When it is finished it remains as it was made for as long as it lasts – or until some later architect designs it afresh, and rebuilds.

A tree by contrast may grow to be tall as a church and yet must be fully functional (apart, perhaps, for the business of reproducing) from the moment it germinates. It must fashion and refashion itself as it grows, for as it increases in size so the stresses alter – the tension and compression on each part. To achieve hugeness and yet be self-building – no scaffold or outside agencies required – and to operate for good measure as an independent living creature through all phases of growth (first as seedling, then sapling, then young tree, then mature tree) is beyond anything that human engineers have achieved. After the tree is cut we see that the wood, of course no longer increasing in size, is the ultimate composite: remarkably complex chemistry (cellulose, lignin, tannins, resins, and often much else besides), minutely structured for maximum strength and functionality; lovely to look upon; and infinitely various. Great human craftspeople from Grinling Gibbons to Henry Moore can create artefacts to show off wood at its best. But the wood itself, on which they work their creativity, is nature's invention.

The terms 'wood' or 'timber', like the word 'tree', tend to be used in different ways. Some define wood loosely, and some more narrowly. Loosely, the word refers to the hard skeletons of conifers and flowering (angiosperm) trees. But some botanists and foresters don't like to think of monocot wood – the kind that comes from palms and bamboo and so on – as 'true' wood because it has a quite different structure. For them, true wood comes only from conifers and broadleaved angiosperms – the 'broadleaves' being all the dicots, from magnolia to oak and teak.

This description focuses on 'true' wood: as in conifers and broadleaves. In both cases, the basic component of the wood, which makes it both functional and strong, is or are the conducting tissues, the basic plumbing. These tissues are of two main kinds. On the inside is the xylem: a mass of tubes that carry water with dissolved minerals up from the roots to the leaves. In broadleaves, most of the xylem tubes are open all the way along; but in conifers they are interrupted by perforated plates (and this is the chief difference between the two types). The second group of conducting tissues form the phloem: strings of cells that carry the products of photosynthesis (organic materials of various kinds, which basically are variations on a theme of sugar) out from the leaves, downwards and outwards to the rest of the plant. The tissues of the phloem are on the outside. Collectively the phloem forms a cylinder, enclosing the solid column of xylem within.

So all in all, you can imagine wood as a close-packed bundle of straws, bound tightly together into a solid whole. But now add one more element to the image. Imagine that swords were thrust into the bundle from the outside, slicing between the bundles – running from the outside towards the middle. The 'medullary rays' run in just this fashion from the centre to the outside. These blades of tissue provide some linkage between the different elements of the xylem and phloem, and also act as a food store for the whole trunk. By carrying material inwards and outwards, the rays enable the trunk to increase in diameter as the tree grows. More generally, they help to ensure that the trunk is itself a larder, to be drawn on as required.

But where does the growth come from? How can the trunk increase in thickness and yet be continuously functional? Here is where the subtlety really begins. Between the xylem and the phloem is a thin layer of tissue known as cambium, which forms a sheath, running from roots to leaves. The cambium is stem-cell tissue; the kind whose job it is to generate more tissue. It generates more xylem vessels on the inside, and more phloem vessels on the outside. So the tree grows thicker year by year – and yet the trunk is always functional. Always, fresh xylem and phloem are coming online. Herbaceous plants and young trees, of course, have some thickness to them from the outset. A tomato stalk grows thicker as the season wears on – more and more cells are produced, all puffed up by water pressure within the cells. But only conifers and broadleaves have the complete sheath of cambium, not far from the surface, that allows the tree to go on getting thicker year by year, perhaps for centuries. This is the phenomenon of 'secondary thickening'. Other trees that are *not* conifers or broadleaves may practise secondary thickening up to a point. Cycads do. The lycophyte tree, *Lepidodendron*, apparently did. (Palms don't. In general, they begin life short and fat and stay at the same thickness until they are 20 or so metres tall.) But no trees apart from conifers and broadleaves have taken secondary thickening to such a peak. It is the final requirement and accomplishment of true treedom (at least up to now).

The cells that form the tubes of the xylem soon die. In fact, in order to become fully functional, they *need* to die. They lose their living cytoplasm: all that is left is the cell wall, cellulose stiffened with lignin. However, as time passes, the cells both of the xylem and phloem not only die, but they lose their function as conducting tissue. Clearly, in any one tree-trunk, the xylem closest to the centre is the oldest: it may have been laid down ten, a hundred, even a thousand years earlier. But xylem that is more than a decade or so old tends to be increasingly blocked, not least with tannins. So the centre of a tree becomes increasingly solid. Not only are the individual cells dead, the whole structure loses its ability to transport water. Phloem is the mirror-image of xylem: its oldest vessels are on the outside, and they become crushed as new phloem tissue is laid down inside them.

But although their days as conducting tissues are over, the very dead xylem in the core of the tree, and the crushed phloem on the outside, do not cease to be functional. The very dead, commonly tannin-soaked xylem within becomes the 'heartwood'; and the newer xylem outside it, still serving as plumbing, forms the 'sapwood' (because indeed it is full of sap). The heartwood truly provides the skeleton of the tree; it is what enables it to become big. The crushed phloem, outside, becomes incorporated into the bark, providing essential protection. 'Bark' in general means everything that lies outside the cambium: the inside layers consist of living phloem, but the layers beyond that are dead. We get a hint of the life that lies just beneath the surface of the tree through the phenomenon of 'cauliflory': the way in which many tropical trees in particular, including cocoa, produce flowers and then fruit straight from the trunk or biggest branches.

In trees that grow seasonally, the addition of xylem and phloem is intermittent. In a typical temperate tree, the new xylem laid down in spring is wide but thin-walled; while the summer xylem is narrower but thicker-walled. The differences can be clear to see, and result in a series of concentric 'growth rings'. Typically there is one growth ring per season, and so the age of the tree can be gauged. In good growing years the growth rings are broad. In bad growing years, they are close together. Thus, knowing the age of the trees, it is possible to work out the climate of past years. If we cut a mature tree in, say, 2004 we can see what the weather must have been like in, say, the 1850s. Some growth rings might be particularly far apart, and some particularly close together. If we have a piece of timber we know was cut sometime in the late nineteenth century, but we don't quite know when, we can see which of its growth rings correspond in width to the ones of the tree felled in 2004 – and which, therefore, correspond to the 1850s. We can then count back and work out when the tree was planted. Then we can overlap that older tree with one that is even older, and so on back. This is the principle of dendrochronology – judging past climates, and the general ages of things, by examining the growth rings of successively older trees. Dendrochronology has provided some remarkable insights in archaeology. (Tropical trees in places where there is a distinct wet and dry season also show growth rings. Trees where the climate is constant do not.)

Many trees have a layer of secondary cambium, outside the principal cambium layer, with the specific job of producing cork. Cork cells (like xylem cells) are born to die: they finish up small, with thick, impermeable cells walls. Cork is wonderful material: it is light; waterproof (hence preventing excessive water loss); it helps to repel pests; and it is relatively fireproof. All trees have some corky cells in their bark, and some have a great deal of it. Trees that are most likely to be exposed to fire tend to have the thickest cork – like, of course, the beautiful cork oaks (*Quercus suber*) of the Mediterranean, and the baobabs of Madagascar, Africa and Australia (which are also used for cork). The one snag from the tree's point of view is that cork is also air-proof, and prevents exchange of gasses. But it tends to be interrupted by passages of only loosely bundled cells, known as lenticels, which let air through.

Bark too, of course, compounded from formerly functional phloem and custom-built cork, is highly evolved and adapted. Of course, much of the variation is not explicable in terms of function; it just is the way things have turned out. It can be used (by experts) to identify species, just as the pattern of the timber itself can. But bark does have many adaptive features. Some, for example, is highly impregnated with tannins to repel pests. The bark of redwood trees is not corky like cork oak, but is fireproof nonetheless: fibrous, and up to 30 centimetres thick. Others, like *Enterolobium ellipticum* (it has no common English name), which must endure periodic fires on the dry forest of the Cerrado in Brazil, have huge ridges of corky bark. I suspect the ridges help to create an up-draught which carries the heat up and away from the trunk.

Many trees shed their bark, sometimes in great swathes, which can be helpful in various ways. Some (especially tropical forest trees) seem to shed it in an attempt to get rid of epiphytes, which can grow in great abundance on their trunks and branches and so weigh the tree down and block its light. The bark of eucalyptus is rich in oils and resins and burns quickly and fiercely. Oddly, this is an anti-fire device. The bark is shed, commonly in shreds, and builds up around the tree as litter. Other plants find it difficult to grow through the chemically rich, dark brew, and so there may be little or no undergrowth. When the bushfires rage they race quickly through the oily, resiny tinder on the ground – and a quick, hot flame is far less damaging than a cooler but slower one. The bark beneath the wisps that are shed is smooth and iron hard, difficult for fire to take hold in. By shedding their bark, London plane trees shrug off the polluting city soot, so they do well as urban trees. This cannot have been an adaptation – the parent species of this hybrid evolved long before cities did – but it is a good example of 'pre-adapation': a feature that evolved earlier in some other circumstance, coming by chance into its own.

Clearly, different species produce different timbers. Some are very light and fast-growing. Some are very dense and on the whole tend to grow more slowly. Quite a few are heavier than water, such as lignum vitae and *Olea laurina* (a heavy-timbered olive). Some timbers are black, some creamy white, some yellow, and some distinctly red.

These are the broad differences. To some extent they seem easily explicable. For instance, pioneer trees – those that invade newly available space quickly – need to grow fast. But they are soon likely to be overtaken by other trees and will then be overwhelmed, so they do not need to be strong enough to endure for a long time. So their timber, typically, is strong and light. A classic pioneer of this type is

Cecropia, whose big, silvery, horse-chestnut-like leaves are such a feature of tropical forest that has been opened up by storms or logging. But nature cannot be second-guessed – we cannot assume that it will always follow our logic. So it is that some pioneer trees endure the later invasions of other species and are extremely long-lived – like redwoods; and some are not only long-lived but also have very hard timber – like mahogany. The baobab tree of Madagascar (and Africa and Australia) on the other hand has extremely soft wood, like a classic pioneer, but commonly lives for 500 years or more. Many other trees begin life in the shade as part of the understorey, grow slowly up to the canopy (or wait for a gap to open) and then endure perhaps for centuries. Their wood is likely to be dense and strong, to enable them to live a long time. Some long-lived trees bend with the wind; others outface it: in Britain, the flexible ash and the resilient oak have become symbols of different life strategies. But other differences – including, perhaps, colour – seem mostly down to chance. The prime requirement is to produce an organism that works. Many of the genes will have odd effects in addition to the ones that seriously contribute to survival. It is hard to see how it matters to a tree whether its timber is black or white or red or a pleasing buff, and the genes that influence colour may be doing something truly useful as well, for example repelling pests. Or they may not. Provided their side effects do no positive harm, then these genes will be passed on through the generations, whatever eccentricities they bring with them.

By the same token, different species have different patterns of grain and 'figure': grain being the narrow stripes that run along the length of the wood and appear when the growth rings are cut across, and figure being the general appearance, whether the growth rings are cut across or not. These variations represent differences in micro-architecture. Clearly, it is vital to a tree that its wood should be functional. Equally clearly, the very fine details of structure do not matter too much – especially in the heartwood, whose only tasks are to provide strength and bulk. We can imagine, then, that trees contain a variety of genes that in some way or other influence grain and figure – to the extent that experts can (usually, and at least in theory) identify the species of any kind of timber from these patterns. The small genetic variations that cause these differences do not matter to the tree.

There can be enormous variation between the different individuals of any one species, too, which again is partly genetic. Grain and figure may vary just like human fingerprints. There may be no specific benefit from such variation. But if there is no great natural selective pressure not to vary, then variations will creep in. But genomes are not commandments, which say exactly what to do come what may. Genes present options. They operate in dialogue with the environment. So the same tree, grown under different circumstances, could grow in very different ways; and the effects of the different circumstances are reflected in the timber.

Thus growing timber responds to stresses and strains and pressures just as the bone of mammals may do. A big horizontal branch puts enormous strain on the point of contact with the trunk. In broadleaf trees, such as oak, you will often see that the base of the branch, where it meets the trunk, is not round. It will be oval: the branch beneath is bolstered by 'compression wood', like a corbel in a cathedral holding up a beam. Conifers adopt the same idea – but use totally opposite physics. In conifers the extra reinforcement is *on top* of the big horizontal branch. The timber added above is 'tension wood': it is acting as a guy rope.

Around the base of tropical trees from many different families you commonly see buttress roots, which take many forms: commonly and bizarrely they are like the fins of a rocket ship, thin vertical triangles of timber protruding from the sides and sometimes extending upwards to 3 metres or more. They can be impressive structures. Yet they are not truly buttresses, for buttresses are under compression: the buttresses of cathedrals support the walls by pushing against them. The buttresses of tall tropical forest trees are again like guy ropes, under tension. More generally, a tree exposed to the wind somehow 'knows' that it is being shaken, and grows thicker.

Heartwood is usually very different from sapwood. Often heartwood is very good at resisting pressure – it has high 'crushing strength' – while the sapwood has high tensile strength. The archers of medieval England made their longbows from the timber of yew, from the particular part of the trunk where heartwood meets sapwood. The former is dark in colour and has great crushing strength; the latter is lighter, and very flexible. With the dark heartwood on the inside and the light sapwood on the outside, the yew bow gave tremendous spring. Result: a very powerful bow indeed. Indeed the English archers made short work of the French knights at Crécy in 1346, and again at Agincourt in 1415. You would think the French knights would have learnt, but apparently not. Unfortunately, the best yews for the purpose came from Spain, with whom the English, at least later, were also intermittently at war. However, 'total war' is a twentieth-century concept, and as late as the eighteenth century the great English navigator James Cook was able to replenish his ships at French-owned ports in the Pacific, even though we were (again) at war with France. So perhaps the English had less trouble buying Spanish yews than might be imagined. Business is business.

Sometimes there is internal tension in a tree that contributes to its strength, in the same way that steel under tension is sometimes used to reinforce concrete. Eucalyptus is often like this. When eucalypts are cut the tensed tissues within them are free to uncoil and the timber may split even as the tree is falling; and when a eucalyptus burns (which it will do when the flames are hot enough, even though eucalypts as a whole are fire-adapted) it may explode, both with the tension and because of the oils trapped in its timber.

The grain makes its way around the bases of branches growing from within; and the cut branch bases form the knots in wood. Some trees, including oaks and redwoods, produce anomalous masses of buds that come to nothing but persist to form burrs. Timber grows around the burrs and then its grain may be all over the place. The grain may go this way and that, too, around the bases of trees, and in parts of the buttress roots. Builders want straight-grained wood, for maximal strength and predictability. But makers of veneers, and turners, interested in decoration, love burr wood, and will pay hugely for it.

Trees grown in forests grow straight and tall, anxious for the light – which on the whole is how builders like it. Trees grown in open spaces may spread themselves like a Persian cat on a feather bed, and take all manner of wondrous forms. Thus the beeches of England's many fine woods tend to be straight and tall as towers while the pampered specimens in Kew Gardens, with no deer and horses to browse their lower branches and armies of gardeners over 200 years to keep competitors out of their light, are spherical as golf balls – albeit 20 metres or so in height. The oaks of ancient windswept Scottish hillsides commonly had bent branches – of particular use to ship-builders, who could fashion the keel and the prow around the natural shape.

Finally, the timber may vary in colour and figure depending on soil and even on infection. Wood may be coloured by minerals – blue or green by nickel, red or black by iron. Some trees, like the zebrawood *Microberlinia* (another of the family Fabaceae), are beautifully striped naturally. Others are striped with colour by fungal infections – red, black, whatever. Infections are not all bad. In the tulip frenzy of seventeenth-century Holland striped blooms were the most highly prized – and the stripes were caused by a virus. (Viruses were not identified as discrete organisms until the twentieth century. But the craftsmen–breeders of earlier centuries had a good working knowledge of disease and knew how to produce striped flowers to order.) Cheeses are beautifully veined by *Penicillium* and other fungi; and wine-makers speak of yeast as 'the noble rot'. The fungus that decorates it from within may rot the wood, to be sure – but when the fungus itself is killed off, it remains in colourfully suspended animation effectively for ever, and again, the results are highly prized by turners.

Thus wood is not only wonderful, it is also endlessly various. If humanity had only one kind of timber to draw upon it could think itself blessed (although we are an ungrateful lot). But in practice we have many thousands – a tree for every job, and for every decorative caprice. If timber is appropriately grown and selected, its qualities can be as tightly specified as steel, and it can be used for the most exacting tasks. Thus in the Second World War the British de Havilland company build enormous numbers of Mosquito light bombers, which incorporated ash, spruce, birch and balsa, each minutely specified. Timber still plays an enormous part in modern aircraft – and of course in ships and boats, even those with fibreglass hulls. Timber, too, could undoubtedly replace much of the steel now used in the biggest buildings – which surely would be friendlier to the planet, since it takes far less energy to prepare a beam of teak, say, than to make a girder. Furthermore, timber is composed primarily of the element carbon, which is derived from the atmosphere in the form of carbon dioxide. It therefore serves as a carbon 'sink': a wooden beam will lock up the carbon of which it is composed for as long as the building that it helps to form continues to stand. After the building has run its course, the timber can be recycled. The prestige buildings of the future, like those of the distant past, could with great advantage be constructed of timber.

On the other hand, if you want a fruit bowl or a bureau that is both unique and beautiful, and does not have to endure the strains of a plane or a boat or a tower, then there are endless capricious twists of pattern and colour to draw upon.

Yet all this benison is merely a bonus; for the real point of wood is to enable plants to grow big, and lift their photosynthesizing leaves high into the air and sunshine, yet keep them bathed in water drawn from the earth. Although we can grow trees, and some people can fashion them in many marvellous ways, we could not have designed such a material with such micro-architecture in a thousand years, or ten thousand, and indeed have not done so; for even our most remarkable modern synthetics do not begin to compete, in versatility and functionality and beauty, with what nature has provided. Such is the power of evolution.

So these are the generalizations. In the next four chapters, I want simply to wallow in the glories: an overview of all the trees (at least, the conifers and angiosperms) that nature has left us with.

II

All the Trees in the World

5

Trees Without Flowers:
The Conifers

Some living bristlecone pines are as old as all written history

Among the conifers are the world's tallest trees (California's coastal redwoods), the oldest (California's bristlecone pines), and some of the most drought-resistant (a cypress in the midst of the Sahara), while various species fill some of the vastest forests in the world's most extreme and dramatic landscapes. Yet the conifers that are left to us are, as botanists are wont to say, 'relicts'.* Conifers first appeared on earth nearly 300 million years ago, in the Permian, long before the dinosaurs, and their heyday lasted until at least 50 million years ago, well into the Tertiary, which zoologists chauvinistically call 'the age of mammals'; and so the earlier (but already ancient) types were browsed by diplodocus and iguanadons, while their descendants saw the world's first elephants and horses and cats, and the world's first squirrels and primates cavorted in their branches. Over all that time the conifers have given rise to scores of families, containing goodness knows how many genera and species. Now only eight families are left to us, with seventy genera – three-quarters of which have only five species apiece, or even fewer (and some are down to one). In all, only about 630 species of conifer are known. Doubtless there are many more still to be found, not least in the uplands of South-East Asia and Venezuela, but they are still vastly outnumbered by the 300,000 or so species of flowering plants. What's left, therefore, is but a shadow of what there has been: 'relicts' is the word. But what's left, nonetheless, is magnificent and endlessly intriguing.

In truth, beginning in the Cretaceous, about 100 million years ago, the conifers have been steadily upstaged by the angiosperms. All conifers are woody. Most are trees, although some are groundhuggers. None live as epiphytes; and just one makes a living as a parasite. There are about fifty times more species of flowering trees than of coniferous trees; yet most flowering plants are herbs, which between them have adopted every known form and way of life – climber, liana, annual, perennial, epiphyte, aquatic plant, and several thousand species that live as parasites.

Broadly speaking, conifers now flourish in conditions that flowering plants find especially difficult. Between them they can hack any kind of climate, from tropical to almost arctic. Of course they will not grow in extreme desert or at the poles – no tree can – but in Siberia the spruces are matched only by birches (sometimes) in the extreme north and in Canada the pines are rivalled (sometimes) only by aspens. On the whole conifers are excellent pioneers, invading soil that has

---

* The ideas in this chapter reflect those of Dr Aljos Farjon of the Royal Botanic Gardens, Kew, who chairs the World Conservation Union's Conifer Specialist Group.

been variously devastated and has not yet built up fertility. But on good or adequate soils and in reliable climates, where growing should be easy, conifers tend to be ousted by angiosperms. So there are no native conifers at all in the vast tropical forests of Central Africa and Amazonia. Yet conifers do thrive in highland tropical rainforest where conditions are somewhat less easy – clambering for example up the hillsides of South-East Asia.

In practice, in the wild, conifers come into their own where the soils are poor or badly drained, or where conditions are in other ways uncertain. Often they succeed by forming mutually helpful relationships with toadstool-like fungi (basidiomycetes); these invade the roots of the trees, but in a benign fashion, and hugely extend their absorptive powers. Such symbiotic associations are called 'mycorrhizae'. Broadleaves too form many such symbioses, but many conifers seem particularly adept at them. Conifers also often thrive in places that are particularly beset by fire. The cones of coastal redwoods and of many pines will not release their seeds unless the cones are first cooked in forest fire (though if it's too hot it burns them up completely).

In general, conifers are light-lovers – an odd thought, as you wander through the green shade of the redwood forest, or peer through the close-set boles of some spruce plantation, or contemplate the long dark winter months of spruce and pine in the sub-boreal forests of the Baltic, or the truly boreal forests of Alaska and Canada, Scandinavia and Russia. So in general they can do well in the company of angiosperms when they are tall enough to overshadow them, but not when they are overshadowed themselves. To this end they have a trick. Many can grow very tall very quickly. Thus one of the biggest of the living giant sequoias is called the 'Boole Tree' (another great tree with a personal name), and is thought to be around 3,000 years old. But when space was cleared around it, other sequoias leapt in – and within a hundred years were just as tall as the mighty Boole. If these newcomers are spared, they will spend the next few thousand years growing thicker.

This, too, is why the conifers of the far north tend to be tall and thin: the sun is always low in the sky so they get most of their light from the side. Their steeple shape (says Dr Farjon) is not primarily a way of keeping off the snow, as is often suggested. Conifers of lower latitudes, where the light comes from overhead, tend to be shorter and flat on top – like the lovely and characteristic stone or umbrella pines of southern Europe (*Pinus pinea*) which turn up in the background of Mediterranean paintings. Perhaps, too, conifers feature on tropical hillsides because the slope gives them a grandstand view of the sun (though as Dr Farjon points out, this would be the case only in the hours after dawn and before dusk). All generalizations are dangerous in biology, however. Some conifers do grow happily as understorey trees in the shade, including that sombre denizen of English graveyards, the yew, and the cypress-like *Thujopsis* of Japan, which grows slowly at first as an understorey tree until it overtops its neighbours.

Despite their limitations, conifers are a huge presence through most of the world – in the Americas, Eurasia, Australasia, and in the past, so the fossils show, in Antarctica. In the north, the greatest centres of conifer diversity are California, Mexico, a slice of eastern China that embraces Sichuan and Yunnan and extends up to the eastern Himalayas, Japan and Taiwan. Taiwan even has a genus named after it, the cypress relative *Taiwania*. In the southern hemisphere, the greatest diversity of conifers is not on the great southern continents but in New Caledonia: an island about the size of Wales or Massachusetts, slap in the midst of the South Pacific, halfway between Australia and Fiji. The websites for New Caledonia dwell primarily on its beaches and nightclubs. Some mention its unique wildlife, but none refers specifically to its fabulous trees (particularly its araucarias). Ah well.

India, however, is strangely deprived of wild, native conifers. Conifers grow very well in India – in plantations. But apart from a few Eurasian types on the Himalayas, the only living native is *Nageia wallichiana* of the southern hemisphere podocarp family in the Western Ghats in the south-west of the country. The reason might be historical. Ancient India was wiped clean about 60 million years ago by the huge Deccan volcanoes, which buried a great part of the subcontinent in lava. The angiosperms, by then well established, seem to have been the first to get back in to the devastated land (although this idea clearly does not chime well with the conifers' reputation as outstanding pioneers).

Conifers are also largely absent from oceanic islands – the kind that arise as volcanoes (such as Hawaii) as opposed to those that are fragments of continents (such as New Caledonia). At least, conifers may be found on volcanic islands that are close to continents, but the furthest from any continental shore is the juniper *Juniperus brevifolia*, on the Azores. The seeds of pines and most other living conifers are winged and wind-blown, and do not generally travel far over oceans. But juniper seed cones are fleshy and tasty (they are the 'berries' that flavour gin) and are eaten and dispersed by birds – which not only travel vast distances but can also control their landing, as wind-blown seeds cannot.

Conifer means 'cone-bearing'. All cones are either male or female: never hermaphrodite, as many flowers are. Many conifers bear both male and female cones on the same individual ('monoecious') while others (like yews and most podocarps) have only one sex per tree ('dioecious'). For conifers, reproduction is often a leisurely affair. Many weeks may pass between the transfer of pollen from the male cone to the female and actual fertilization, when the pollen tube grows into the ovule. The female cone, when fertilized, may take several years to mature. Some conifers shed their mature cones, and some retain them on the tree. In some, like the knobcone pine (*Pinus attenuata*) and the Monterey pine (*Pinus radiata*), the growing branches may envelop and eventually encase the old cones (which makes for some interesting patterns when the timber is sawn across, again much favoured by turners). Firs, pines, cedars and so on have the classical cones we all admire and many like to collect; but in others, like the yews, the lower part of the seed coat grows up around the developing seed to form a fleshy 'aril', superficially like a fruit. In junipers, the cone scales fuse and become succulent or pulpy, imitating a berry. In *Podocarpus*, the basal parts of the cone below the developing seed swell up to form a brightly coloured, fleshy receptacle. The fleshy 'fruits' of yews, junipers and podocarps could well have evolved early in conifer history, to be dispersed by birds and sometimes by mammals. Mammals are ancient, after all – dating from the Triassic, which again is pre-dinosaur; and birds date from the Jurassic.

In the timber trade, conifers are lumped together as 'softwoods', while the broadleaves are 'hardwoods'. This is rough and ready, for some conifer timbers (like yew) are far harder than some angiosperm timbers (like balsa) – but no conifer timber is as hard as the hardest broadleaves, some of which can be worked sharp as steel, like oak and mahogany, and some of which are too hard to be worked at all except with tungsten and diamond tools that are too expensive to be worth the trouble (some hardwoods are even spiked with silica, which makes them even more difficult).

WHO'S WHO AMONG THE CONIFERS

The classification of conifers has wavered a little these past few decades. A common traditional taxonomy recognized eight families: Araucariaceae, Cephalotaxaceae, Cupressaceae, Pinaceae, Podocarpaceae, Sciadopityaceae, Taxaceae, and Taxodiaceae. Most modern taxonomists, however (including Aljos Farjon) merge the Taxodiaceae with the Cupressaceae, reducing the list to seven. But some (including Dr Farjon) divide the Podocarpaceae into two – splitting off the 'celery pines' into the Phyllocladaceae. So now we are back to eight.

Note, in the following account, that three of the eight families contain only one genus, and only three families contain more than ten genera: the Cupressaceae with thirty, the Pinaceae with eleven, and the Podocarpaceae with eighteen. In all Dr Farjon now recognizes seventy genera – many with only one or a very few species. This is typical of ancient groups that have found just a few niches in the modern world – 'few' being a relative term that in practice means extensive and various.

Of the eight families, the Podocarpaceae live mainly in the southern hemisphere, and the Araucariaceae is exclusively southern. The rest are primarily or exclusively northern (although of course human beings have introduced trees from all families to just about everywhere). In the deep past all the world's landmasses were grouped in two vast supercontinents, Gondwana to the south and Laurasia to the north. So it is tempting to speculate that the Podocarpaceae and the Araucariaceae originated in Gondwana, and the other families in Laurasia. But their places of origin are far from certain. In the past, as we will see, there were araucarians in the north as well as the south

– the south is where they just happen to have survived. The modern cypress family, Cupressaceae, occurs all over the world. I *like* the idea that the Araucariaceae and Podocarpaceae are Gondwanan, and perhaps the Cupressaceae, and that the rest are Laurasian. But I have sometimes been forced to acknowledge that some of the things I would like to be the case, aren't.

The relationships between living conifers, and between the living and the extinct, are not easy to pin down. Morphology (structure) is the main guide to relationships, but the various conifers seem to have too few distinctive features, and sometimes the most important structures – like the cones of podocarps – are too reduced to make much sense of. Often it is hard to decide whether any one particular feature is 'shared derived', and signifies close relationship, or is merely 'primitive', and common to everybody. Molecular studies should help to clarify things, but as yet there seem to be too few. So it is not yet entirely clear whether the various conifer families, as now recognized, all form truly coherent groups (true clades); and it is far from clear how and to what extent the various recognized living families relate to each other. Thus some show the Pinaceae as outliers – the sister group of all the rest. But others group the Pinaceae with the Podocarpaceae and Taxaceae. I have described the families in no special order.

Kauris, the Monkey-puzzle and the Long-lost *Wollemia*: FAMILY ARAUCARIACEAE

The Araucariaceae were very various in dinosaur times (245 to 65 million years ago), and they lived all over the world. Now there are just forty-one species left, in three genera – *Agathis, Araucaria and Wollemia* – all in the southern hemisphere. The greatest variety of Araucariaceae is on the magical Pacific island of New Caledonia, perhaps the most pristine remaining fragment of ancient Gondwana.

Among the twenty-one species in the genus *Agathis* is one of the mightiest trees of all: *Agathis australis*, the kauri of New Zealand's North Island. The grandest of the grand is Tane Mahuta: it is 51.5 metres tall, its lowest branches are nearly 18 metres above the ground, and its trunk is 13.77 metres in girth – nearly 4.5 metres in diameter

– which means it would touch all four walls if planted in an average suburban living room. I have stood at the huge buttressed feet of Tane Mahuta. It is surrounded by other enormous trees, but it makes them seem ordinary. Its trunk rises out of the gloom like an iceberg in the Southern Ocean. The mass of epiphytes it holds aloft in its great spreading boughs is a fantastical, floating garden. It must have supported entire dynasties of lizards and invertebrates who never went anywhere else and must have thought, if they could think at all, that Tane Mahuta was the whole world.

Tane Mahuta is reckoned to be around 1,500 years old. Yet until 1886, when it was destroyed by fire, there lived another kauri known as Kairaru that was more than 20 metres in girth, and was thought to be at least 4,000 years old. Small wonder that the Maoris revered the kauri tree – ranking it second only to the totara, of which more later. They never felled one without a ceremony in which they asked forgiveness – although, as a Maori lawyer remarked to me somewhat wryly, 'They must have held an awful lot of ceremonies', for they used vast quantities of kauri timber for houses, boats and carving, chewed its gum and used it for starting fires: they felled whole forests. Fossil kauri gum is still there to be dug up: a form of amber. The Europeans were even more rapacious and reduced the kauri's range from around 1.2 million hectares to around 80,000. Now New Zealand is protecting its native trees. To walk through a natural New Zealand forest with its spooky battalions of understorey ferns, and with little fantail birds to lead the way is one of life's delights – and a very accessible one, since the New Zealanders have built so many easy paths, with wooden bridges to span the protruding roots and raised causeways to avoid crushing the wet places and for the kiwis to pass underneath. (Kiwis, flightless and indeed wingless, prowl by night. They hunt for worms and other invertebrates not by sight, as most birds do, but by scent.)

There are *Agathis* in Borneo, too, in tropical rainforest; and in New Caledonia there are five species, all endemic. By conifer standards *Agathis* is a fairly recent genus: the oldest date from about 65 million years ago.

The genus *Araucaria*, for which the whole family is named, has nineteen species of which no fewer than thirteen are endemic to New Caledonia – meaning they occur nowhere else. New Guinea has several

Araucaria, too, and so does Australia. Well known in Europe and America – though not too far north (except as an unusual houseplant) because it's tender – is the beautiful Norfolk Island 'pine', *Araucaria heterophylla*, that grows straight as a Christmas tree but with branches that curl up at the ends; much favoured in the prestige gardens of embassies and smart hotels in warm countries the world over. But the araucarias best known to northerners are the only two species that are native to South America. The monkey-puzzle tree (aka Chile pine), *Araucaria araucana*, from Chile and Argentina, was once a favourite in suburban gardens (and is still hanging on there; many only just coming in to their pomp). The leaves of the monkey-puzzle are leathery and spiky, and cling closely to the stems. Perhaps they do give monkeys pause for thought – but monkeys did not appear in South America until about 30 million years ago, and the genus *Araucaria*, apparently far older than *Agathis*, was around about 120 million years before America had any monkeys at all. Perhaps the monkey-puzzle had no thought of monkeys but evolved its daunting leaves to deter dinosaurs. (So, at least, some botanists have speculated. Certainly, trees are incomprehensible until and unless we consider their past.)

The monkey-puzzle has poor timber, but the Parana pine (*Araucaria angustifolia*) is much loved by do-it-yourselfers for its lovely variable colours, from smooth creamy white to chestnut brown and rich streaky red, tough but not too tough to work with tools of ordinary steel. The Parana pine grows mainly in Parana, Brazil, and also in Paraguay and Argentina, as a flat-topped tree up to 40 metres high with a straight clear bole around 1.2 metres in diameter. It is sometimes known as 'Brazilian pine', and is Brazil's chief timber export.

Wollemia, the third remaining genus of this once great family is the archetypal relict – to be ranked with the coelacanth, the ancient lobe-finned fish that was found in the ocean depths near Madagascar in the 1930s. For until 1994 *Wollemia* was known only from fossils, dating from 120 million years ago. Then a group of thirty or so turned up at the bottom of a canyon in the Blue Mountains of New South Wales, Australia. They were growing alongside flowering trees by a stream. They don't seem to be holding their own in the wild and must now be actively conserved. Finding *Wollemia* was not quite like finding *Tyrannosaurus rex*, but it is conceptually similar.

Nowdays there are no Araucariaceae in Africa, although the fossils show they were there in the past; or of course in Antarctica, where they once abounded; or in the northern hemisphere as a whole, though they were once widespread there. Truly they are a relict group, and we should be grateful for the survivors.

The Plum Yew and Other East Asians:
FAMILY CEPHALOTAXACEAE

Here is another archetypal relict family with only one genus – *Cephalotaxus* – and just eleven species. It grows as an understorey tree – a shade lover – in temperate forest on mountains, mixed with flowering trees, from the eastern Himalayas through China, Japan, Taiwan, Thailand, Vietnam and Malaysia. *Cephalotaxus* are vaguely yew-like to look at (and in less sophisticated days were sometimes classed with the yews). They are known to Western gardeners by a variety of names that include 'cow's-tail pine', because everything with dark-green needle leaves tends to be called a 'pine' sooner or later; and 'plum yew', because the female cones of *Cephalotaxus* give way to a single, soft-skinned seed that looks roughly like an olive – or an unripe yew 'berry'.

Cypresses, Junipers, Swamp Cypresses and Redwoods: FAMILY CUPRESSACEAE

The Cupressaceae is the only conifer family that occurs all over the world, in all continents (except Antarctica) and both hemispheres – it is impossible even reasonably to guess whether the family arose in the north (Laurasia) or the south (Gondwana). Cupressaceae also has the most genera of all living families of conifers – thirty – although not the most species (just 133). Yet its relict status shines through, for eighteen of those genera contain only one species. In each of them, the once possibly startling diversity hangs by a thread.

The family Cupressaceae has been extended in recent years. In its earlier form, it included only the cypresses, junipers, the Australian

Callitris and the thujas, which look like cypresses. But botanists had suspected for many a decade that there is no clear distinction between the Cupressaceae and the trees that were then placed in the Taxodiaceae – the swamp cypresses and the redwoods. Their cones are similar in significant details (not least in the way they develop), and the bark of a big cypress – thick, soft and stringy – is indistinguishable from that of a redwood.

Now it is clear that the old-style Taxodiaceae is not a coherent grouping. Really it's just a group of genera that look roughly similar because they share primitive features – not because they have any special, close relationship. In fact the various members of the Taxodiaceae are no more closely related to each other than some of them are to the old-style, narrowly defined Cupressaceae. The newly discovered *Metasequoia* seems particularly close to the cypresses. Thus the traditional Taxodiaceae family may be compared to reptiles or the bryophytes – not a true group (a clade) but a 'grade': a collection of creatures with similar general features. So the old-style Taxodiaceae are now combined with the old-style Cupressaceae to form a new, expanded Cupressaceae family. However, you will still find the traditional name 'Taxodiaceae' on labels in botanic gardens. These things take time to catch up.

Between them the trees of the newly expanded Cupressaceae family live in an extraordinary variety of places. Some, like *Chamaecyparis*, *Fitzroya*, *Sequoia* and *Thuja*, prefer very wet and tall forests on coasts. Others (*Chamaecyparis*, *Cupressus cashmeriana*, *Taiwania*) live in monsoon cloud forest high in mountains. Some (*Callitris*, *Juniperus*) thrive at the edge of deserts. One – *Cupressus dupreziana* – survives right in the heart of the Sahara where there is virtually no rainfall, drawing its water from a fossil aquifer far beneath the surface. Some junipers live at the edge of the Greenland ice cap, in permanent snow. The various trees in their various habitats are correspondingly various in form: from the very squat, like *Microbiota*, which hugs the ground for survival in the Russian Far East; to the giant redwood (*Sequoia sempervirens*), which basks in the mists of coastal California, and is the tallest tree of all. The Cupressaceae family also includes some of the world's oldest living organisms – in the genera *Fitzroya* and *Juniperus*. Like the pines the species of the cypress-redwood family are generally happy in poor soil – and some grow from crevices in rocks, apparently without soil at all.

The old-style Cupressaceae family included twenty-two genera. Most of them contain only very few species, and some only one. *Callitris* is one of only three genera in the family that contain more than ten species. In fact it has fifteen, of which thirteen live in Australia (two in Tasmania) and two (inevitably it seems) in New Caledonia. In general *Callitris* favour upland semi-aridity, and grow alongside eucalyptus as 'fire climax' species: the kind that thrive when everything else is burnt out. *Callitris preissii* (aka *C. robusta*) finds favour as a garden tree sometimes known as 'cypress pine'.

The sixteen species of *Cupressus* are the 'true cypresses'. Between them they span the northern hemisphere, happy on temperate, moist coasts or in desert or high mountain. In south-west North America one species extends down into Honduras. *Cupressus dupreziana*, as we have seen, hacks it out in the Sahara. The 'classic' cypress of the Mediterranean and the Middle East is *Cupressus sempervirens* ('always living'). Some *Cupressus* are native to the Himalayas and western China. But in the Mediterranean and Asia in particular it is hard to decide which are truly native, because so many have been moved and replanted since Roman times, or perhaps even earlier. *Calocedrus decurrens* from western North America has the nostalgic niff of school classrooms (at least as they were in my day), for its soft, fine-textured timber furnishes 75 per cent of the world's pencils.

The genus *Cupressus* should probably be extended to include the six species of *Chamaecyparis*; either that, or the two should first be combined and then split up again in new ways. The features that are commonly used to tell the two genera apart – notably the arrangement of the leaves – do not seem to carry enough weight, when taken with other characters. Species of *Chamaecyparis* grow in North America and eastern Asia as tall trees in temperate mixed or all-coniferous forests, from sea level to the mountains. Best known is the garden favourite *Chamaecyparis lawsoniana*, Lawson's cypress (aka 'Port Orford Cedar') of south-west Oregon and north-west California. In Britain it is often grown as a hedge. In America (though not in Britain) it is acknowledged as a fine timber tree for furniture, ships, oars, paddles for canoes, and church organs. Leyland cypress or leylandii (x *Cupressocyparis leylandii*), which grows so fast and casts the shadows that cause so much suburban strife, is a hybrid of *Chamaecyparis nootkatensis* (aka 'Alaska cedar', originally from the coast of north California and Alaska) and *Cupressus macrocarpus*. Apparently the hybrid first arose in a garden in Montgomeryshire, on the English–Welsh border, at the end of the nineteenth century. The fact that these two hybridized so readily is another reason for thinking that *Chamaecyparis* and *Cupressus* should not be treated as separate genera. But as a further complication, Aljos Farjon and others have proposed that *Chamaecyparis nootkatensis*, together with a newly discovered conifer from Vietnam, should be placed in a new genus, *Xanthocyparis*.

Fitzroya now contains only one species – the cypress-like *F. cupressoides* – but it's a great one. It is native to the coast and up into the foothills of the Andes of southern Chile and Argentina (a wonderful place to see trees) and given time, it becomes massive. But it needs a great deal of time: the oldest living *Fitzroya* that can be aged are at least 3,600 years old – but there are hollow ones, too, even bigger and undoubtedly older, although (because they are hollow) their age cannot be directly measured. Sometimes *Fitzroya* grows with its own kind, in groves, and sometimes intermingled with the southern beech, *Nothofagus*.

Juniperus is the biggest genus of the Cupressaceae: its fifty-three species account for nearly 40 per cent of the whole family. Apparently (like some groups of pines) the junipers have radiated to form many new species in (geologically) recent times. Some live to several thousand years. Between them they span the northern hemisphere – and *J. procera* is found south of the equator in east and south tropical Africa. They seem to tolerate almost anything from subarctic tundra to semi-desert, taking all forms from ground-hugging shrubs to tall trees: virtually all are drought-resistant. On mountains, some junipers grow to the topmost limit of the treeline. *J. brevifolia* is endemic to the Azores – and is the only conifer established on any mid-oceanic volcanic island. Its juicy 'berries' must have been taken there by birds. *J. communis* is the most widespread conifer species of all, and has even put in an appearance as one of Britain's three (possible) native conifers. *Juniperus* as defined here includes *Sabina*, a name that still

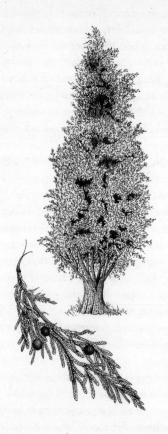

The most widespread of all conifers: one of the many junipers

features in many texts and may turn up on botanic labels but does not seem to be distinct enough to warrant generic status.

Thuja are long-lived and cypress-like. The grandest of all the five species is *Thuja plicata*, known confusingly as the western red cedar – the 'cedar' of the timber trade: fabulous all-weather wood that makes fine garden furniture and shingle roofs that will last a lifetime without further preservation. For good measure too its timber (as in most species of *Thuja*) is aromatic, and its leaves when crushed smell of pineapple. In its native Oregon, Washington and British Columbia, western red cedar grows to 60 metres, with great buttressed bases that can be 10 metres across. The native Americans of the north-west coast salvaged its long-dead trunks from the swamps and hollowed them into vast canoes and totem poles; and the Haida Indians of the shores of southern Alaska carved arrows from the stems, fashioned the tough knots into fish-hooks, and wove the fibres from the bark into ropes, baskets, mats, clothes and hats. In more humble guise, western red cedar features in suburban gardens, filling the same kind of role as Lawson's cypress. But the first *Thuja* to make it to Europe – to Paris, in the sixteenth century – was America's other native: the smaller *T. occidentalis* from the eastern states.

Three more species of *Thuja* live in north-east China, Korea, Japan and Taiwan. This is a common pattern of distribution: many trees of many kinds – including oaks – are distributed both in North America and East Asia, notably China. They seem for some reason to have found it easy to straddle the Pacific. In any case, trees do not respect political boundaries and indeed reveal how arbitrary the lines that we draw on maps really are. *Thuja* in general like it cool and moist, and grow from the coast to the hills. Though most are tall there is one, *T. koraiensis* of north-east China and Korea, that grows on exposed mountain ridges as a twisted shrub; showing again that any one group of organisms may essay a great variety of body forms. *T. sutchuensis* is one of those conifers that was presumed to be extinct but then turned up. To be sure, it didn't go missing for quite as long as *Metasequoia* or *Wollemia*, but it was thought to be long deceased until the late twentieth century when some were found alive and well in the Daban Shan mountains of northern China.

The genus *Thujopsis* has only one species, from Japan, which can look curiously like a plastic imitation of a cypress. It again favours cool moist places from the coast to the mountains, and is another conifer that grows slowly at first in shade but eventually overtops its neighbours. *Thujopsis* comes in various cultivated varieties and is much favoured in gardens.

The four species of *Widdringtonia* are among the few conifers that put in an appearance in sub-Saharan Africa: in South Africa, Malawi, Mozambique and Zimbabwe. *Widdringtonia* are fire-adapted: they do well when the surrounding shrubs are cleared by fire, as happens regularly in the African summer. The genus has been sadly depleted by felling, however. The forests of 'Mulanje cedar' on the steep slopes of Malawi's Mount Mulanje are the only substantial stands of *Widdringtonia* left to us. They are on my wish list of trees to be seen in the wild.

Two other genera of the old-style Cupressaceae are worth particular mention. Both have only one species each, both of which can grow into very large trees. They are *Platycladus* of east and north-east China, Korea and the Russian Far East; and *Taiwania*, which of course occurs in Taiwan and also in China (Yunnan), Myanmar (Burma), and Vietnam – where it was discovered only in this century. It grows up to 70 metres. The businesspeople and diplomats who flock to Taiwan probably feel they have little time for trees. A pity. Perhaps they should make time.

Of the eight genera that once formed the Taxodiaceae family, five have only one species each: *Cryptomeria*, *Glyptostrobus*, *Metasequoia*, *Sequoia* and *Sequoiadendron*. Of the other three genera, *Athrotaxis* has three species, while *Cunninghamia* and *Taxodium* have two apiece. This gives a grand total of twelve species. Here is a bunch of relicts indeed – magnificent, most of them – but not much left from a group that once bestrode the northern hemisphere.

Cryptomeria japonica, the 'Japanese cedar', known in the trade as 'sugi', accounts for much of the remaining forest in Japan. Yet those forests are probably not natural: rather, they represent the remains of some of the oldest forestry plantations in the world. The timber turns dark green when buried in the ground to give *jindai-sugi*, which serves as a semi-precious 'stone'.

Three genera of redwoods are left to us: the coastal redwood (*Sequoia*); the giant sequoia (*Sequoiadendron*); and the dawn redwood (*Metasequoia*). Truly the remaining redwoods are 'relicts', for 100 million years ago, when the climate was much milder and flowering plants were first coming into their own, there were a dozen more species of redwood throughout western North America, Europe and Asia. There was even one in Australia. Various species of *Metasequoia* were widespread in the Tertiary. In the Eocene, around 45 million years ago, when all the world was wonderfully warm, they grew far up in what is now Arctic Canada, only 10 to 15 degrees from the

Pole. But then the world began to cool in what has been called an 'icebox effect' – prompted by a steady diminution of atmospheric carbon dioxide – and by the time of the Pleistocene (around 2 million years ago), the genus almost went extinct. Now, only the dawn redwood is left to us: *Metasequoia glyptostroboides*. Even this was presumed to be extinct, until a few turned up in central China in the 1940s. It's hard to say exactly how the last dawn redwoods in the wild have been scratching a living since the time they went missing, because the area around their present habitat has been so cultivated this past few hundred years; but in general they seem to prefer the same kind of niche (wet) as the swamp cypress, *Taxodium*.

The genus *Sequoia* is also reduced to a single species: *S. sempervirens* – the coastal redwood of western, lowland California and Oregon. Yet the genus was once present in three continents (of which one was Australia). *S. sempervirens* needs the coast. It gets about a third of its water from the fogs that rise almost daily from the cold currents of the north Pacific and condense against its dark green leathery feathery leaves. Some coastal redwoods are managed for timber on a 100-year cycle: their rich reddish brown wood provides everything from telegraph poles to coffins and organ pipes while its fireproof bark, up to 20 centimetres thick, supplies fibre for fibreboard.

There is only one species of *Sequoiadendron* left to us, too: the giant sequoia (*S. giganteum*) that survives in groves on the western slopes of the Sierra Nevada in California, sometimes in pure stands and sometimes with other conifers. Like *Sequoia*, *Sequoiadendron* once included several species, widespread over North America. Like *Sequoia* too, it is adapted to fire, and in general regenerates only after fires, so that there are successive waves of new recruits, each following a fire. It grows quickly when young, soon overtops its rivals, and may live for thousands of years – not as tall as the coastal redwood, but sturdier. On the face of things *Sequoiadendron* seems a fine survivor. Yet it is down to one species, in only one small restricted area. Nature is unpredictable.

The old-style Taxodiaceae family takes its name from *Taxodium*, the genus of the swamp cypresses. They generally live as their name suggests in swamps and on the waterlogged fringes of lakes; and their roots have vertical outgrowths that project above the ground or the surface of the water which apparently act like the ventilator pipes on the decks of ocean liners, bringing air with its much-needed oxygen down to the roots. The broadleaved trees of the mangrove swamps have a comparable arrangement. In Mexico, however, swamp cypresses sometimes grow well above the water-table, sometimes in the company of broadleaved trees. Again, a once-various genus is now reduced to two species: *T. distichum* of the south-eastern United States, which is deciduous; and *T. mucronatum*, of Guatemala, Mexico and southern Texas, which is reported to be evergreen.

Finally, closely related to *Taxodium*, is its Asian equivalent *Glyptostrobus*: this time again reduced to a single species along streams and other damp places, including river deltas, in southern China and Indochina. Again we see the North American–Chinese connection.

Pines, Firs, Spruces, True Cedars, Larches and Hemlocks: FAMILY PINACEAE

Pinaceae is not the biggest family when measured in genera – a mere eleven – but it does have the most known species, with 225. This may simply be because of all the conifers, the Pinaceae are the best described. This is partly because the three biggest genera include the most economically valuable trees of all – the pines *Pinus*, the firs (*Abies*) and the spruces (*Picea*) – and partly because all but one species of Pinaceae are native to the northern hemisphere, where most scientists do their work. The only one of the family that has strayed south of the equator is *Pinus merkusii*, from northern Sumatra (note the similarity to junipers, which also have just one southern species). Even the fossils of the Pinaceae are exclusively northern. It is unclear, as we have seen, where most of the conifer families originated, but the Pinaceae family seems emphatically Laurasian.

The Pinaceae as a whole live in many habitats, but when it's very dry they tend to be replaced by Cupressaceae – all except some pines, which can take extreme aridity. Firs, larches, spruces and pines are among the most extreme northerners among trees – found along the northernmost treeline in Eurasia and North America, and climbing mountains as high as any tree will go. Sometimes they grow alongside broadleaves – especially pines, which are pioneers, and firs and hemlocks (*Tsuga*), which tend to creep in later.

The biggest natural forests of Pinaceae are in the extreme north – North America, Scandinavia, Russia. Yet these boreal forests contain very few different species. As with most groups of creatures, the greatest diversity is towards the equator, where the growing season is longer and there is abundant seasonal rain. There are also mountains towards the equator, which create boundaries between populations which, when thus isolated, evolve into new species.

In practice the various Pinaceae are focused mainly in four great centres of diversity (although we might with justice add a fifth). Two of these centres are in Asia, which has the greatest variety of genera; and two (or three) are in North America, which has fewer genera but has the greater number of species, though mainly from the genus *Pinus*.

The greatest of the four centres – with species from all eleven genera – runs from China through to the Himalayas (Sichuan and Yunnan to Nepal). Most various among them are the firs and the spruces. Several genera are endemic to China: *Cathaya*, *Nothotsuga* and *Pseudolarix*. Japan and Taiwan form another, separate centre, with species from ten of the eleven genera. The one lacking is *Cedrus*, the genus of the true cedars.

Of the North American centres, California has five genera: *Abies*, *Picea*, *Tsuga*, *Pseudotsuga* and *Pinus*. Mexico, a separate centre, has the lion's share of *Pinus*, with forty-three species; and also has *Abies*, *Picea* and *Pseudotsuga*. Perhaps Mexico has so many species of *Pinus* because it has so many natural fires, and pines are good fire-resisters – indeed are fire dependent: they cannot release or germinate their seeds without it. Finally, the Atlantic plain of the south-east United States really forms yet another centre, with an array of pines that is largely different from those of California and Mexico.

With 109 known species, *Pinus* is the largest genus of all the living conifers (with *Podocarpus* running a close second). It also seems to be the oldest known genus of Pinaceae, known from the Cretaceous of Europe and North America: all the other genera first appear early in the Tertiary, about 60 million years ago. Many *Pinus* species were more widespread in former times than they are today, yet they have not retreated as the sequoias or the araucarias have done. Features that distinguish *Pinus* include the papery sheath that surrounds the base of their needle leaves – leaves that sometimes are born singly, but more often in clusters of up to five. Yet their overall form is highly various. Most grow tall with a central trunk; some are more spreading like cedars; and some are multi-stemmed shrubs.

Pinus is also the most versatile of the Pinaceae. Its various species extend from the tundra treeline in Eurasia, up to the alpine treeline in Europe and the western United States, through the salt-spray Pacific coast of North America, and down to the tropical coastal savannahs of North America. Some form big, open forests with just one species; others grow on mountains with other conifers; many populate desert shrub land; and many are particularly fire-adapted and grow in savannahs and northern forests that are prone to fire. Indeed, *Pinus* dominates most where fires are frequent. In the lowland tropics of Central America and South-East Asia, too, species of *Pinus* flourish largely because they recover from hurricanes more quickly than most. Most *Pinus* are adapted to poor soils, their roots extended by mycorrhizae. Indeed, the vast pine (and birch, spruce and alder) forests of Latvia seem largely to be rooted in sand. The first Europeans to settle on Cape Cod in the wake of the Pilgrim Fathers cut down the pine forest to expose what they assumed would be rich soil beneath and found only sand dunes that are most unsuitable for wheat – and, since they didn't have a fish-hook between them to pull the teeming cod from the sea, they nearly starved. In Spain pines grow half buried by the dunes.

Economically, *Pinus* is the most important of all the genera of trees. There are vast plantations worldwide of several species, in the southern hemisphere as well as the northern, like those of Caribbean pine (*Pinus caribaea*) near Brasilia and Monterey pine (*Pinus radiata*) just about everywhere.

The genus *Abies* are the firs – all forty-eight species of them: mostly denizens of uplands up to sub-alpine altitudes, from temperate to extreme northern cold. They grow in North Africa, throughout Europe and Asia south to North Vietnam, and in North and Central America (Honduras): sometimes in pure fir forests, sometimes with other conifers or broadleaves. Unlike most of the Pinaceae, they prefer rich soils. Most firs grow as spires like Christmas trees and can be very tall: the tallest are the grand firs (*A. grandis*) which in Vancouver Island approach 90 metres. Needles from the balsam fir (*A. balsamea*) of Canada and the lake states of the US, yield the scent of 'pine' soap, while resin from its trunk becomes 'Canada balsam', the finest cement for optical instruments, much favoured in microscopy for sticking cover slips to slides. The female cones of firs stand tall and upright along the upper branches: those of the noble fir (*A. procera*) of Washington and Oregon are magnificent, up to 25 centimetres long.

Cathaya is worth a passing note – it consists of a single species (*C. argyrophylla*) indigenous to central China, scattered on limestone or up on mountain slopes among the deciduous broadleaves. It was not described until 1958. The Chinese guard it jealously. No part may be collected, and it may not be cultivated from seed outside China. But, says Aljos Farjon, 'it is neither particularly rare nor threatened, nor does it appear to be of high economic value, either in forestry or horticulture'. But perhaps, thus guarded, it is one for the connoisseur to seek out; like some rare icon in some remote Greek monastery.

The four species of *Cedrus* might be called the true cedars (although of course since common names are largely arbitrary it's a moot point what is 'true' and what isn't). They are scattered along the Mediterranean from the Atlas Mountains of North Africa to Cyprus, Lebanon, Syria and Turkey, with one species – somewhat different from the rest – in the western Himalayas. They love cool mountains, alongside other conifers, where there is plenty of snow in winter. As befits a conifer of low latitudes, they hold their branches and their leaves horizontally, layered like a cake stand. *C. atlantica* is the Atlantic cedar, with the lovely blue variety *C. atlantica* var. *glauca*, much favoured in gardens. *C. indica* alias *C. deodara* is the deodar from the western Himalayas. *C. libani* is the cedar of Lebanon, which indeed is from Lebanon and also from south-west Turkey. It provided the timber for Solomon's temple at Jerusalem. The timber is much too scarce now for everyday use; when large logs become available they are cut radially to provide veneers with a sinuous grain.

Larches (*Larix*), eleven species of them, are among that minority of conifers that are deciduous: shedding their leaves in winter to stand characteristically skeletal with a straight central trunk and thin simple branches held more or less horizontally. They grow widely through the boreal forest of Eurasia and North America, while in lower latitudes, as in the Himalayas, they tend to prefer mountainsides. *L. deciduas* of Europe provides all-purpose timber: for pit-props, ships (still used for trawlers in Scotland), and much favoured on housing estates for gates, posts and fences. The northernmost tree in east Siberia – at latitude 73 degrees – is *L. gmelinii*. The most northerly are extremely stunted.

But in North America the northernmost conifer is a spruce, *Picea glauca*. In all, there are thirty-four species of *Picea*. In the extreme north, in the boreal zone of North America and Eurasia, they often form vast single-species forests. Further south they prefer mountains and often grow in mixed forests with other conifers. Western China and the eastern Himalayas again have the greatest diversity. Spruces are beautiful trees – tall and steeple-like like firs; Europe's tallest native tree is a spruce – Norway spruce (*P. abies*) from Scandinavia, Eastern Europe and the Alps. The virgin cones of some species, still to be pollinated, are beautiful reds and yellows. Spruces are useful too. They provide light timber (known as 'deal') and are the archetypal Christmas tree: the name 'spruce' is related to 'Pruce', which derives from 'Prussia', which is where Christmas trees originally came from. After the First World War Britain planted vast plantations of sitka spruce, *P. sitchensis*. It is a lovely tree that in its native north-west America may grow to 80 metres. But in the fashion of those post-war, no-nonsense days it was planted in military lines and largely at the expense of native species and familiar landscape, and got itself a bad name. Now, at least sometimes, commercial planting is more sensitive.

Tsuga is the genus of the hemlocks: nine species, native to North America (two in the east and one in the west) and Asia – spreading up to 3,000 metres in the Himalayas and through China to Japan and Taiwan. Again, trees reflect the historical link between North America and Northern Asia – the Pacific athletically bridged. Several of the hemlocks are most people's idea of what a conifer should look like: tall, dark and needle-leaved like spruces, although some like *T. canadensis* of the eastern United States are smaller, slow growing, and often cut into hedges. Hemlocks are also among the minority of conifers that tolerate shade, and in Asia in particular grow in forests alongside broadleaves, when again they tend to have more rounded crowns.

The four living species of *Pseudotsuga* again show the historical link of North America to Eastern Asia: there are two in America (one endemic to California) and two more through China, Japan and Taiwan. The Douglas fir, *P. menziesii*, grows along California's Pacific coast right up to British Columbia, and also in the Rockies from Canada to Mexico, and was first brought to Europe by the great Scottish naturalist-explorer David Douglas. It is the biggest of the Pinaceae – indeed among the biggest trees of all: there is a flagpole in Kew Gardens more than 60 metres tall, sawn straight as a die from a single trunk. Douglas fir timber is among the strongest and most rigid of softwoods, and is used for everything from heavy construction to carvings; it also supplies more veneer and plywood than any other species. The Asian *Pseudotsuga* are in general more modest, mostly growing among broadleaved trees in deciduous mountain forests.

Rimu, Totara, Kahikatea – the Most Various Conifers of All: FAMILY PODOCARPACEAE

The Podocarpaceae family currently has eighteen genera, but these are under-researched and difficult to classify not least because their cones are much reduced and hard to decipher (in most conifers it is the cones which are most informative). With more study (particularly of DNA), the Podocarpaceae might be split further. Already, however, they are the second largest family of conifers (after Pinaceae), with 185 species.

All of the genera are found in the southern hemisphere, but only half of them venture into the north. In the northern hemisphere podocarps extend from the Andes into Central America and the Venezuelan highlands, but they are most widespread in Malaysia, Indonesia, Indochina, and subtropical China north into southern Japan. It is not clear that the Podocarpaceae family as a whole originated in Gondwana (though this is the best bet) but its biggest genus, *Podocarpus*, almost certainly did. Fossils of this family are known from the Jurassic onwards (about 140 million years ago), all from the south.

Both in form and ecology the podocarps are impressively various.

Most are trees, scattered in moist tropical or subtropical forest. Some help to form the understorey, others join the canopy, and some tower above the rest as 'emergents'. Some grow in mossy forests in the highest tropical mountains. Some – especially in the far south – form low shrubs well above the treeline. Many grow on poor soils, including swampy peat, but others compete with angiosperms where nutrients are more abundant. Many, including *Podocarpus*, have broad ever-green leaves that are unlike most living conifers, and fleshy cones of various shapes and colours that again are much under-researched but are evidently dispersed by mammals and birds. Thus *Podocarpus* plays the angiosperms at their own game. The extraordinary island of New Caledonia as usual has the greatest oddities. There, *Retrophyllum minus* grows in running water – which is a most unusual way for a tree to behave. *Parasitaxus usta* from New Caledonia is the only parasitic conifer: it taps into the roots of another podocarp, *Falcatifolium taxoides*.

Podocarpus includes 107 living species, though more are likely to be found in remote tropical forests. On the other hand, says Dr Farjon, 'revision of the genus is long overdue', which means that it could well be divided (and *Podocarpus* itself will be reduced). *Podocarpus* is the only genus of podocarps to occur on most of the southern continents and major islands, especially Borneo with thirteen species and New Guinea with fifteen. Six are known from the Venezuelan highlands – but there is much exploration yet to be done.

Podocarps are key players in New Zealand's ecology, and were vital to the religion and economy of the Maoris, and later prized by the Europeans. The tallest of all New Zealand's trees is the kahikatea, *Dacrycarpus dacrydioides*, at nearly 60 metres. In the early nineteenth century, soon after they decided that God had obviously made New Zealand for their personal use, the British coveted the fluted, flakey-barked trunks of kahikatea as masts for their swelling fleet: warships to counter Napoleon; cutters for fast reliable trade overseas. But the timber of kahikatea is weak. It was a great disappointment. Nowadays, relationships between the Maoris and the Europeans, though far from disastrous, continue to be edgy. Recently air traffic control demanded that a lofty row of kahikateas near Rotorua airfield should be trimmed. The Maoris said no way. Rotorua is definitely

Maori country. The last I heard the airfield may have to be moved.

Though it disappointed the British navy, the kahikatea is a fine tree. As in many plants, the juvenile and mature leaves are different: the former superficially yew-like in two rows, the latter fleshy and roughly cypress-like. When the female cones mature the receptacles grow red and berry-like, each with a purple seed at its tip, so the two together are like Russian dolls.

The matai is better for timber – known to the timber trade as 'black pine', and to botanists as *Prumnopitys taxifolia*. *P. ferruginea* is the miro, alias 'brown pine'. The 'mountain pine' and the 'yellow pine' belong in yet another genus, *Halocarpus*, while the 'silver pine' is *Lagarostrobus* (and the 'yellow silver pine' is *Lepdothamnus*). But perhaps the greatest of these misnamed 'pines' is the rimu, aka 'red pine', *Dacrydium cupressinum*. It is no longer cut, but existing planks and joists are rescued for furniture and household goods of all kinds, like bowls and salad servers.

To the Maoris, however, the greatest of all trees was the totara, *Podocarpus totara* (with the smaller upland version, *P. cunninghamii*). *P. totara* grows on both islands up to 40 metres, with trunks 2 metres across. The Maoris revered it, not simply for its general magnificence but for the red of its timber, the colour of royalty. They also hollowed out the trunks to make from a single piece of timber canoes to be paddled by a hundred men.

The Celery Pines: FAMILY PHYLLOCLADACEAE

The single genus *Phyllocladus*, with just four species, is commonly included within the Podocarpaceae. These are the 'celery pines', which grow in Malaysia, Indonesia, New Zealand and Tasmania, usually as large canopy trees, though stunted up towards the treeline in mountain cloud forests. They are called celery pines because their leaves are celery-like, though fleshy; not what you would associate with a conifer. Botanically speaking, however, these are not leaves at all. They are 'phyllodes': flattened green stems that do the job of leaves. The true leaves have gone missing. But such vegetative features are not usually considered of huge taxonomic significance. True evolutionary relationships are more reliably revealed by sexual characters, which with conifers means cones; and these, and the fruit-like bodies that develop from them after fertilization, are very like those of podocarps. But *Phyllocladus* was first placed in its own family in the 1960s; and although many taxonomists do not favour this separation (including Judd), Aljos Farjon feels that the phyllodes are such a striking feature that they justify separation; he also points out that the celery pines have a different number of chromosomes from the Podocarpaceae; and their mechanism for pollination is also distinct. The discussions will doubtless continue. More data are clearly needed, not least from DNA.

A One-off from Southern Japan:
FAMILY SCIADOPITYACEAE

Here is another very small family – indeed with just one species: *Sciadopitys verticella*. It's an evergreen conifer endemic to southern Japan, where it grows on steep slopes and ridges, sometimes in clumps and sometimes mixed with broadleaved trees. It competes because it can tolerate such poor soils.

The Sciadopityaceae are an ancient group – fossils that possibly belong to this family are known from the upper Triassic, around 200 million years ago, before the dinosaurs got fully into their stride. Traditionally it has been included in the Taxodiaceae (which means it would now be in the Cupressaceae). But its needle-like leaves are not leaves at all – again, they turn out to be phyllodes. Also, studies of structure and of DNA support the notion that the remarkable *Sciadopitys* indeed belongs in a family of its own.

The Yews: FAMILY TAXACEAE

In the family Taxaceae are five genera, with twenty-three species between them. Mostly these are in the northern hemisphere, but there are a few in the south including – inevitably it seems – in New Caledonia. The only widespread genus is *Taxus*, the yew, with ten species found throughout North America and down into Honduras, and in Eurasia down into the southern hemisphere in Malaysia– Indonesia. The British are used to common yew (*Taxus baccata*) lowering gloomily in dank churchyards. Yet it lives in the tropics too, albeit confined to mountainsides. *Toreya* is less widespread, but also occurs in Asia as well as North America. Trees of the Taxaceae are slow growing and long-lived – hence the somewhat speculative though not to say outlandish claim in the preface, that Pontius Pilate might have dozed beneath one that is still growing in Scotland. Taxaceae seem doleful by nature: adapted to the shady understorey in coniferous or mixed forests. The fruit-like arils, of various colours, are dispersed by birds. Like the podocarps, the group as a whole has not been studied enough. In the past, because of the peculiar and much-reduced female cones, yews were thought to form a group on their own, separate from the conifers. But closer study suggests that their cones have evolved from more complex seed cones and this, plus evidence from DNA, confirms yews as bona fide conifers.

The wood of yew is valuable in many ways. In particular, it provided the medieval English with their longbows.

It would be good to devote this entire book to conifers. They are so various and wondrous and in many ways, in many contexts, they have turned the course of human history. Yet they are at least matched and in some ways for outstripped by the flowering trees that will occupy the next five chapters.

6

Trees With Flowers: Magnolias and Other Primitives

Beautiful but simple: is magnolia the most ancient flowering tree?

There is more to flowers than meets the eye. Of course they may be beautiful. They are also, beautiful or not, superb essays in engineering, wonderfully efficient first at achieving fertilization, and then at producing and dispersing their seeds. The archetypal flower is supported by a circle of sepals, usually green, that make the calyx; inside that is a circle of petals, which between them form the corolla; and in the middle are the sexual parts – the male stamens, with the anthers at the tips, containing pollen; and the female carpels. Each carpel has an ovary with its ovule (or ovules) inside, a projecting style, and is tipped by the stigma which receives the pollen. Immediately we see a key difference between angiosperms and all other seed plants. In the angiosperm the ovule is completely enclosed within the ovary, and the male gamete (reduced to a nucleus) is carried to it via a pollen tube that must burrow through the full length of the tissue of the style. In conifers and all other seed plants, the ovule is not completely enclosed. The pollen tubes do not have to burrow through living tissue. In cycads and ginkgoes, motile sperms do the fertilizing.

Then there is another key difference – much less obvious. Uniquely, angiosperms practise 'double fertilization'. As noted earlier, the pollen contains the male gamete – and other cells too. So too does the ovule – a true egg cell, and also subsidiary cells. In all but the most primitive angiosperms, the male gamete fuses with the egg to form a new embryo, as in all other organisms that practise sex. But a second cell in the pollen fuses with two of the cells in the ovule to form a combined cell with three sets of chromosomes; and this peculiar triploid cell then multiplies to form a food store, rich in carbohydrates, protein and often fat, that surrounds the embryo. Double fertilization is a very neat trick – and unique to angiosperms.

The unique ability of angiosperms to take so many different forms – mighty trees and creepers, tiny floating duckweeds, and everything in between – seems at first sight to have nothing to do with the innovations of flowers and seeds and fertilization. But it happens anyway. Perhaps the more sophisticated devices of sexual reproduction simply allow new ways of living.

Flowers are immensely variable. Some are huge and showy, others cryptic to the point of invisibility. Some have all the standard parts – sepals, petals, stamens and carpels – but others have abandoned one or several of the basic components, and some incorporate various bracts (modified leaves) and other structures. In some the petals are green like sepals, and in others the sepals are coloured like petals, and in some the two kinds of structure are more or less indistinguishable. Most flowers are hermaphrodite (like those of magnolias) while others

are single sex; and some 'monoecious' angiosperms keep both sexes of flowers on the same plant (like oaks) while other 'dioecious' types have only one sex per plant (like holly). The sexual organs themselves take many different forms (as is also true of animals) and are key features for identification.

Many flowering plants are pollinated by various kinds of animals – not just insects, but also bats, birds (such as hummingbirds) and sometimes mammals, from fruit bats to giraffes (or so it is said). Best known among the insect pollinators are bees, butterflies and moths, but flies and beetles are important too. Beetles were probably the first insect pollinators of all – and probably developed the trick in association with cycads. The brash new angiosperms lured at least some of them away.

But many flowering plants, including most temperate trees, are pollinated by wind. Some – probably more than is yet appreciated – make some use both of animals and of wind. In a few, like the sea-grasses, which flower under water, and perhaps some plants of the mangroves, the pollen is conveyed by water. In general, the animal-pollinated flowers are showy and the wind-pollinated ones much less so. But flies, for example, hunt largely by scent and pollinate many very inconspicuous flowers (like those of the garden shrub *Fatsia*, a close relative of ivy), while wind-pollinated flowers often take the form of catkins, which can be very spectacular indeed.

In the plants that are considered to be most primitive, all the petals are much the same, all the sepals are much the same, and so on; and all the parts of the plant are separate. So the flower consists simply of repeated modules, arranged in spirals like the scales of a cone (although flowers are not, just to emphasize, botanically related to cones). The whole is a kind of Kiddicraft flower. Magnolias and waterlilies are of this primitive type. Such flowers are radially symmetrical: symmetrical whichever way you look at them. In other flowers the different parts may be fused and extended in all kinds of shapes, and are commonly designed to attract very particular kinds of insects, and then to ensure that the visitors are well furnished with pollen. In such types (and many wind-pollinated types) the flowers are bilaterally symmetrical, like a face, or indeed are asymmetrical. Orchids are perhaps the most famous floral elaborators; but many other families have elaborate, specialist flowers too, including the pea family Fabaceae (Leguminosae), which includes many of the world's most significant trees. But primitive flowers can also trap insects in very clever ways. 'Primitive', after all, is not a pejorative term. It merely means 'close to the ancestral state'. Daisies, incidentally, may *look* primitive (radially symmetrical, repeated parts). But in truth their flowers are complex inflorescences (collections of flowers), each of which is highly modified. So the daisy family, the Asteraceae (formerly called the Compositae) is commonly considered to be the most 'advanced' of all. Perhaps unsurprisingly, though in truth for different reasons, the Orchidaceae and the Asteraceae both include a spectacular number of species – far more than any other family. Of the two, however, only the Asteraceae has trees – though not many; and little to write home about. The family with the most trees (as well as many of the grandest), is the Fabaceae.

The fruits of flowering plants, containing the seeds, are also immensely variable. Some are big, bright and fleshy and are intended to be spread by animals. Some (notably orchids) are tiny and wind-blown. Some wind-blown types are bigger but are fitted with wings. This is the case in many trees – such as the familiar sycamore and ashes, but perhaps most spectacularly in the dipterocarps, the great forest trees of South-East Asia, which South-East Asians at least consider to be the most important tropical trees of all. Other airborne seeds are fitted with cottony extensions that serve as parachutes. We all know dandelions and groundsel. In Yorkshire I have driven through blizzards of migrating thistle seeds that seemed to go on for miles. Cotton, too, of course (a relative of the hollyhock) produces its fibre as a way of spreading its seeds. Among trees, perhaps the most spectacular and significant parachute seeds are those of the various kapoks – the cotton used to stuff mattresses and parkas. Some fruits, notably the coconut and the giant coco de mer, are expressly designed to float.

So various are flowering plants, it has often been suggested that they cannot be a single group: not a true clade, with just one common ancestor. But they clearly are. All but the most primitive practise double fertilization, and that is so weird, and so complicated, that it surely could not have evolved more than once. So flowering plants, for all their marvellous variety, *are* a single invention, all derived from one common ancestor, which arose about 145 million years ago. Truly they are one of nature's greatest inventions.

But it is not clear who the common ancestor was, or what it looked like. There are two ways to go about finding out. One is to look at the fossil record, to try to see what the very first flowering plants looked like. The other approach is to look at living plants, decide which are the most primitive, and then assume that the very first flowering plants must have looked roughly like the still-existing primitives. (The latter exercise is helped by applying cladistic principles to the DNA, and inferring which plant's DNA is the most basic, with the most shared primitive features. But the details need not delay us.)

Both approaches are necessary, but both are riddled with traps. The fossil record is notoriously patchy (or 'spotty', as the palaeontologists tend to say). We are most unlikely to find the first organisms in a new lineage because, obviously, new lineages begin with just a few individuals – and rare organisms have scant chance of being fossilized, and then recovered. More generally, if we fail to find what we are looking for, that does not mean that our quarry did not exist. But at least if we do find something, that shows that it did exist. In reality, some of the earliest known angiosperm fossils are waterlilies, which seems to fit very nicely with expectation, since waterlilies have a simple kind of flower. But waterlilies are herbs. The problem with this is that the timber of flowering or broad leaf trees closely resembles the timber of conifers. That structure is very complex, and is not likely to have evolved more than once. This and other evidence suggests that conifers and angiosperms shared a common ancestor. If flowering plants had first arisen as herbs, then they would have to have reinvented timber that was very like that of conifers, and this seems most unlikely. So it seems that the very first flowering plants must have been trees – and thus for all their primitive flowers, waterlilies are highly evolved specialists, devoid of wood. They could not have been the first. The first angiosperm must have been a tree. Magnolias are reasonable candidates, but the oldest-known magnolias are not quite old enough, and it is hard to imagine that the first ever angiosperm took quite such spectacular form.

Then again, among living angiosperms, two quite opposite kinds of flower have been mooted as the most primitive. One is the big showy kind, as in waterlilies and magnolias. The other is a simple but small and modest kind, as found in pepper vines (the plants that produce peppercorns; not sweet or chilli peppers). But the first ancestor of all the flowering plants *either* had magnolia-like flowers *or* pepper-like flowers. It could not have had both. If it had some compromise form, able to evolve either way, then we simply do not know what it was. The same kind of dilemma applies to fruits. Are the most primitive the small and scrutty kinds, or the big and fleshy kinds, packed with seeds, like custard apples?

So the origin of flowering plants remains mysterious, but we do know that they are now immensely various, and their variety is reflected in their taxonomy. Roughly speaking, there seem to be around 300,000 species. Different botanists differ in the way they divide these species into families: recent published papers recognize anywhere between 387 and 589 families. The specialists who form the Angiosperm Phylogeny Group have, for the time being, plumped for 462. Only specialists can get their heads around so many, and so it is useful to group the families into orders. Again, different taxonomists recognize different numbers of orders, but the authorities I am following in this chapter (who I refer to collectively as 'Judd') divide them into forty-nine, most of which contain trees. Forty-nine is still too many for most people to bother with, but most of the trees are contained within about thirty of the orders, and that is manageable.

One last word before the catalogue begins – on the idea that was first mooted by John Ray in the seventeenth century, and was developed by Antoine Laurent de Jussieu at the end of the eighteenth. Ray (you will remember from Chapter 2), divided flowering plants into those with narrow leaves and those with broad leaves; and Jussieu found that the narrow-leaved types were 'monocots' (with a single cotyledon) while the broadleaved types were 'dicots' (with two cotyledons). So the distinction has stood for the better part of 200 years. In rough and ready terms, the distinction still stands. But those interested in phylogeny – the true history of plants – as opposed merely to convenience, must now make a serious modification. For it is clear that the first ever flowering plants were dicots. Some of those early, primitive types are still with us – including the magnolias, the waterlilies and the peppers.

After some time, there arose from among the ranks of the primitives a new group that went off in novel directions. Some of these avant-garde types evolved into the monocots – which thus emerge as comparative latecomers. Some evolved into a quite new kind of dicot, now known collectively as the 'eudicots' ('eu' meaning 'good'). The monocots are a true clade – all deriving from a common ancestor. The endicots are also a true clade. It is also possible, indeed quite likely, that the ancestor of the monocots was also the ancestor of the eudicots – in which case the monocots and the eudicots together form a true clade.

But the primitives (magnolias, peppers, etc.) remain out on their own: not members of either of the two modern clades (although flowering plants as a whole form a true clade).

So in effect we now have three great groups of flowering plants. First there is a mixed bag of 'primitive dicots': not a clade; just the kinds that seem to have retained many of the main features of the first ancestor. Then there are the monocots, which are a true clade. Then there are the eudicots – another true clade, this time of derived dicots. The primitives are discussed in the rest of this chapter; the monocots have the next chapter to themselves; and the eudicots are spread over the two chapters after that (because they are too diverse to be accommodated comfortably in one).

The Mixed Bag of Primitive Dicots

As you can see from the chart, the primitive dicots include seven orders. The three extreme outliers need not delay us too long. At least, the Amborellales does include one tree – which, like so much of botanic interest, lives in New Caledonia. It is aromatic, but it is only small. The Nymphaeales is the order of the waterlilies: of huge interest botanically and ecologically, but emphatically non-tree-like. The Australobaileyales are interesting as the possible sister group of all the other flowering plants – and because the order contains the family Illiciaceae, which includes the genus *Illicium*. From the bark of various species of *Illicium*, comes the spice star anise.

The remaining four orders collectively form a group that botanists refer to as the 'magnoliid' clade. All of them contain trees, and three of them contain some very significant trees indeed.

Magnolias, Custard Apples and Nutmeg:
ORDER MAGNOLIALES

The Magnoliales order contains around 2,840 species in six families, of which the three outlined here – Magnoliaceae, Annonaceae and Myristaceae – include highly significant trees.

Some traditional classifications[1] list up to twelve genera in the Magnoliaceae. But Judd combines eleven of the traditional types into *Magnolia*; the other genus is *Liriodendron*. Between them they include 218 species, and they have the kind of distribution that we have already seen among the conifers: many in South-East Asia, from the Himalayas out to Japan, and many more in the south-eastern United States and Central America, with some in South America. Why are they spread out this way? Did they at some point leap the Pacific? Perhaps. But there are fossil Magnoliaceae in Europe and even in Greenland – which suggests that the family may once have spread more or less continuously from South-East Asia to the Americas before the two continents drifted apart, and that they have simply died out in the middle of their range.

Magnolia is of course magnificent, with some of the most stunning flowers of any tree – sometimes star-like, sometimes huge like water-lilies. Their greatest value for humanity is in horticulture; and I commend to you the magnificent botanic gardens in Yunnan, south-west China, which has a hundred or so species. Some magnolias give serviceable timber. The bark and flower buds of M. *officinalis* of China are used medicinally, and are a valuable export. (China is also one of the world's greatest centres of biodiversity and its potential for ecotourism is unsurpassable. It is up there with Africa, Madagascar and Amazonia).

Liriodendron includes the two species of tulip trees – one Chinese (*L. chinense*) and one North American (*L. tulipifera*). Their leaves are strange and absolutely characteristic – like glossy, dark green versions of maple leaves, but with the pointed tip cut out. Their flowers are tuliplike, although difficult to appreciate since tulip trees can be big (the tallest in England is 36 metres) and the flowers are born aloft. The American tulip tree yields a creamy timber, with streaks of olive green, black, pinkish brown or even steely blue, from growing in mineral soils, which is much valued for carving, and for doors and suchlike. As a timber tree, *Liriodendron* is sometimes known as 'yellow poplar' and is sold as 'American whitewood'. Such names do it scant justice.

The 2,300 species of Annonaceae, arranged in 128 genera, form a glorious family widely spread through lowland tropical and subtrop-

You can't mistake the leaves of the tulip tree, Liriodendron

ical forest. In many ways they seem wonderfully primitive. Their flowers are pollinated by beetles – which they have evolved to encourage: they have a fruity odour, and they reward their visitors with thick fleshy petals for feeding on, and extra fleshy tissue that serves no purpose except to provide beetle food. Some flowers of the genus *Annona* are able to heat up – a fairly common trick among several plant families. This encourages the beetles to stay inside the flowers overnight and mate, so becoming covered in pollen. Various genera, notably *Annona* and *Rollinia*, provide marvellous fruits of the kind that seem primitive and in Cretaceous times doubtless were food for dinosaurs: big and pulpy, with many big seeds. Custard apple (*Annona*) with its grey, tessellated skin, is the best known of these fruits in the West. Others include the cherimoya, soursop and sweetsop. The fruits of *Monodora myristica* are sometimes used in place of nutmeg, and some Annonaceae with their thick, fibrous bark are grown as ornamentals, at least in warm countries.

The third important family of the Magnoliales order is the Myristicaceae. It includes 370 known species in seventeen genera, which occur across the tropics: in South and Central America, across equatorial Africa, through south India and South-East Asia and into Queensland, Australia. The trees are usually dioecious (only one sex per tree) and although their flowers are small and inconspicuous they are pollinated by insects – beetles and thrips. Clearly they are very different from those of the Annonaceae or Magnoliaceae – illustrating that although flowers are one of the main guides to classification, they may nonetheless differ enormously even between closely related families.

The biggest genus is *Myristica* with 125 species, centred on New Guinea. *Myristica fragrans* from the Molucca islands of Indonesia is the most economically important of all the family: its big seeds are nutmeg; the fleshy coating of the seed (the 'aril') is scarlet while it remains on the tree (although concealed inside a thick fruit with a pale green skin) but is a pale buff-pink after drying, and is the stuff of mace. The seeds of the Brazilian genus *Virola* are ground to make snuff that is hallucinogenic. (Perhaps the world needs more hallucinogens rather than less. The puritanical attitude of the West might more properly be seen as an offence against nature. More in Chapter 12.) *V. surinamensis*, whose seeds are waxy, are used to make 'butter' for eating and for candles. So too are those of *Gymnacranthera farquhariana* from India.

Greenheart, Stinkwood and the Green Bay Tree:
ORDER LAURALES

The Laurales order includes around 3,400 species in seven families, of which the most significant is the Lauraceae (named, incidentally, by Antoine Laurent de Jussieu, who first made clear the monocot–dicot distinction).

The members of the Lauraceae family are mostly trees or shrubs – in all there are around 2,500 species in fifty genera, a huge presence in tropical wet forests worldwide, and in the subtropics. They serve humanity in many ways: extraordinarily nutritious fruits, many fine timbers, and a host of medicines and other drugs. The Lauraceae illustrate beautifully how biochemistry runs in botanical families.

Best known of all Lauraceae fruits is the avocado, *Persea americana*, native to Central America. It has more protein than any other fruit, and is 25 per cent fat. It also has a wonderful strategy to prevent inbreeding. As with other Lauraceae, it is pollinated by insects. It has two kinds of flowers, inventively dubbed A and B: some individuals have A flowers and some have B flowers. The stigmas of A flowers are receptive to pollen only on one particular morning, while the anthers of A flowers do not release their pollen until the afternoon of the same day. In B flowers, the stigmas are receptive on one afternoon, while pollen is not released until the following morning. So A flowers can be pollinated only by B flowers, and B flowers can be pollinated only by A flowers.

Many Lauraceae have oil cavities in their leaves and elsewhere, and many are aromatic. Thus the family includes the bay trees (*Laurus nobilis*), sassafras (*Sassafras albidum*), cinnamon (*Cinnamomum verum*) and camphor (*C. camphora*), once much used in mothballs (though modern mothballs are of naphthalene) – one of many examples of a tree's own insect repellent turned to human use.

Among the many valuable timber trees in the Lauraceae family is the Queensland 'walnut', *Endiandra palmerstonii*, which grows to a magnificent 40 metres or more and whose timber resembles European walnut (which comes from the quite different Juglandaceae family). It has pinkish sapwood and pale to dark-brown heartwood, streaked with pink or purple-black – much prized for furniture of the board-room kind. The mangeao of New Zealand's North Island (*Litsea calicaris*) is another giant (40 metres or more), giving cream to pale-brown timber that is favoured for everything from turnery to dance-hall floors to pit props – and also for excellent veneers for export. Imbuia (*Phoebe porosa*) from Brazil is yet another forest giant (up to 40 metres) with a dark-brown, fine-grained, lustrous timber that again is much coveted for high-class joinery.

The many fine trees of the genus *Ocotea* are prized both for their beautiful timber and for their broader biochemistry. South Africa's stinkwood (*O. bullata*) is 18 to 24 metres tall in the forest, yielding a dark timber with a fine grain that is indeed malodorous when freshly worked but settles down when dried. *Ocotea usambarensis* from Kenya is camphorwood. It is up to 45 metres tall, yields a greenish-brown timber that matures to deep brown, smells of camphor and is much favoured for making wardrobes, effectively with the mothballs built in. (I have found, in travels with tropical loggers, that many trees are surprisingly smelly when first cut, and not always pleasantly so.)

But the most famous of all the Lauraceae timbers comes from the greenheart (*O. rodiaei*), the pride of Guyana. Again it reaches up to 40 metres, with a long cylindrical trunk that may be 25 metres tall and a metre in diameter. The sapwood is pale yellowy-green, the heartwood light olive to dark brown, often streaked with black. Greenheart again is highly versatile: much favoured in maritime circles for jetties and groynes, and in ships for planking and stern posts, but also for turnery and the butts of billiard cues. It is used to make longbows, too, the technology of which has come on apace since the English first juxtaposed the heartwood and sapwood of yews. Modern longbows are laminated and greenheart often forms the central layer.

In addition, the nut of the greenheart yields material called 'tipir' which the native people of Guyana have long employed as a medicine. The Wapishana tribe grate the nuts and use the extract to stop haemor-rhages, prevent infections, and as a contraceptive. In the late 1990s, however, an American entrepreneur, after spending time with the Wapishana, tried to patent tipir as an antipyretic useful in preventing flare-ups of malaria, and perhaps for treating cancer and Aids. The tribe accuse him of theft: 'biopiracy'. The dispute rumbles on. Meanwhile greenheart, precisely because it is wonderful in so many ways, is being seriously overlogged. But still it is outstanding even among the distinguished family of Lauraceae, which are huge players in tropical forests and economies worldwide.

Winter's Bark and White Cinnamon:
ORDER CANELLALES

There are only two families in the Canellales order. The Winteraceae family includes up to 120 species (no one is quite sure) in seven or eight genera, largely from around the coasts of South and Central America, eastern Australia and New Guinea, plus one in Madagascar. Several have peppery leaves and bark that are said to be medicinal: best known is winter's bark (*Drimys winteri*), which was once used to prevent scurvy. The wood of the Winteraceae is strange: the conducting tissue of the xylem contains only tracheids – rows of cells with perforated ends – as in a conifer. They do not lose the cell walls at the ends to form the continuous tube-like vessels that are more typical of broadleaves. But it is not clear whether this arrangement is primitive or secondary: the ancestors of the modern Winteraceae may have had vessels, which may subsequently have been lost. As we have seen, evolution often leads to simplification, and several other groups of flowering trees, unrelated to the Winteraceae, have wood with tracheids rather than vessels.

The second family, the Canallaceae, includes sixteen or seventeen species of aromatic trees, in five genera. The bark of *Canella winterana* is also known as 'white cinnamon', and is a tonic and a condiment; it is also used to poison fish in Puerto Rico (so they can be scooped out of the water). The aromatic wood of *Cinnamosma fragrans* of Madagascar is exported via Zanzibar to Bombay, where it is used in religious ceremonies. The bark of *Warburgia ugandensis* of Uganda serves as a purgative, while its leaves are used in curries and its resin is handy for mending tools.

Black Peppers, White Peppers and Dutchman's Pipe:
ORDER PIPERALES

The last order of primitive dicots is the Piperales. As defined by Judd it includes five families, of which the most significant is the Piperaceae, mainly of rainforests throughout the tropics. Its 2000 or so species

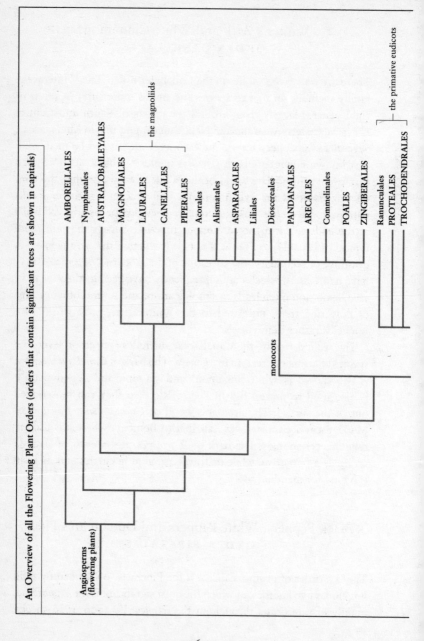

An Overview of all the Flowering Plant Orders (orders that contain significant trees are shown in capitals)

Angiosperms (flowering plants)

AMBORELLALES
Nymphaeales
AUSTROLOBAILEYALES
MAGNOLIALES
LAURALES
CANELLALES
PIPERALES
} the magnoliids
Acorales
Alismatales
ASPARAGALES
Liliales
Dioscereales
PANDANALES
ARECALES
Commelinales
POALES
ZINGIBERALES
monocots
Ranunculales
PROTEALES
TROCHODENDRALES
} the primative eudicots

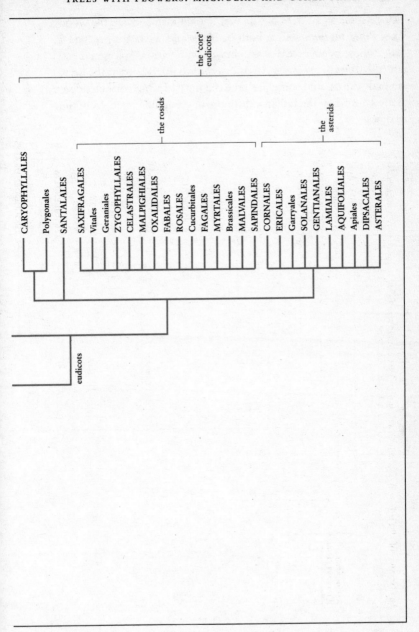

include some small trees but its best known members are the woody vines *Piper nigrum*, source both of black and white pepper, and *P. betle*, whose peppery leaves are chewed with various spices and betel nut (the fruits of the palm *Areca catechu*) to provide what the herbal manuals call 'a mild stimulant', the red juice of which dribbles down many a beard on the Indian subcontinent (including mine, from time to time).

7

From Palms and Screw Pines
to Yuccas and Bamboos:
The Monocot Trees

An arborescent relative of asparagus: the dragon tree

If some earthly firm of engineers had designed the magnolia or the bay tree, they would be pretty pleased with themselves. Here, they would conclude, is the finished article: root, trunk, branches, leaves, fruits and well-protected seeds; everything taken care of, the whole structure beautifully integrated, beyond improvement; the apotheosis of the plant.

But nature is never satisfied. However fine its inventions may be, evolution powers on. Some time after it had produced the first, magnificent, magnolia-like tree, the mighty laurels and the ancestral peppers, nature came up with something completely different: the monocots. They are easily discerned (usually) by the kinds of features that John Ray noted. They tend to have long narrow leaves. The veins in the leaves run in parallel from base to tip whereas those of dicots generally form a branching network. The parts of the flowers (petals, sepals, stamens, carpels) are typically arranged in multiples of three, while those of dicots (both primitives and eudicots) more usually occur in multiplies of four or five. But the differences run far deeper. The monocots represent a new and different way of being a plant.

What really matters, what is truly profoundly different, is the way the monocot grows, particularly as manifest in the leaves and roots. Notably, the leaf of a dicot grows out from the edges. The youngest bits are furthest from the twig. The strap-like leaf of the monocot grows from the bottom up; typically from a bud at the tip of a stem, known as an apical bud. The youngest part of the monocot leaf is at the bottom and the oldest at the top: so that grass leaves die from the tip downwards, and an onion leaf is white and immature at the base and green or even senescent at the top. (The great tactic of the grasses is to keep their growing tip, the apical bud, below the surface of the ground so that it is not destroyed by grazing animals – and in fact grasses, in contrast to almost all other plants, positively gain from being grazed and grow rank if they are left alone. To find a way of *benefiting* from the attentions of the predators who come to eat you is a trick indeed; and explains why the world's 'grasslands', basically created by members of the single family Poaceae, are almost as extensive worldwide as the world's forests, which contain many thousands of species in scores of families.) The roots in monocots are different too: in general they are much more likely to grow straight from the stem (the technical term for this is 'adventitious'), rather than from other roots.

As you can see from the chart, there are ten orders of monocots – five of which contain significant trees, and five of which are predominantly herby. This is in contrast to the dicots (both the primitives and the eudicots), in which most of the orders contain trees.

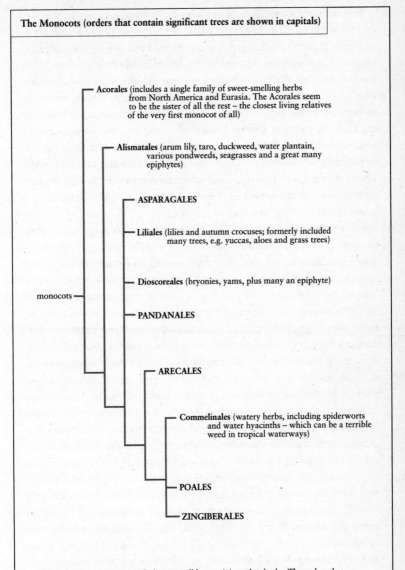

The Monocots (orders that contain significant trees are shown in capitals)

Acorales (includes a single family of sweet-smelling herbs from North America and Eurasia. The Acorales seem to be the sister of all the rest – the closest living relatives of the very first monocot of all)

Alismatales (arum lily, taro, duckweed, water plantain, various pondweeds, seagrasses and a great many epiphytes)

ASPARAGALES

Liliales (lilies and autumn crocuses; formerly included many trees, e.g. yuccas, aloes and grass trees)

Dioscoreales (bryonies, yams, plus many an epiphyte)

PANDANALES

monocots

ARECALES

Commelinales (watery herbs, including spiderworts and water hyacinths – which can be a terrible weed in tropical waterways)

POALES

ZINGIBERALES

The class of monocots as a whole may well have originated as herbs. The orders that contain trees may each have reinvented the form of the tree independently, as discussed in the main text. But the other five monocot orders have remained herby.

This difference can be explained on evolutionary grounds. We can assume that the first flowering plants of all were primitive dicots – and that these ancestral types were trees. Then we merely have to suggest that the dicots that are herby, like dandelions and waterlilies, have simply lost their woodiness and their arborescence. But it seems very likely that the very first monocot was itself a herb. So each modern order of monocots that contains trees must have reinvented the form of the tree afresh. Dicots as a whole seem to have stayed with the timber of the original angiosperm ancestor. All their timber is basically very similar – and similar to that of the conifers, with whom they probably have shared a common ancestor for almost 300 million years. But the timber of monocot trees is highly variable, and in general is nothing like that of dicots at all.

Most obviously, most monocot trees do not undergo secondary thickening of any kind. Palms, which are the most various and ecologically significant of all, may be very big – up to 60 metres tall – and their trunks may be up to 2 metres in diameter. But in general they are just as thick when the tree is young as they are when it is at full height (although the stems of some of them do sometimes thicken, sometimes along only part of the length as in bottle palms; but they do this just by accumulating more tissue – there is no regular secondary thickening from a sheath-like cambium). Some other monocot trees, such as the dragon tree, *Dracaena*, do undergo some secondary thickening. But the mechanism is quite different from what we see in oaks or magnolias (or indeed in pines). In particular, there is no continuous sheath of cambium, regularly turning out new tissue. The dragon tree's form of secondary thickening is another reinvention.

I will not dwell on the five monocot orders that do not contain significant trees, but a brief mention is called for just to provide context. The Acorales are sweet-smelling herbs found both in North America and Eurasia. They may be the most primitive of all the living monocots: closest to the ancestral form. The Alismatales include some extremely interesting plants of huge ecological importance, including pondweeds, seagrasses (mentioned again in the context of mangroves), a great many epiphytes (of great significance in tropical forests) and the important staple food crop, taro. The Liliales are of course the lilies, and the autumn crocuses. You will find trees included in the

Liliales in some traditional classifications, including those yucca and aloe: but these two genera have now been repositioned, as will become apparent. The Dioscereales are the brionies; and the Commelinales are more water-plants, including the spider-worts and the water hyacinth, which has become such a pest in many tropical waterways. The five monocot orders that do include trees – including some extremely significant trees – are as follows.

Joshua Trees and Dragon Trees:
ORDER ASPARAGALES

This order was of course named after the asparagus, in the Asparagaceae family – which also includes some shrubby species, and a few woody vines. Also in the Asparagales are the families of the daffodils, the hyacinths, the irises, the onions and the orchids. But more directly to the point is the Agavaceae family. The genus for which the family is named, *Agave*, includes about 300 species of prickly succulents like pineapple tops that are native to the Americas but feature in warm gardens worldwide. Several are big with woody trunks: bona fide trees. *A. americana* provides the Mexicans with pulque, which is sometimes distilled to make mescal. *A. sisalana* and *A. fourcroydes* provide strong fibres – sisal hemp – for ropes and fishing nets. *A. americana* was imported to Europe soon after Europeans became aware that the New World exists, and is now grown everywhere that is not too cold. But the genus of Agavaceae most familiar to gardeners in temperate regions is *Hosta*; another homely link to a more exotic world.

The Aspholadaceae family includes the genera *Aloe* and *Yucca*. Both include significant trees. The aloes of Africa, Arabia and Madagascar may be seen as the Old World equivalent of the American agaves: they look very similar, and turn up in the same gardens worldwide. (There are many pleasing parallels between American and Old World flora and fauna. Animal examples include the iguanid lizards, boas and condors of the Americas versus the agamid lizards, pythons and vultures of Africa and Eurasia.) The yuccas include the wonderful Joshua tree, *Yucca brevifolia*, 10 metres tall, with simple curving branches which have prickly mop-like tops: they enhance and haunt the semi-desert of the south-west United States and Mexico. *Aloe* and *Yucca* (unlike the palms) have reinvented a form of secondary thickening though without the continuous cylinder of cambium that the conifers and dicot trees have.

The Ruscaceae is the family of the butcher's broom – but also of the magnificent and extraordinary dragon trees in the genus *Dracaena*. In general the species of *Dracaena* resemble a yucca or an agave, but some of them are big – up to 20 metres – and hold their own in tropical forests. Some, including *D. cinnabari* and *D. draco* (the dragon tree) yield a red resin, known as dragon's blood. *Dracaena*, too, has reinvented a novel form of secondary thickening.

Finally – worth mentioning at least as a curiosity – there are the extraordinary grass trees, *Xanthorrhoea* and their relatives from Australia, in the Xanthorrhoeaceae family. You may well come across them – there are several species in the beds at Kew – and very strange they look: like little wooden posts with a big tuft of 'grass' at the top. They are not hugely important, perhaps. But they represent yet another way of being a tree, and if a few historical coins had flipped differently who knows how they might have taken off? It might have been that when we now think of trees we would think not of oaks and beeches but of grass trees, no doubt further evolved into all shapes and sizes.

The 'Screw Pines': ORDER PANDANALES

So to the Pandanales, with the single family of the Pandanaceae, colloquially known as screw pines. The 'screw' is understandable, since the tops of the stems are twisted and the long narrow leaves, which in reality are in three rows, seem to form a spiral. The 'pine' is incomprehensible. 'Palm' would have been closer, since many of the Pandanaceae resemble palms – although they are branched, which palms usually are not, and are typically supported by long stilt roots, which is not usual in palms. But screw pines are not closely related to palms, and indeed have no close relatives at all. Not all the screw pines are trees – some are climbers – but members of the biggest genus,

Pandanus certainly are: up to 30 metres tall. The arborescent forms mostly grow around the sea or in marshes – indeed, says Corner, 'there are tracts of rivers where the pandans prevail in serried and impenetrable ranks'. Where screw pines flourish, palms can't get a look in. Some screw pines, however, prefer mountains, at the top of the treeline. They live mostly around the Indian Ocean and the west Pacific. Some have economic importance, not least as ornamentals. It is worth keeping an eye out for them – they turn up in the oddest of places. Some provide good food, too – notably *P. leram*, whose big spherical fruits are boiled to a mealy mass known as Nicobar bread-fruit; and pandanus cake features as a dessert (a delectable one, I am told) on the menus of Thai restaurants. It's odd that such a significant group of plants – some of them big trees – are so little known outside botanic circles. People probably mistake them for palms.

The Palms: ORDER ARECALES

There is only one family in the Arecales order – that of the palms, the Arecaceae (formerly known as the Palmae). But that family includes more than 2,600 species in more than 200 genera, and they are among the wonders of the world.

Most palms are from Asia (especially the south-east), the Americas and Australia, with surprisingly few species in Africa although they grow widely there. In the rainforest of Amazonia palms more than hold their own among the dicots, as major players in the canopy. Most are tropical and subtropical, and on the whole they are tender, presumably because the all-important growing tip (the apical bud) is vulnerable. In most species (though not all) the bud is irreplaceable, and once it is killed (by cold or because someone harvests it as a 'palm heart') the whole tree dies. But the bud is typically protected by fibres and leaf bases that presumably help keep out the cold. So it is that some, like the European fan palm (*Chamaerops humilis*), grow naturally around the Mediterranean, where it can be very cold. Several have drifted north into the United States, including the petticoat palm, *Washingtonia filifera*. Some flourish in mountains – in tropical latitudes, to be sure, but at altitudes where the dicot trees are stunted.

The giant fishtail *Caryota* in the uplands of Malaysia may be 40 metres tall, towering above the oaks and laurels all around. The wax palm, *Ceroxylon*, at more than 60 metres is the tallest palm of all – yet it grows high on the Andes, at 3,000 to 4,000 metres. The Chusan or windmill palm of China, *Trachycarpus fortunei*, can cope with snow and ice.

Most palms look as we all imagine palms should: a tall straight stem with a crown of leaves on the top. But others deviate startlingly from our imagined archetype. Some keep their stems underground, so their leaves seem to shoot straight from the soil surface. The underground stems do not grow horizontally, like the rhizome of an iris. The trunk of *Attalea*, for example, first grows downwards and then, as if realizing its mistake, grows upwards again, to form a U. *Attalea* is also the genus of the American oil palm and is a relative of the coconut – although the coconut palm itself, *Cocos*, grows tall and straight, though often leaning over. There are many so-called 'stemless' palms in the Cerrado, the dry forest of Brazil. Other palms, notably in the genus *Calamus*, are climbers: indeed *Calamus* provides rattan cane, the stuff of colonial screens and chairs that creak on the veranda. Yet there are around 370 species of *Calamus* and many are bona fide trees, suggesting that climbing is a late evolutionary departure, and illustrating once more the extreme versatility of nature.

In many palms the trunk is smooth; in others, patterned with the scars of earlier leaves. Many retain the ragged bases of old leaves, like the African oil palm, *Elaeis*; or indeed like America's *Washingtonia filifera*, which Hugh Johnson in *The International Book of Trees* compares unkindly to 'an Alpine haystack', even though Californians are so fond of it. Many palms are valued for their timber – like palmyra (*Borassus*), much favoured in India. Palm wood, says Corner, can be 'as hard as steel'. Yet the structure is quite different from that of dicots or conifers. The xylem and phloem run from roots to leaves in scattered strands, for there is no cambium to divide the two into neat concentric cylinders. So there can be no secondary thickening. Instead, the stem starts stubby and simply grows taller – although in bottle palms the trunk may swell at intervals, through the general proliferation of tissue and enlargement of cells. These swellings may in part serve as food stores. Bottle palms are not a discrete group:

there are various kinds, in unrelated genera. Best known and most magnificent among them is the royal palm, *Roystonea*, whose tall, pale-grey trunks are swollen for much of their length and topped by a huge shaggy crown of dark-green pinnate leaves. *Roystonea* came originally from Cuba, but is grown as a street tree throughout the affluent tropics. I am told there is a most amazing avenue of royal palms in the Botanic Garden at Rio. (Memo: Always check out botanic gardens!)

Many palms have vicious spikes. Those on the trunk may be like bodkins, up to a foot in length. Some that grow from near the base originate bizarrely from the tips of adventitious roots, rising directly from the trunk. I have grabbed the leaves of palms to keep my balance along Amazonian forest paths and come close to shredding my hands on barbs as sharp as fishhooks, ranked along the midrib. Not nice. Yet on islands in the Indian Ocean that lack insouciant ramblers and rapacious herbivores the palms are spikeless – just as birds on islands that traditionally had no humans or cats tend to be tame (and were wont to perch on the rim of sailors' cooking pots). Armaments, and the emotions of fear and aggression, require energy. With no threat, an air of relaxed innocence can prevail. There is an obvious moral in here.

The roots of palms are peculiar. Firstly, they are adventitious, growing directly from the trunk; generally from the base, of course, but also from higher up, occasionally forming significant stilt roots as in the screw pines, so that the whole tree can look like a rocket about to take off (and sometimes remaining dwarf but with hard and sharp tips to the roots, which form some of the most murderous spines). Big dicot trees have extremely thick roots – but they become thick in the same way as dicot trunks do, by secondary thickening. Palm roots and screw pines have no secondary thickening, but have instead the thickest primary roots seen in nature. They can also be, as Corner says, 'exceedingly numerous' – so that 8,000 roots a centimetre thick may emerge from the base of a coconut tree or an African oil palm, and sometimes up to 13,000. The root system begins normally enough in the seed – in the dicot manner, from the radicle (the primary root) of the embryo. But that first effort soon dies, and the adventitious roots take over. Because there is no secondary thickening, the roots remain cylindrical, though they taper off towards the end and finish with a stout root cap. Seldom are there root hairs of the kind that dicots generally rely upon: water is absorbed from a short region just behind the root cap. Palm roots (like screw-pine roots) are very rich in xylem. They are highly efficient conduits.

Palm roots are lignified through the middle and have huge tensile strength. The roots of the coconut spread up to eight metres in any direction, branching more or less at right-angles two to four times as they advance. Their spread far exceeds that of the crown, and they provide prodigious anchorage. Coconuts grow on windblown islands by the edge of the sea. The waves dig out the sandy soil around them until the trunks collapse to the horizontal, held out like the bowsprit of a ship to a length of 20 metres, with a huge mop of leaves and a couple of score of coconuts at the end. The strain is fierce. Yet the trees remain stable. As Corner says, nothing less than a hurricane can dislodge them. Perhaps it's the roots of palms above all that have made them so successful. They can hang on through storms where dicots and conifers are toppled. Perhaps palms established themselves first as swamp and riverside trees and then spread to the surrounding forests.

The apical bud at the top of the palm, from which the crown arises, is the biggest bud in all of nature. The buds are harvested from many species as 'palm heart' or 'palm cabbage', common on Brazilian menus. Few palms can grow a replacement bud, so harvesting kills the tree: a huge loss if they are taken carelessly from the wild. But some species, like the Brazilian *Euterpe*, can be coppiced like a European chestnut or a hazel. When a stem is cut, more stems grow to replace it. *Euterpe* also yields a fine fruit with the common name of acai (pronounced 'ass – ay – ee'). Its deep-purple juice is a delightful drink, makes excellent ice cream and is one of the most valuable of the 'non-timber forest products' from Amazonia (though of course behind rubber, cocoa and brazil nuts). For small farmers, a copse of *Euterpe*, constantly harvested but then regenerating, can be a fine source of income. For good measure, it has lovely, fernlike leaves.

In general the leaves of palms are either pinnate (feather-like) as in the coconut or palmate (hand-shaped), as in the European fan palm. For some reason the palmate types include most of the hardiest species. The flowers may be of one sex or both. If they are unisexual, the tree itself may be of one sex only (dioecious), as in the American oil palm; or carry both sexes (monoecious). The inflorescences may be huge – up to 250,000 separate flowers. In some species, like the date palm (*Phoenix*), the inflorescence grows from between the leaves and hangs out from the side of the tree, typically below the leaves. In others the inflorescence erupts from the top like a slow firework. Sometimes this is the tree's last fling, for some apical flowerers die after reproduction. The flowers are mostly pollinated by insects – commonly beetles, bees or flies. They often have nectar to reward the pollinators.

The fruits of palms are immensely variable. Most are 'drupes' – a fleshy fruit with a stone, enclosing one or two seeds. Often they are small and berry-like. But often too they are huge, sometimes warty and sometimes fibrous – like the coconut. The seeds of *Lodoicea*, endemic to the Seychelles, are the biggest of any plant: like a great pair of brown wooden buttocks, and also known as the double coconut, occasionally as the bum seed and more felicitously as coco de mer. *Lodoicea* seeds are dispersed by water (like those of coconut and nipa palms) and can stay at sea for many months.

Many palm fruits and seeds are rich in fat, including oil and wax. The coconut is a staple in many countries and a source of valuable oil. The oil palm (*Elaeis guineensis*) from West Africa provides an oil that's used for many purposes, including soap. Oil palm has long been Malaysia's second biggest earner after rubber, and is now being grown on a larger and larger scale throughout the tropics in the get-rich-quick manner of modern agribusiness, to the huge detriment of the traditional farming that actually feeds people. *Attalea* is the American oil palm. *Copernicia cerifera* provides carnauba wax.

Other palm fruit, rich in sugar, are delectable. Besides acai and date, there is the betel nut (from the betel palm, *Areca catechu*). The 200 species of *Bactris* include the peach palms. The pulp of the palmyra fruit yields a pleasant jelly; in India the germinated nuts are eaten as vegetables, and its sap, drawn by cutting through the great bract that envelops the base of the inflorescence, yields a sugar called 'jaggery' and is also distilled to make arrack or toddy. The Chilean wine palm (*Jubaea chilensis*) is the hardiest of the pinnate leaved palms and perhaps the sturdiest of all palms, with a trunk a metre thick. From South America too comes the ivory palm (*Phytelephas*,

The coco de mer *is the world's largest seed and a prodigious mariner*

literally elephant plant): its big and immensely hard seeds are 'vegetable ivory', once favoured for billiard balls.

Palms yield many fibres. The coir of coconuts, which wraps around the nut, is immensely valuable for ropes, matting and potting compost. It helps the wild seeds to float, as they set out across the oceans from island to island. But the cultivated types have less fibre and more fleshy nut, and are more inclined to sink. The leaves of many palms are used to make mats or walls. Those of *Borassus* and of the thatch palms, *Thrinax* and *Coccothrinax*, are used for thatching. Raffia (*Raphia ruffia*) is used for weaving mats and baskets. Rattan (*Calamus*) is stiffer, and is used to make furniture.

Then there are many ornamentals, raised in botanic gardens almost worldwide (including the west of Scotland, anomalously warmed by the Gulf Stream), and along main streets in rich, warm cities from Florida to the Mediterranean to Melbourne. These include the fishtail palm (*Caryota*), the European fan palm, the cabbage palm (*Sabal*) and the queen palm (*Syagrus*). The ugly-attractive *Washingtonia*, sometimes known as the petticoat palm, is the California fan palm. There are the hundred different parlour palms (*Chamaedorea* species),

Royal palms may grow to 30 metres. This is a young one

the needle palm (*Raphidophyllum*), the foxtail palm (*Wodyetia*) and the graceful, fan-leaved *Livistona*, from Australia and Asia. The king palm of Australia (*Archonthophoenix alexandrae*) is magnificent, with its 20 metre tall trunk. Vast, too, is the Indian talipot (*Corypha*), with a trunk up to a metre across.

Finally, out on its own, an eccentric even among the bohemian palm family, is the nipa palm (*Nypa*). Molecular studies suggest that it may be the sister of all other palms – closest to the common ancestor. The most ancient palm fossils known are nipa palms, which lends some support to that notion: they date from the (fairly) early Cretaceous, around 112 million years ago. There are many more from the early Tertiary, around 60 million years ago, not least from the depths of London clay (for London from time to time has been tropical and swampy, and with global warming might soon be again).

But nowadays the nipa palm grows among the mangroves of Asia and the western Pacific, sometimes as the dominant species – with its roots in the sea (or at least in very salty water) and its stems prostrate (meaning not upright) and, unusually among true palms, branching. It holds its feathery leaves vertically. Many palms that may live with their feet in water, including the coconut, the oil palm, and the date palm have loosely spaced cells through the core of their roots that allow air to circulate. But the nipa palm goes one step further. It has large air spaces within its roots which connect with cavities in the base of the floating leaves – rather as the lungs of birds connect with further spaces in their bones. Corner suggests that 'the rise and fall of tide may with slow strokes pump the air around' inside the nipa palm. It's as if it breathes, but using tidal energy – a common trick among mangrove trees. Sometimes entire clusters of nipa palms float away on little islands, perhaps to take root somewhere else, sometimes simply to perish at sea. Sometimes, doubtless, animals ride on these floating islands and populate new continents. The present distribution of animals suggests that such events have taken place many times in the past. Nipa seeds, too, like coconuts and cocos de mer are dispersed by ocean currents. Truly, says Corner, the nipa is 'the swamp palm par excellence'.

All authorities agree that far more research is needed on palms, but it is not easy. A detail, though an important one, is that botanists rely quite heavily on material kept in herbaria – but it is extremely difficult to store the often enormous leaves and inflorescences of palms. Thus, ubiquitous as they are, the palms remain largely elusive. Yet if humanity were allowed to retain only one family of trees from all the several hundreds, the Arecaceae would surely be on the shortlist.

Pineapples, Sedges and Grasses – Including Bamboos: ORDER POALES

Some of the eighteen families within the Poales order are hugely important to our story. The Bromeliaceae includes fifty-one genera and 1,520 species and is best known for the pineapple. But it does have some tree-like forms (including some relatives of the pineapple) and a great many epiphytes – which again include relatives of the pineapple and also *Tillandsia* the so-called 'Spanish moss', which festoons the trees in the swamps of the southern United States, and is very much part of its scenery and folklore. No movie from the Deep South is complete without it. The Cyperaceae, too, the reeds and sedges, include the paper reed or papyrus (*Cyperus papyrus*) which, growing along the banks of the Nile, is distinctly tree-like.

But the family that is most germane to our story, and indeed to our entire existence, is the Poaceae, formerly known as the Gramineae – the 650 genera, with nearly 10,000 species, of grasses. Poaceae is the most successful plant family on earth, alone responsible for vast biotopes and ecosystems (thanks to their hidden apical bud). It includes the cereals – just three of which, wheat, rice and maize, provide humanity with half of all our calories and more than half of our protein, while the more fleshy grasses feed most of our cattle and sheep. But also, which is what matters here, the Poaceae include the bamboos, in the subfamily Bambusoideae. Truly they are trees, often prominent and sometimes dominant in tropical forests throughout the world. I have gazed up at them in Chinese forests – and up and up and up: some grow to 40 metres, taller than most tropical forest trees. One of the world's favourite animals, the panda, has turned itself from a perfectly good carnivore (pandas are basically bears) into a dedicated scoffer of bamboo. (Though if you want to catch a panda, lure it into a (bamboo) cage with roast pork. It's the same with all vegetarians.) Bamboos, like palms, are in many ways eccentric: for example, many flower only at extremely long intervals – every ten to eighty years – and when they do, they all flower simultaneously. How do they manage this? Are they responding to some quirk of light or climate, or to some cryptic message passed between them? And what

Bamboos too may compete in the forest as canopy trees

prompts the message? Some species die after reproducing, sometimes leaving the creatures that depend upon them stranded (like the giant pandas). Binge flowering produces a boom population of young plants over the following years. Many other trees (and animals, such as wildebeest and zebras) practise the same tactic: they produce so many offspring, all at once, that their predators cannot possibly catch them all. A steadier output would produce a steady kill. Perhaps, too, the sudden demise of the adult bamboo plants before the next generation gets going helps to reduce the population of pandas.

For the traditional peoples of Asia, bamboos are among their greatest assets. They lend themselves to every purpose, from brushes and pens to food, pots, cutlery, furniture and musical instruments of all kinds – percussion, wind and strings. They have created an entire aesthetic of painting and architecture – the swishy calligraphy, the great sagging roofs of palaces and temples. Virtually the whole interior of the headquarters of the International Network for Bamboo and Rattan in Beijing is made from, or veneered with, bamboo. Whole books are devoted to bamboos. They deserve them.

Bananas and the Traveller's Palm:
ORDER ZINGIBERALES

The Zingiberales order is named for the family Zingiberaceae, which includes ginger, cardamom and turmeric – beautiful, wonderful, and valuable, but not trees. But two of the other seven families in the order certainly do include trees. The Musaceae are the bananas. They are giant herbs, rather than true trees, for their stems are not woody, but fleshy and fibrous. But their stalks are thick and tough enough to give them the form of a tree, and they live like trees. The banana genus, *Musa*, with thirty to forty species, originates in Asia, particularly in Myanmar and New Guinea: Heywood says 'they are essentially jungle weeds of disturbed habitats'. Bananas are now grown worldwide, and they grow as wayside weeds throughout the tropics. The cultivated banana originated as a hybrid of two species, *M. acuminata* and *M. balbisiana*. Extraordinarily, the modern cultivars are triploid – meaning they have three sets of chromosomes, which also means that they are sexually sterile (since the parent cells from which gametes are produced must contain an even number of chromosome sets). So the fruits of the cultivated types are seedless and sterile – all form and no reproductive substance. Thus a bunch of bananas emerges botanically as an infructescence of monocot triploid parthenocarps (and you never know when such knowledge might come in useful). Growers reproduce the plant itself by cloning from suckers. Other species of *Musa* provide 'Manila hemp', for ropes. Africa has its corresponding genus, *Ensete*, with six species. The exotically decorative *Heliconia* also belong to the Musaceae.

The Strelitziaceae are exotic and alluring: they include the bird-of-paradise flower, *Strelitzia reginae*, which is not a tree, and also the traveller's palm, *Ravenala*, of Madagascar, which very definitely is. This is an extraordinary-looking plant, again beloved of botanic gardens, with its near half-circle of ragged, banana-like leaves, all springing from the top of a long straight trunk that may be 30 metres high. The flowers, encased in tough bracts, produce an abundance of nectar – which attracts the black-and-white ruffed lemur, to whom the prodigious trunk presents no obstacle at all. As the lemurs feed on the nectar they are coated with pollen. Here we see lemur qua hummingbird, bee or butterfly: the key pollinator.

That ends our rapid recce of the monocot trees. The remaining families of broadleaves all belong to the clade of the modern dicots – known as the eudicots.

Known as 'traveller's palm' – but Ravenala *is related to ginger*

8

Thoroughly Modern Broadleaves

Trees? Why not? Big cacti form veritable desert forests

Just to recap: Flowering plants (angiosperms) are traditionally divided into two great groups – the dicots and the monocots. The trees among the dicots are commonly known as 'broadleaves', to distinguish them from the monocots, like palms and grasses, which have narrow leaves. That is: broadleaved trees are dicot trees.

But as outlined in Chapter 6, the dicots are no longer perceived as a single, coherent group. They include a mixed bag of primitive types, such as magnolias, peppers and waterlilies, which are presumed to resemble the ancestors of all flowering plants; and they also include a particular, more modern group, a true clade, known as the 'eudicots'. What I am now calling 'thoroughly modern broadleaves' are the eudicot trees.

The eudicots are the most varied plants of all. They range from tiny floating duckweeds, through a host of herbs and scramblers and climbers, to some of the world's mightiest trees. To look at them, you would see no reason to suppose that they all arose from the same ancestor, and that they indeed form a true clade. But as is so often the case in taxonomy, it's the small, cryptic features that betray true relationships. All the eudicots have a characteristic kind of pollen which has three slits in it, and is known as 'tricolpate'. The eudicots have other features in common too, of course; and their DNA confirms their general relationship. But the feature that pins them down, shows them all to be of the same broad lineage, is the tricolpate pollen. Different authorities divide up the angiosperms into different numbers of families but most agree that there are about 450, and that most of these belong among the eudicots. So there are at least several hundred eudicot families which, you may well feel, is too many to keep track of. Fortunately, however, these families are further grouped into orders of which Judd recognizes thirty-one. This, I hope you will agree, is a manageable number.

As you can see from the figure on p. 159, fifteen of the thirty-one eudicot orders form a true clade, known as 'the Rosids': not that they are all particularly rose-like, but they do include the order of the roses, the Rosales, and the grouping has to be called something. Another ten form the group known as 'the Asterids': this time named after the Asterales, which include the family Asteraceae, which were formerly known as the Compositae, and include the daisies. But six of the thirty-one eudicot orders belong neither to the Rosids nor to the Asterids, and again can reasonably be seen as primitive outliers ('primitive' being a relative term: they are not primitive relative to magnolias, but they are primitive relative to daisies). This chapter deals with the six primitive orders of eudicots. The next chapter looks at the rosids, and the one after that is a rapid survey of the asterids.

Of the six outlying orders (neither rosids nor asterids), two contain

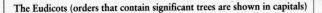

The Eudicots (orders that contain significant trees are shown in capitals)

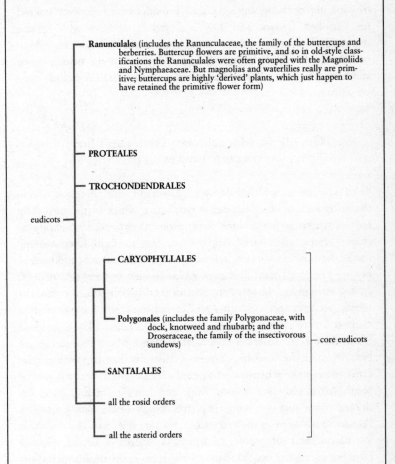

As you can see, three orders among the eudicots can be seen as 'outliers' – primitive types whose relationship to each other and to the rest of the eudicots is not clear. The remaining orders form the 'core eudicots'. The groups 'rosids' and 'asterids' can logically be ranked as subclasses. The individual orders within them are discussed in the next two chapters.

no significant trees. The Ranunculales is named after the buttercups. It does include some shrubs – for example the berberries – and a few trees among the moonseed family, but nothing that need delay us (though one of the moonseed family provides curare, much favoured for poisoned arrows, and now deployed in medicine as a muscle relaxant). The other order without trees is the Polygonales, which include the families of rhubarb and dock, and of the insectivorous sundews. The four orders of primitive eudicots that do include trees are as follows.

Grevilleas, Maçadamias, Planes and Box:
ORDER PROTEALES

The Proteales, as now defined by Judd, is highly intriguing. Of course the order includes the Proteaceae family, for which is it named. The root of the name is the same as in 'protean'; it implies an ability to change shape, and indeed young Proteaceae typically have juvenile leaves that are markedly different in shape than those of the mature plants. The family contains over 1,000 species, in sixty-two genera, spread throughout the southern continents, with occasional encroachments into the north: South America (whence they have spread to Central America); sub-Saharan Africa; eastern India; all of China and South-East Asia; Australia and New Zealand. Many Proteaceae are beautiful, like the bottlebrush trees of Australia that drip with nectar: *Banksia*, and the pincushion-flowered *Hakea*. The *Protea* flowers of South Africa's upland *fynbos*, huge and flamboyant and good for drying, alone justify the journey (the beauty of the landscape is a bonus). Quite a few of the Proteaceae, too, are fine trees. Among them are *Macadamia integrifolia* of Australia, source of the egregiously hard but excellent nuts, which are a staple for many traditional Aborigines and are also the only native Australian food plants of significance to world markets. The Rewarara of New Zealand (*Knightia excelsa*) which grows up to 40 metres, has juvenile leaves that are long and thin and mature leaves like a chestnut's, though thicker and glossier. Australia's *Grevillea robusta* is a magnificent timber tree confusingly known as the silk oak and sometimes even as the golden pine (while other trees are also sold as 'silk oak'). It is also widely grown in India as a shade tree in plantations of tea and coffee, both of which grow best when not too exposed; and its leafy branches when trimmed make fine cattle fodder.

It is a taxonomic novelty to find the plane family, the Platanaceae, in the Proteales order; yet DNA studies show this family to be close to the Proteaceae. There is only one genus, *Platanus*, with about ten species, but it's enough to leave fine scope for confusion; for to the British *Platanus* are the plane trees, while the Americans call them sycamores (though some are also called buttonwoods). To the British, the sycamore is *Acer pseudoplatanus* – meaning 'maple false-plane'. (Just to stir the pot a little further, the 'sycamore' referred to in the Bible is a species of fig, in the genus *Ficus*.) In any case planes make fine ornamental trees especially in cities (they shed their bark and the soot with it) and in North America their timber has been used for everything from barrels and butchers' blocks to fine veneers; while the Native Americans of the east used planes for dug-out canoes, one of which was reputed to be 20 metres long and weighed four tonnes. Planes grow wild throughout the south-eastern United States, and also in the eastern Mediterranean, north India and China. Given that Platanaceae is closely related to Proteaceae, and Proteaceae is obviously Gondwanan, the southern supercontinent), where did the Proteales as a whole originate?

The family of the box trees, Buxaceae, has not been easy to place. It has at times been linked to the rubber tree within the Euphorbiales order – although its members do not have latex, as the euphorbias do. Now it seems to have fetched up in the Proteales. Its 100 or so species (in four to six genera) stretch throughout all the tropics (apart from Australia) and through Europe as far as Scandinavia. Common box, *Buxus sempervirens*, is one of only two evergreen hardwood trees that seem to be native to Britain. (The other is the holly.) But I say 'seem' because botanists have been arguing for 150 years about whether box really is a native, or was simply introduced a long time ago. The box tree is not big – up to 10 metres – but it is highly prized, partly for high-class hedges (the stuff of medieval knot gardens) but also for its close-grained, pale-yellow timber: favoured for rulers and also for wood engraving. Traditionally woodcuts were made with blocks of fruit wood (usually pear), cutting along the grain (as 'twere into a plank), while finer work – engraving – needed metal blocks. But then in the late eighteenth century the English artist Thomas Bewick showed that boxwood, cut across the grain, gave results comparable with metal: and boxwood is now the wood engraver's staple fare (and extremely expensive).

An Oddball from Japan and Taiwan:
ORDER TROCHODENDRALES

There is just one family (the Trochondraceae) in this order, which has just one species, *Trochodendron aralioides*. It comes from Japan and Taiwan, and is large, with a trunk up to 1.5 metres thick. I regret I have not seen a *Trochodendron* and can throw no further light on it (but it provides yet another reason for getting into the countryside in Japan and Taiwan and checking out the extraordinary flora there).

Prickly Trees from Madagascar and 'Tree' Cacti:
ORDER CARYOPHYLLALES

A big grouping, with 8,600 species in eighteen families, two of them including trees (the others range from carnations to sugar beet). First we have the extraordinary family of the Didiereaceae. They are unique – endemic – to semi-desert in Madagascar: that long-isolated island, effectively a small continent, that has so many unique creatures including all the present-day lemurs (and other 'prosimians' like the aye-aye); the fossa, a civet that looks for all the world like a slinky cat; and, probably until about the fifth century AD, the heaviest bird of modern times, the elephant bird – which is probably the roc of Arabian legend. The elephant bird laid the biggest eggs of any known creature, big as an over-inflated rugby football (fragments of shell can still be found). For their part, the Didiereaceae offer yet another remarkable example of convergent evolution, for most of their eleven species (in four genera) might reasonably be compared to the swollen baobab tree, which is also at its most various in Madagascar, though also is widespread in Africa. But more strikingly the trees of the Didiereaceae resemble columnar cacti, with green, swollen, prickly trunks. One of them, however, *Alluaudia procera*, Heywood describes as 'a bent and thorny telegraph pole up to 15 metres high'. It is possible, I suppose, that if the historical coin had flipped differently then Europe and North America might have finished up with Didiereaceae, and Madagascar with oaks and ashes; and then we would think that oaks were weird. Incidentally, Judd no longer recognizes Didiereaceae as a discrete family, but includes it within the purslane family, The Portulacaceae. More work has still to be done on the DNA, however. Watch this space.

Also within the Caryophyllales order are the cacti themselves, the Cactaceae. Although they are close relatives of the Didiereaceae, they surely evolved the cactus-like form independently. Almost all cacti are American (though many – notably the prickly pears in the genus *Opuntia* – have been naturalized in warm dry countries everywhere, and indeed were once a serious pest in Australia). Of course, most of the 1,400 or so species of cacti (in ninety-three genera) are not tree-like. Many horticultural favourites like *Mammillaria* and *Notocactus* are for the most part prickly orbs. But some, including *Opuntia* and many of the columnar *Cereus* and *Carnegia* have woody trunks, and form veritable desert forests. Most tree-like of all are the extraordinary *Pereskias*, almost certainly the most primitive cacti of all, which have green, swollen branches but also have leaves. Within the cacti, indeed, we can see how a family that began as trees (similar to *Pereskia*, and resembling Didiereaceae) has evolved several times, independently, into forms that are very non-tree-like: including the spherical types and the various epiphytes.

All in all, phylogeny never fails to intrigue. It is pleasant to contemplate that cacti and carnations, which seem such poles apart, are not too distant cousins; or that the supremely exotic Didiereaceae and the purslanes that grow as wayside weeds (and feature in traditional European salads), are of the same (extended) family. What does this say of their evolutionary history, and the forces and happenstance that took virtual siblings in such different directions?

Sandalwood and the Mistletoes:
ORDER SANTALALES

The Santalales are a strange lot. Many are epiphytes (plants that grow on other plants – usually trees); and different groups have evidently evolved the epiphyte mode independently. Many are parasites or semi-parasites at least for part of their lives – not simply hitching a ride on trees but tapping into their stems or roots. In the parasitic kinds, the roots are replaced by penetrating organs known as 'haustoria'. But some (including some of the parasites!) are valuable trees – and among them is sandalwood, one of the most treasured of all. Sandalwood shows another common quality of the Santalales: some are exquisitely scented.

The order has proved hard to classify. Many traditional taxonomies recognize four families, although modern work on DNA suggests they should be split into seven – but at least two of those seven (says Judd) probably need regrouping. So the present classification is a little untidy but Santalales as a whole seems to be a good grouping, a true clade, which is what really matters.

Two families, the Loranthaceae and the Viscaceae (traditionally often combined into one), are those of the mistletoes. They are green and do their own photosynthesizing, but their haustoria tap in to the xylem of the trees on which they grow, typically high on small branches, and because they exert more osmotic pressure than the tree itself they suck out the mineral-rich sap. The mistletoes are quaint. Some of the Loranthaceae have bright flowers and are pollinated by birds. The Viscaceae are the 'Christmas mistletoes': favoured as Christmas decorations and charms: *Viscum album* in Europe and *Phoradendron leucarpon* in North America. Christmas mistletoes were sacred to the Druids, symbols of immortality – the obsession of Asterix's chum and Mentor, Getafix; and (so legend has it) provided the spear that killed the Norse god Balder. But they also do immense damage to the trees that they favour – mainly members of the Rosaceae and Tiliaceae families, but others too. They create general water stress and ruin the timber by inducing big knots and causing their hosts to produce masses of twigs known as 'witch's brooms' (though there are other causes of witch's broom as well, including the bacteria which commonly affect birches). *Arceuthobium* of the western United States is the dwarf mistletoe: a major pest of conifers.

The Misodendraceae family contains just one genus with about eleven species of shrubs that live as parasites on the trunks and branches of the southern beech, *Nothofagus*. As we will see shortly, *Nothofagus* is a great Gondwanan genus, related to the northern oaks and beeches (although now given its own family), which extends all through Australasia and South America; but the particular parasite that plagues it is confined to southern Chile and Tierra del Fuego. I never cease to be amazed by the number of sub-plots in nature. To the Chilean *Nothofagus* and the parasites of the Misodendraceae their contest is a huge drama – of which virtually all other species, including most of our own, are completely unaware.

Among the twenty-five genera of the loosely defined tropical family Olacaceae (including the Opiliaceae and Schoepficaceae) are shrubs and climbers. There are also some bona fide trees, including some that are locally valuable. *Scorodocarpus* of Asia smells strongly of garlic (its name means 'garlic fruit') and its timber, though smelly, is very strong and used in heavy construction. Tallow wood or hog plum (*Ximenia americana*) is used as a substitute for sandalwood in South America: its timber is hard and yellowy-pink (and its fruits, laden with prussic acid, are extremely bitter). The African walnut, *Coula edulis* is another very strong timber for building.

But the family that makes the Santalales order so important in human affairs is the Santalaceae, whose 400 or so species in thirty-five or so genera include *Santalum album*, the sandalwood tree (from the Hindi, *sandal*). Its heartwood is the stuff of incense and is beautiful: smooth and pale sandy-coloured, cutting like wax, carved throughout Asia into a million artefacts (endlessly intricate boxes, elephants inside other elephants, and so on). Entire trunks are smoothed and used as scented pillars in many an eastern temple. In Mysore it was a royal tree. Apparently sandalwood is endemic to Timor and its neighbouring islands, but was brought to India more than 2,000 years ago: there are texts that apparently refer to it in the Pali *Milanda-panha* (150 BC) and the *Mahabharata*. Its small, purple-black fruits are dispersed by birds, who presumably helped it to spread throughout India. All in all it is immensely valuable commercially and culturally.

Yet the sandalwood tree is a parasite too: at least when young it taps into the roots of a variety of trees. Among its many favoured hosts are the strychnine tree, *Strychnos nux-vomica* (a relative of the buddleia). Another host is the pestilential *Lantana* – a shrub imported into India by the British from South America as an ornamental, which now seems to grow along every wayside and in every wood throughout the tropics. Dr Sas Biswas, of the Forestry Research Institute in Dehra Dun, north India, tells a charming story of sandalwood trees he once found growing in a dead straight row in the middle of nowhere. Why were they there? Who had planted them so carefully and then abandoned them? No one, is the answer. But in the past there had been a garden; and around the garden was a fence; and along the fence grew the inevitable *Lantana*; and sandalwood had grown as parasites from its roots. Now the garden and the fence are long gone and the *Lantana* with it, but the sandalwood remains. Dr Biswas is adept at reading the history of landscapes from the trees that are left in it: the ambitions of the people who lived there, and the time and chance that in the end reduced their ambitions to nonsense. But nature is opportunist, and lives on.

Finally, members of this whole order are particular pests of some species of acacia trees and other leguminous trees, of which more later. On the other hand, opportunist foresters seeking to grow sandalwood trees in plantations, often use acacias as hosts to start them off. It's an ill wind.

9

From Oaks to Mangoes: The Glorious Inventory of Rose-like Eudicots

Big, hollow baobabs may double up as funeral parlours and cafés

Fifteen of the eudicot orders are grouped together to form what might be called a sub-class, known informally as 'the rosids': the rose-like eudicots. In truth, most of them are not literally rose-like. What really links them all together is not any obvious feature of their flowers, but details of their DNA. But it seems to be enough. At least in the present state of knowledge, the rosids do seem to form a coherent group. If future scholars decide that they are not so closely related as it now seems – well: that's the way science goes. Nothing is ever absolutely certain.

Of the fifteen rosid orders, four do not contain significant trees. The Vitales are vines, including the families of grapes and of Virginia creeper. The Geraniales include the cranesbills and the pelargoniums, parched and sooty-potted on their suburban windowsills. The Cucurbitales includes the family of the melons, cucumbers, squashes, marrows, gherkins, gourds and pumpkins. They are singularly untree-like – and yet the Cucurbitales seems to be closely related to the Fagales, which contains only trees, including some of the mightiest, like the oaks and beeches. The Brassicales includes the family of cabbages and wallflowers. Fine plants all. But the rest of this chapter focuses on the eleven rosid orders that do contain trees – including most of the world's finest and most valued.

Witch Hazels, Katsura and Sweet-gums:
ORDER SAXIFRAGALES

The Saxifragales are a difficult group, still being sorted out. One reason they are tricky is that within the Rosid subclass are many plants that are clearly somewhat primitive – close to the presumed ancestral state. The Saxifragales may be the most primitive of all (the 'sister' group to all the other Rosid orders). Plants with lots of primitive features and few truly distinctive features are always hard to classify – and modern studies, based on cladistic analysis of DNA, often differ markedly from more traditional treatments based solely on anatomy.[1] Judd now includes thirteen families within the Saxifragales, totalling about 2,470 species. Among them are many pleasant and well known herbs and shrubs, including the families of the saxifrages, the stonecrops and houseleeks, and the peonies.

But there are also a few families with some most interesting trees. The Hamamelidaceae includes about eighty species in twenty-five genera of which the most pertinent here is *Hamamelis*, the witch hazels. *Hamamelis virginiana* from the eastern United States has hazel-like leaves and yields a lotion widely used as an astringent and for soothing cuts and bruises; and diviners, at least in the US, favour its

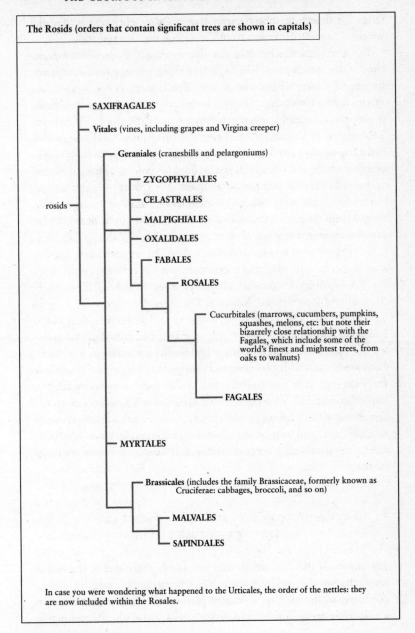

The Rosids (orders that contain significant trees are shown in capitals)

rosids
— SAXIFRAGALES
— Vitales (vines, including grapes and Virgina creeper)
— Geraniales (cranesbills and pelargoniums)
— ZYGOPHYLLALES
— CELASTRALES
— MALPIGHIALES
— OXALIDALES
— FABALES
— ROSALES
— Cucurbitales (marrows, cucumbers, pumpkins, squashes, melons, etc: but note their bizarrely close relationship with the Fagales, which include some of the world's finest and mightest trees, from oaks to walnuts)
— FAGALES
— MYRTALES
— Brassicales (includes the family Brassicaceae, formerly known as Cruciferae: cabbages, broccoli, and so on)
— MALVALES
— SAPINDALES

In case you were wondering what happened to the Urticales, the order of the nettles: they are now included within the Rosales.

twigs for detecting subterranean water. Other *Hamamelis*, like the winter-flowering *H. mollis* from China, are fine ornamentals.

The Cercidiphyllaceae has just one species, *Cerciphyllum japonicum*[2] – the much-valued katsura tree from the northern temperate forests of Japan, China and Korea. The katsura is big – up to 30 metres high with a trunk up to 1.2 metres in diameter – and sometimes it has several trunks, deeply furrowed and often spirally twisted. The timber of katsura is excellent: not too heavy, straight-grained, lustrous, much prized for wood-carving, delicate mouldings, high-class furniture and veneers. It is also used for pencils and cigar boxes, and to make traditional Japanese shoes.

Finally, the Altingiaceae family includes rasamala (*Altingia excelsa*) from Assam through South-East Asia, valued both for its heavy timber and for its fragrant gum, used in perfumery; and also *Liquidambar*, which grows both in Asia and North America. Superficially, *Liquidambar* species look like maple. Best known is *Liquidambar styraciflua*: the American sweet gum which grows from New England through Mexico and into Central America. The trees are often huge – up to 46 metres tall and 1 metre in diameter. Sweet-gum heartwood is brownish and pinkish, often darkly streaked, and is sold as 'red gum'; while its sapwood is creamy white and is sold separately as 'sap gum'. Both are favoured for furniture and veneers (and also more mundanely for packing cases and pallets). But also, when sweet-gum bark is wounded it exudes a vanilla-scented resin known as storax or styrax which is used in perfumery and also as a medicine (as an expectorant and inhalant, and to treat skin diseases). Like maple, *Liquidambar* is famed for its autumn colours and several kinds are grown as ornamentals.

Creosote Bushes and the Wood of Life:
ORDER ZYGOPHYLLALES

The plants of the Zygophyllaceae family, the only one in the order, are herbs, shrubs and trees with showy flowers pollinated by insects, that tend to favour dry and salty places and often predominate in scrub. They are also rich biochemically and so are oily and aromatic, and are favoured as medicines. Among them is the creosote bush, *Larrea* to be found out in semi-deserts and in gardens as ornamentals. Among them too is the genus *Guaiacum* from the Caribbean and Central and South America: small trees (up to around 9 metres tall and 40 centimetres thick), known collectively as 'lignum vitae'.

Lignum vitae of course means 'wood of life', for in the sixteenth century the tree was believed to offer a cure for syphilis. But in addition, the timber, sometimes greeny-brown and sometimes almost black, is one of the heaviest of all timbers, and enormously strong, with prodigious crushing strength. So it finds favour for sculpting and turning (although it is fearsomely hard on the tools), and for example for mallet heads: ideal for croquet. Because the wood is innately oily, too, it is self-lubricating, and so is favoured for rollers and wheels in pulleys and various machines – especially for parts that are hard to get at and lubricate. The related American genus *Bulsnia* is also sold as lignum vitae, but is not so good. But *Guiacum* is becoming rare, and indeed is now covered by CITES (the Convention on International Trade in Endangered Species), which at least gives some legal control over harvesting.

Spindle trees and Khat: ORDER CELASTRALES

The Celastraceae family is named after *Celastra*: a climber known as bittersweet. *C. scandens* is the native North American species, much admired. But it is now largely replaced by *C. orbiculatus*, introduced from East Asia – an aggressive weed that grows over everything, often smothering other plants, and is pertinent here not as a tree but as a spoiler of trees (at least when out of its native lands).

But the fifty-five genera of Celastraceae also include the genus *Euonymaeus*. Among these is *Euonymous europaeus*, the spindle tree, whose fine-grained wood is favoured for carving and turnery, and whose seeds yield an oil for soap and a yellow dye for colouring butter. *Euonymous hians* from Japan is also used in turning, and for making printing blocks. Other *Euonymous*, including some that are shrubby rather than arboreal, yield a rubber-like latex and some feature in Native American medicine. The pharmacological propensities of the Celastraceae family are also evident in the khat tree, *Catha edulis*, whose leaves are chewed in the Middle East as a stimulant. The kokoon tree from Sri Lanka (*Kokoona zeylanica*) yields a useful oil.

Rubber Trees, Mangroves, Willows, Poplars, and Some Truly Prodigious Hardwoods:
ORDER MALPIGHIALES

The thirty-five families of the order Malpighiales include a great variety of truly remarkable trees both tropical and temperate: the rubber tree, the cactus-like euphorbias, the red mangroves, the willows and poplars, and some formidable American and African hardwoods. Some are of huge ecological significance, some are the basis of important industries, and some are just very strange. Three families in particular are outstanding.

The family Euphorbiaceae is truly extraordinary. It includes the herbs known as spurges, common as wayside flowers and beloved of gardeners. It also includes the finest range of cactus look-alikes: some are an almost perfect imitation of the *Cereus*-style cacti, having dark-green succulent columns with spiky ridges – almost uniquely goat-proof, and much favoured for hedges in Africa. Cassava or manioc, *Manihot esculenta*, is a major, starchy staple throughout the tropics. The shrub *Ricinus* provides extremely valuable castor oil. The oil of *Jatrophus* is so pure that it will run a diesel tractor without refining – pressing and filtering are all that's needed. Among the bona fide trees – providing more valuable oils – are the candlenut tree, *Aleuritis moluccana*, and the tung tree, *A. fordii*, and the Chinese tallow tree, *Sapium sebiferum*. Most important of all, however, by far, is the rubber tree, *Hevea brasiliensis*, a native of Brazil but now grown also in Africa and Asia, and most prodigiously in Malaysia. (Note, though, that the archetypal house plant known as the 'rubber plant' is a species of *Ficus* or fig.)

Rubber in its raw form is latex, a creamy gum that plants exude when wounded. Why they do this is a mystery: none of the many explanations (most obviously, that latex is intended to heal the wound) seems to stand up to scrutiny. But a great many plants do it, from quite a few families. Several have been tried out as a source of rubber. The Russians have produced rubber from their native *Guayale*. Even dandelions produce latex. Notoriously, in the late nineteenth and early twentieth centuries, the Belgians enslaved many thousands of Congolese people and sent them off into the jungle to tap latex from the vine *Landolphia* (a relative of the periwinkles in the family Apocynaceae). The endeavour produced some 60,000 tonnes of rubber, at an estimated cost of one human life for every 4 kilograms.

But *Hevea* is the best source by far. Various species are widespread through the American tropics, a few trees per hectare, living in the wild for a hundred years or so and growing to 40 metres. The native people had known their properties for centuries, and made play-balls and religious figures from their latex. Europeans came across them and the people who made use of them in the eighteenth century. In England, the chemist and theologian Joseph Priestley found that balls of latex would erase pencil marks, and so coined the deprecatory term 'rubber'. Many other languages prefer more dignified variations on the beautiful native word *cachuchu*: 'weeping wood'.

Rubber first took off commercially in the mid nineteenth century, with new technologies to extend its use. First technologists learned to shape it. Then in 1839 (patented in 1844) the American Charles Goodyear developed 'vulcanization': hardening the rubber by combining it with sulphur. Rubber made an enormous impression at Prince Albert's Great Exhibition in London in 1851: Goodyear built an entire 'Vulcanized Court', with rubber walls, ceiling and furniture. The turnaround came with John Dunlop's invention of the rubber tyre in 1888. The rise of the motor car from the late nineteenth century onwards, and particularly Henry Ford's mass production, were the final flourishes.

For a time, Brazil did very well out of rubber. Indeed it seemed set to transform the country's economy. The focus and symbol of the trade was and is Manaus: a teeming city built a thousand miles up the Amazon. In truth this is not quite so bizarre as it might seem since the Amazon is broad and access is easy (relatively speaking), and once the trees had been cleared around the place where Manaus now stands, the ground proved solid enough. (New York was once a forest too, and London was largely swamp. Venice, in effect, still is.) Manaus blossomed in the early decades of the twentieth century when Henry Ford was turning out the first true 'people's cars' in the form of the Model T. In a fit of what seemed at the time to be perfectly justified bravado, Manaus built itself an opera house, and very splendid it still is. Caruso sang there and Pavlova was due to dance but could not face the final leg of the journey and sent her apologies from Belém, at the mouth of the Amazon. Whether the megastars who favoured Manaus with their glittering presence were aware of the horrendous cruelties inherent in old-style rubber production (torture, rape, enslavement and murder), I have no idea.

But the seeds of destruction had already been sown, literally, before Manaus was even built. In the 1860s Sir Clement Markham of Britain's India Office, who had already organized the introduction of *Cinchona* (for quinine) from tropical America to plantations in India, sought to do the same with *Hevea*. Botanists from Kew soon established that *Hevea brasiliensis* was the best of the genus. Eventually, in March 1876, Henry Wickham brought 70,000 seeds of *Hevea brasiliensis* from Brazil to Kew, where just over 2,000 of them germinated. From Kew young plants were sent to Ceylon (now Sri Lanka) and Malaya (now Malaysia). Soon the Asian plantations were the world's major producers. Their only real setback since then has come from the rise of synthetic rubbers, first developed in America during the Second World War when the Japanese occupied the principal rubber plantations of the East: the US synthetic rubber programme was second only to that of the atomic bomb. By the 1980s, natural rubber filled only 30 per cent of the world market but by 2002 it had sprung back to 40 per cent. Perhaps this is part of a general world shift from industrial chemistry to biotechnology. But natural rubber still fills various special niches, for example in aeroplane tyres and in condoms.

Brazil now produces far less rubber than Malaysia: for one thing, its plantations are beleaguered by an untreatable fungus called South American leaf blight, which those of Asia have escaped. But for Brazilians, rubber is still of immense cultural importance. The rubber tappers' movement in Acre in the 1980s did much to draw the attention of the West to the plight of the Amazonian forest. Overall, the

Brazilians continue to smart at what they see as the theft of their inheritance. Some argue that Henry Wickham did nothing underhand. At least, he declared his cargo to the Brazilian customs officers as 'exceedingly delicate botanical specimens specially designate for Her Majesty's own Royal Gardens at Kew': disingenuous to be sure, but not inaccurate. Others see the entire episode as biopiracy. All in all it does seem most unjust, but this is a tricky area. After all, the Brazilians are doing very well out of eucalyptus, which Europeans brought from Australia in 1828 (though it has not always been planted wisely and in places is a serious pest). Brazil is also pushing ahead vigorously with teak, from India. Its biggest agricultural export these days is soya – a Chinese plant. Brazil's cattle came originally from Europe (and to some extent from India). For their part, the Chinese grow enormous quantities of potatoes and maize, which are American, and wheat, from the Middle East. Deciding who has a right to what is a key issue of present-day politics and globalized commerce. (More in Chapter 12.)

On more practical matters: plantation rubber trees generally start producing useful amounts of latex at around seven years; are producing maximally by age fifteen; and are generally chopped down and replanted after thirty years (when they are about 20 metres tall). So the grower has about twenty-three productive years out of thirty. In the past the timber was simply burned, but it is a pleasing reddish brown, and strong, and in recent decades has become a major crop. Malaysia and Thailand now export nearly $1.5 billion worth of rubberwood furniture. South-East Asia could harvest more than 6.5 million cubic metres of rubberwood per year – almost as much as the entire timber harvest of Central America.

In Malaysia I found the rubber plantations in many ways attractive: green and aromatic shade in a colonnade of trunks, and all as cool as Sussex. But the work – slashing the bark every day and replacing the receptive cups – is desperately tedious, and recruitment of labour is difficult. My solution would be to make the work more interesting – by integrating smaller plantations with mixed farming, as I have seen in China. Indeed, rubber trees do lend themselves to agroforestry. Other crops (including valuable herbs) may be grown among them and their shade is good for livestock. As a bonus, they produce big oily seeds – which are thrown spectacularly for several metres as the fruit dries and splits. In the wild, in Amazonia, these seeds are dispersed by large river fish. In plantations, I would have thought they would be ideal for turkeys. I would hate to be a traditional rubber-tapper, and feel sorry for those who are (and their daughters in Malaysia take off for the electronics factories). But a mixed exercise in agroforestry, with other crops and livestock, is a different proposition all together. A grand challenge: most interesting.

Then there are the Rhizophoraceae – the family of *Rhizophora*, the red mangrove trees. Mangrove forests grow at the edge of the sea throughout the tropics: in northern Australia, South-East Asia, West Africa, around the Red Sea, along the north coast of South America, on both coasts of Central America and in the Caribbean. Although mangroves occupy only 180,000 or so square kilometres of the earth's surface they are hugely important ecologically and economically. Around their roots breed a host of sea creatures, including many ocean fish. Local people take as much from the mangrove forest as forest people take everywhere – fuel, timber, fruits – and fish, too. Offshore from the mangroves, typically, are the seagrasses – food for fish, molluscs, manatees and marine iguanas – and beyond them lies coral reef, which in diversity of wildlife is second only to tropical forest. The mangroves, if left intact, protect the seagrasses and the reefs. Since the mangroves, seagrasses and corals are all nurseries for marine creatures, the consequences of mangrove destruction extend through all the oceans. Yet they are being destroyed – to make way for marinas and promenades, for lagoons to raise tropical shrimp for Western supermarkets, or even, it sometimes seems, just for the sake of it. In Panama in 2003 I was shown a mangrove that had been filled in with rubble to provide a park for containers, of the kind used for ocean-going cargo ships. The local government bought the idea from some entrepreneur on the grounds that it would provide employment. The only employee when I was there was a man with a gun, to keep people off. The entrepreneur, having pocketed the taxpayers' money and destroyed the mangrove forest (and the wild creatures, and the livelihoods of the people who were living there) was about to sell his barren dump to the Chinese. That's business, apparently.

To return to the natural and saner world: about eighty species of trees have mastered the adaptations needed to live in the intertidal zone. Of these, thirty or forty are core species, which turn up in most mangrove forests; and of these the most important overall are seven or eight species of *Rhizophora*.

The third great arborescent family of Malpighiales is the Salicaceae. It is named for the genus *Salix*, the willows. There are about 400 species – although as outlined in Chapter 1, they are hard to identify and very prone to hybridize, so it will always be effectively impossible to say exactly how many there really are. They range from small shrubs to big trees, and from the tropics to the extreme north. Up cold and windy mountains, and on the edge of glaciers up towards the Arctic, they are often the chief of the woody species. In such territory, they send out underground stems to form vast clones: a wood that in effect is a single plant. Thus the creeping willow, *S. repens*, colonizes marshland and begins its transition into forest. Most willows like the edges of rivers, where they are commonly planted to stabilize the banks. They may serve, as reedbeds do, to purify the water. Some are ornamental: the original weeping willow in particular, *S. babylonica*, originated in China, drooping languidly over lakes and lazy rivers as if specifically intended for patterned tea sets.

Willows belong in the northern hemisphere, although one (*S. muc-ronata*) crosses the equator in Kenya. Many kinds live in western China, which is truly one of the world's great centres of diversity; but, like so many trees, these Chinese willows have not yet been properly studied. Willows are nearly all dioecious (males and females on separate plants) and produce catkins that are usually pollinated by insects, although wind probably plays a part. Their seeds are tufted, and float on the wind. The different forms of willow find many traditional uses: the thinner twigs (especially of osiers, *S. viminalis*) for baskets, coracles and hurdles; the bigger timbers for construction. Female clones of *S. alba* var. *caerulea* are the sole source of wood for cricket bats (though sadly threatened from time to time by the bacterium *Erwinia salicis*). The finished cricket bat is a botanical extravaganza: white willow for the blade, bamboo and rubber for the handle, plus twine to bind the handle and glue to hold it in the blade, both of which may come from plants, and linseed to keep the bat supple. Willow is also an important fuel wood, now much vaunted as a source of biomass to supply energy without exacerbating global warming. Finally, the bark of willow is particularly rich in salicin, the root molecule of salicylic acid, the stuff of aspirin; of proven use as an analgesic and anti-inflammatory, and now favoured to reduce blood clotting and guard against thrombosis.

Also in the Salicaceae and closely related to willow is poplar (*Populus*), the twenty-nine or so species of which include the aspens, like North America's quaking aspen, *P. tremuloides*. Poplars are hugely favoured for plantations worldwide, for matchsticks and paper pulp. They also serve as windbreaks and (like eucalypts) help to dry out wetlands, acting as they do like wicks. A major task now is to conserve their genetic diversity, as the wet river banks that they favour are drained and contained. The levees of the Mississippi have hugely reduced the natural regeneration of the native *P. deltoides*. Attempts are afoot worldwide to conserve the diversity in arboretums. In Europe, EUGORGEN, the European Forest Genetic Resources Programme, is intended to conserve the natural diversity of the black poplar, *P. nigra*, and holds nearly 2,800 clones from nineteen countries. In the Pacific north-west, GreenWood Resources holds 100 stands of *P. trichocarpa*, to counteract the losses along the Columbia and Willamette rivers and their tributaries. Efforts to conserve poplars in situ include the Tarim River nature reserve in the Xinziang Autonomous Region of China, largely intended to conserve the remaining third or so of the original *P. euphratica*; and a plan supported by the International Poplar Commission and the United Nations to conserve the native variety of *P. ciliata* in the Himalayan foothills of India. Such efforts are heartening. *Populus*, however, is only one among many thousands of genera of trees, and the vast majority are receiving no help at all. Even if they escape extinction, few will escape severe genetic diminishment.

There are some other notable Salicaceae too – including some that are traditionally placed in the Flacourtiaceae, which Judd enfolds within the Salicaceae. Among them are the Maracaibo boxwood, *Gossypiospermum praecox*, from Cuba, the Dominican Republic, Colombia and Venezuela – highly favoured for specialist tasks, not least for the working parts of pianos; and odoko from West Africa, species of *Scotellia*, hard and tough and excellent for floors.

Several other families in the Malpighiales are worth a passing mention. The Violaceae in temperate regions manifest as herbs including violets; but in the tropics produce some fairly mighty trees. The Malpighiaceae, for whom the order is named, include the Barbados cherry. The Clusiaceae are best known as the family of St John's wort, which finds favour as an antidepressant. But it also includes some handy trees – not the least being the 200 or so species of *Garcinia* which include the mangosteen, *G. mangostana*, a native of Malaysia with fruits the size of a ping-pong ball, whose brownish purple leathery skins enclose treacly and truly delicious (I can vouch) creamy-white segments. An excellent fruit, though the trees are slow-growing and not easy to cultivate.

The Ochnaceae family includes the ekki tree from West Africa, *Lophira alata*, from whose timber (for some reason) the tracks of the Paris metro are made. The metro trains have rubber tyres. I venture that few who ride the Paris metro realize their debt to the trees of the Malpighiales.

Star fruit and Coachwood: ORDER OXALIDALES

The Oxalidales is yet another order that has been subject to serious reclassification. The family for which the order is named, the Oxalidaceae, is primarily tropical and subtropical and is known to northerners primarily through wood sorrel (*Oxalis acetosella*). It does include a few small trees, however – one of which is the star fruit or carambola, *Averrhoa carambola*. The fruit is juicy, sometimes sweet and sometimes acid, is deeply ribbed and thus star-shaped in cross section, and has lately become fashionable outside its native Indonesia. Modern DNA studies also place the Cunoniaceae within the Oxalidales: and these include *Ceratopetalum apetalem*, a tall (18–24 metre) and valuable timber tree from New South Wales, known as coachwood (or lightwood or scented satinwood). Its browny-pink, fragrant timber is used for lots of things, especially joinery and mouldings, but also for gunstocks, shoe-heels and musical instruments. Thus the wood sorrel emerges as yet another homely herb with wildly exotic relatives.

Trees for Fodder, Fuel, Flowers and Beautiful Timbers: ORDER FABALES

Within the Fabales order are four families but the only one that need delay us is the Fabaceae – and they should delay us plenty because, together with the grass family, Poaceae, the Fabaceae is the most important plant family of all, both ecologically and economically. It is also the third largest with 18,860 or so known species in 630 genera, but the inventory is still rising fast. Only the orchids and daisies have more species. Fabaceae used to be called Leguminosae – the pods are called 'legumes'. The old family name is now officially defunct but the adjective 'leguminous' and the informal noun 'legume' (applied to the whole plant as well as to the pods) live on.

Fabaceae take every form: herbs such as vetches, clovers and alfalfa; vines like peas and runner beans, which climb by twining and/or with tendrils; woody climbers like *Wisteria*; shrubs like gorse and broom. But in addition, many of the tropical genera in particular make fine trees that are of huge significance worldwide in forests and savannah, while providing every kind of service for humanity and our domestic animals.

The special trick of the Fabaceae – although not all do it – is to retain colonies of nitrogen-fixing bacteria of the genus *Rhizobium* in special nodules within their roots. The nitrogen-fixers are to a large extent self-nourishing, and provide nitrogen-rich leaves and seeds even in poor soils – and 'nitrogen-rich' in practice generally mean 'protein-rich'. Thus many legumes are especially valued for food – including all the pulse crops (peas, beans, lentils, chick peas, ground-nuts (alias peanuts) and many a tree); and even more provide outstanding fodder, whether growing wild or cultivated (when the trees also provide shade for the grazing animals). Even when not eaten, the nitrogen-rich leaves are often dug or ploughed into the soil to make green manure. Equally to the point, nitrogen-fixing plants are generous with their nitrogen: they enrich the soil around them. So clover and alfalfa and vetches have long been used to enrich pastures – nourishing the grass that grows alongside; and pulse crops (beans, peas, lentils, chick peas) complement cereals (which of course are also grasses); and leguminous trees of many kinds are the greatest of all the candidates for agroforestry, which offers one of the principal hopes for a sustainable world.

Many genera are outstanding but perhaps the most important, ecologically and economically, are the acacias. The 1,300 or so species grow almost throughout the tropics and subtropics: more than 950 in Australia, where they are known as wattles; another 230 or so in the New World; 135 in Africa – mainly out on the savannah where, flat-topped, they are often the only source of vital shade; eighteen more in India; and a few others dotted around Asia and endemic to odd islands. Not all in the genus *Acacia* are trees (some are shrubs or woody climbers), but a great many are.

Some acacias thrive in the wet – some in the American tropics live in rainforest; and some, like *A. xanthophloea*, survive periodic flooding. But most thrive in harsh, dry environments, and have many adaptations to extreme aridity. Some, like *A. eriloba* of Africa, have extremely long taproots, stretching down to aquifers as much as 12 metres below the surface. Some have very small leaves – or have replaced their leaves with flattened leaf-stalks (petioles) known as 'phyllodes' (as in the celery pines, described in Chapter 3). Generally acacias shed their leaves when it's very dry, sometimes all at once, sometimes progressively as the aridity increases – never having more than the conditions will support; but some desert kinds produce fresh leaves before the rains return, to the delight and benefit of camels, antelopes, giraffes, and the nomadic tribes of Africa who need fodder for their cattle, sheep and goats.

In general acacias do well in soils that are poor and disturbed – and so they are excellent colonizers: for example the Australian black-wood, *A. melanoxylon*. Some, like Australia's *A. auriculiformis*, tolerate toxic or highly acid soils. Many acacias are adapted to fire, including most of those in Australia: in some, fire stimulates germination; in others (including some from Africa), it promotes coppicing (regeneration of shoots). On the other hand, some dryland kinds withstand freezing. In some the seeds are known to remain viable in the ground for up to sixty years. Some reproduce by apomixis (a form of parthenogenesis: the new tree grows from an unfertilized ovule). Some spread themselves by suckers as many trees do (for example willows, poplars, elms and redwoods).

Their pioneering hardiness is both an asset and a menace. It is good for land reclamation – and so in Australia *A. auriculiformis* is used to colonize acid mine tips. But it also means acacias make excellent weeds. We see the worst and the best of them when foresters or gardeners take them from one continent to another – for all nature is unpredictable and nothing more so than the behaviour of 'exotics'. Most plants or animals die when taken to new places, which they are not adapted to. Some settle in and become naturalized, and whatever the native wild species may think of the invaders they can be economically valuable – so it is that Australia's *A. mangium*, for example, has become a valued timber tree in India. But some become rampant and are hugely destructive – and so Australia has its rabbits, cats and foxes and also *A. nilotica*, from Africa, plus others from America. Australia has got its own back, however, with exports of immensely destructive acacias to Africa, Portugal and Chile (and of course of eucalypts to absolutely everywhere, and possums to New Zealand).

Like various other members of the Fabaceae, acacias have formed some close symbiotic ('mutualistic') relationships with ants. In fact different acacias have clearly formed such relationships independently, more than once. Thus many acacias have thorns, typically at the bases of their leaf stalks; and some species in Central America have 'swollen' thorns that are hollow, and accommodate colonies of ants. *A. melanoceros* houses ants of the genus *Pseudomyrmex*. In Africa, whistle-thorn acacias such as *A. seyal* have resident colonies of *Crematogaster*. Ant acacias often provide board to go with the lodging, in the form of protein-rich food stores. The ants, in turn, rid their hosts of pests – not only insects but also, presumably, browsers: for few would risk ants up their snouts. (I have had ants up my arm in India, picked up from an epiphyte, where they create airy chambers by sewing the edges of the leaves together. I wouldn't fancy them up my snout either.)

Taken all in all, acacias are wonderfully integrated socially. Below ground many (though not all) house nitrogen-fixing bacteria to aid with nourishment. Typically, too, they also form mycorrhizae in association with fungi, which further increases their nutritional efficiency. Many harbour ants for housekeeping. They employ a variety of insects – flies and beetles but mostly bees – and sometimes birds to pollinate their flowers; and Africa's *A. nigrescens* may be pollinated at least in part by giraffes. In some species a variety of animals help to spread their seeds: some have brightly coloured arils (fleshy exteriors) around their seeds to attract birds; others increase the attraction by suspending their seeds beneath the pods – in some the seeds are dispersed by antelopes and elephants, passing through their guts. Thus an acacia tree is a veritable hotel; or perhaps it should be seen as the ultimate networker, with a host of mutually beneficial associations with representatives from just about every other class of organism.

As we will see, too, in Chapter 13, acacias also team up with each other, issuing chemical warnings to their fellows that giraffes are on the prowl. Clearly this is necessary. In recent years giraffes have been introduced to places in South Africa where giraffes do not naturally live; and they have all but wiped out the native *A. davyi* at least where the trees are accessible, because, apparently, these acacias are not well adapted to giraffes. Here again we see the menace of introduced species, and also a clash of conservation aims: do we prefer big mammals or native trees? Here is one more reason why so many trees of all kinds are endangered – including thirty-five of the acacias (which is almost certainly an underestimate).

Many acacias are cultivated for many purposes. *A. auriculiformis*, *A. mangium* and the Australian blackwood, *A. melanoxylon*, are the most favoured timber trees. The Australian blackwood grows wild in Queensland and New South Wales to a height of up to 30 metres; nonetheless it grows as an understorey tree, beneath the giant eucalypts known as mountain ash, which may grow to nearly 100 metres. The dark, black-flecked timber of Australian blackwood is highly valued for everything from boats to billiard tables. Other acacias are grown for chipping and for pulp. As noted above, too, various members of the parasitic families of Loranthaceae and Santalaceae favour acacias of Africa and Australia as hosts; and these acacias, accordingly, are grown as host trees in sandalwood plantations, proving that foresters can be opportunist too. Some acacia seeds are highly nutritious: *A. colei* and *A. tumida* were introduced to the Sahel for firewood and shade but are now showing promise for human food. Various acacia seeds in Australia are finding favour as fashionable 'bush food'. Acacias throughout their range provide hugely important browse for wild mammals from antelope to elephants – and fodder for domestic livestock. Some provide valuable gums and medicines. Some are used in perfumery. Some, however, are highly toxic – both seeds and leaves. Several are valued ornamentals.

Then there is the genus *Leucaena*. The twenty-two known species grow wild in the Americas from Peru to Texas, from sea level to 3,000 metres. Native people eat the edible pods for their garlic flavour. Many can be grown as shrubs and are valued for fodder – but they do raise a problem since they contain exotic amino acids that do not normally form part of animal proteins, and which when eaten may lead to the loss of hair and hooves. *Leucaena* also provides some of the world's fastest-growing trees. Best known is *L. leucocephala*, which was first brought out of the Americas four centuries ago and is now grown worldwide for fodder but also for timber. *L. diversifolia*, originally from highland Mexico, is now used widely to provide shade in coffee plantations (the best-flavoured coffee is 'shade' coffee; it grows more slowly than in open sunlight, and slow is good) and for firewood and green manure. *L. esculenta* also comes from highland Mexico and has edible pods.

But foresters are not content with the wild species, and *Leucaena* is one of a fairly long shortlist of trees (it's a long list, but short relative to the total number of species) that have been intensively cultivated and bred. 'Breeding' means selecting the best, and crossing different species to produce hybrids that (with luck) combine the best qualities of both. Many of the hybrids between different species of *Leucaena* are fertile and a few can be reproduced by cuttings (so it does not matter if they are fertile or not). Thus hundreds of crosses made between sixteen different species of *Leucaena* at the University of Hawaii have produced some highly desirable hybrids. One, between *L. leucocephala* and *L. diversifolia*, known as *L. x spontanea*, takes just twelve years to grow into a tree with a trunk 40 centimetres in diameter: massive. Such trees, grown in appropriate plantations, help to take the burden of human needs and ambitions from the wild forests, and if burnt for fuel they are 'carbon neutral', and so do not contribute (in net) to global warming. On the other hand, *Leucaena* can be very nasty weeds. *L. leucephala* in Hawaii is one such. This particular species may well be a hybrid that arose in Mexico – a hybridization mediated by human hand.

Then there are the 200 or so known species of *Dalbergia*. They include shrubs and climbers – but also provide some of the world's most prestigious timbers, valued for xylophones, piano keys and billiard tables. Sadly, many *Dalbergia* species are endangered in the wild through deforestation; but some are widely cultivated. Indian rosewood, source of fabulous veneers, is *D. latifolia* (although Burmese rosewood, equally fabulous and sometimes known as narra, is a different legume, *Pterocarpus indicus*). *D. sissoo*, known as the sissoo or shisham, is native to the gravelly foothills of the Indian Himalayas. It grows slowly, and crooked, but is amazingly resistant to searing temperatures, drought and frost; and is hugely valued locally for fodder, fuel, charcoal and medicines, while its flowers provide bees with nectar for honey. Sissoo also provides beautiful, dark-brown timber. The Forestry Research Institute at Dehra Dun has a gun carriage of sissoo: I can see it in my mind's eye thundering across the maidan, pulled by frantic horses, urged through the dust by equally frantic soldiers in scarlet and brass, some bursting with glory and others cursing their luck. The African blackwood is *D. melanoxylon*. The Brazilian tulipwood (not to be confused with the tulip tree, *Liriodendron*) is *D. decipularis*. In France it was known as *bois de rose* and made some of the finest furniture for Louis XV and Louis XVI.

The reddish-brown, dark-etched Rhodesian 'teak' is *Baikiaea plurijuga*, greatly favoured for turnery. The boldly striped zebrano from West Africa comes from various species of *Microberlinia*: again favoured for carving. Purpleheart from South America, figured like fine tweed, is a species of *Peltogyne*: used to make apparatus for the gymnasium, skis and billiard-cue butts. Among the thirty or so species of *Albizia* from Africa are some that produce timber for big ships and jetties, floors and veneers. Various species of *Pterocarpus* provide hard, dark timbers including *P. indicus*, which we have already met. *Instia palembanica* is cultivated as 'Borneo teak'. In Brazil, various leguminous genera (and a few that are not legumes) provide the valued timber collectively known as 'angelim' (as outlined in Chapter 2). The entire vast country of Brazil is named after one of its own leguminous trees: brazilwood, *Caesalpinia echinata*.

Robinia was apparently tropical in origin but just four closely related species survive, not in the tropics but in North America. *R. hispida* is the ornamental 'rose acacia'. Best known is *R. pseudacacia*, otherwise known as the false acacia or black locust, which was introduced to Europe in the 1700s and selected for the navy – as 'shipmast locusts'. These have been cultivated intensely (and apparently 250,000 hectares of them are planted in Hungary). *Tipuana tipu*, sole member of its genus, is the pride of Bolivia; also known as a rosewood, it grows up to 20 metres as a street tree, as a windbreak and for fodder, both in Bolivia and Argentina.

Many legumes are grown as ornamentals. The suburban favourite is *Laburnum*. The tropical American *Enterolobium*, of which the huge-leaved *E. cyclocarpum* is known as 'monkey ears' or sometimes as 'elephant's ear'. The round-leaved Judas tree is *Cercis siliquastrum*. The forty or so species in the genus *Parkia* that grow widely in the tropics are glorious umbrella-shaped forest trees which hold out their flowers to be pollinated by bats and dangle their bright fruits to be dispersed by birds. The Malaysians eat the seeds of *Parkia* as petai, which has a strong flavour of garlic (which persists in the urine). The 400 or so species of *Mimosa* provide great benison, from thorn fences to their much-admired pom-pom flowers. The rain tree of India is *Samanea saman*. For reasons best known to itself, the rain tree seems to encourage epiphytes – which most trees seem to go to some lengths to get rid of. Indeed it is often grown for its epiphytes. It also harbours the lac insect, which periodically sprays the ground with water (or so it seems); and this, presumably, gives rise to the tree's common name. The asoka, *Sarraca indica*, is planted around Buddhist and Hindu temples, where its yellowy-red blossoms are religious offerings. Asokas are said to blossom more vigorously when given a good kicking by young women. Don't we all.

Many other leguminous trees serve many more workaday functions. The forty-four species of *Prosopis* are mostly generally resistant to drought and salt – which makes them promising candidates for the many million hectares worldwide now spoiled by salinity, brought about by over-zealous irrigation. *Prosopis* includes the mesquites: *P. glandulosa* is the North American kind, favoured for its aromatic charcoal that adds flavours to the barbecue. *P. cineraria* of tropical

India also provides charcoal, plus firewood, fodder, green manure and goat-proof thorny fences. The extraordinary *P. tamarugo* of Chile is slow-growing but widely planted for its resistance both to salt and very low rainfall.

Many legumes are toxic and many are medicinal (these are two sides of the same coin). One of several 'ordeal' toxins used in Africa in various kinds of initiation is the famous red bark from *Erythrophleum*. The fifty-two species in the genus *Sophora* provide hardwoods but also toxins and medicines: the Japanese pagoda tree, *S. japonica*, has been cultivated in China for more than 3,000 years for its beauty, and for dye and medicines.

Many leguminous trees provide food. Several species from the huge tropical genus *Inga* have edible fruits and seeds, including the ice-cream bean, *I. laurina*. (The genus *Inga* is huge. Exactly how many kinds there really are, at least in tropical America, is the subject of research discussed in Chapter 12.) Tamarind is the fleshy fruit of *Tamarindus indica*, which adds astringency to curries and pickles, of which the Indians tell a charming story. Thus: a man set out on a long journey, but his wife didn't want him to go. So she asked the local guru how she might hasten his return. 'Make him promise,' said the guru, 'to sleep every night under a tamarind tree on the outward journey; and to lodge beneath a neem tree every night on the way home.' The man kept his promise. But tamarind trees exude toxic vapours (or so it is claimed) and make you feel ill; while neem trees are restorative. So the further the man travelled, the worse he felt; and as he got nearer to home again, he felt better and better. There are many morals in this tale, beyond doubt. (The neem or nim is from the mahogany family, Meliaceae, of which more later.)

It would be easy to fill several volumes with leguminous trees. But we should move on.

Apples, Plums, Elms, Figs and Cecropias:
ORDER ROSALES

The Rosales have been radically reorganized these past few years, thanks mainly to molecular studies. Judd recognizes seven families – all of which have interesting trees, and deserve discussion. But the relationships between those families, and the plants that they include, are often surprising – and the modern taxonomy differs enormously from most traditional treatments of only a decade or so ago. The position is still fluid because in many ways the Rosales are especially tricky, with enormous morphological variety on the one hand, and a strong tendency to hybridize on the other, which sometimes makes it hard to tell where one group ends and the next begins. But the following reflects the state of play.

Fittingly, the most primitive of all the Rosales families, apparently the sister to all the rest, is the Rosaceae, with its simple round flowers evolved to attract generalist insect pollinators – flies for the smaller kinds, bees for the bigger ones, and long-tongued moths for the biggest. Among the 3,000 species of the Rosaceae are many lovely and useful herbs, like cinquefoil and strawberry. But most (three-quarters) of the eighty-five genera include woody plants. Some are mainly shrubby, as in the roses (*Rosa*), or the blackberries, raspberries and loganberries (*Rubus*). But many are very significant trees.[3]

All of the most important temperate tree fruits are from the Rosaceae family. The apples are the genus *Malus*. All the hundreds (actually, thousands) of varieties that are generally eaten are variations on a theme of *M. domestica* which, so recent studies in Oxford confirm, has been selected over many a century from the wild Asian *M. sieversii*. The various wild ancestors had small fruits, like modern crab apples, but as cultivation spread westwards so they became bigger until by the Middle Ages we had recognizably modern fruit. Domestic varieties of apples stay constant from generation to generation (or century to century) because they are reproduced as cuttings and grown by grafting desirable varieties on to robust rootstock. Thus all the Cox's Orange Pippins in the world, for instance, are a clone of the first ever Cox that was bred in the nineteenth century. Several other

Malus species are kept as ornamentals. The seventy-six species of pears (Pyrus), are close relatives of apples: some species grown for fruit, some for ornament (some have pleasantly silvery foliage for instance, sometimes weeping), and valuable smooth pale-golden timber for much-prized kitchen furniture and for woodcuts. Pear trees can grow big: 18 metres tall, 5 metres in girth (about 1.5 metres in diameter). *Cydonia* is the quince and *Eriobotrya* is the loquat.

The genus *Prunus* abounds with good things. *P. dulcis* is the almond. *P. armeniaca* is apricot. *P. avium* is the sweet cherry and *P. cerasus* the sour cherry. *P. persica* is the peach and *P. domestica* is the plum. Many *Prunus* too are grown as ornamentals – notably the Japanese flowering cherries – and several are noted for their fine reddish timber. In the wild, *Prunus* and *Crataegus*, the hawthorn, are early to put in an appearance as new forests establish themselves, although *P. serotina*, the black cherry, grows on in mature deciduous forest. *P. avium*, the bird cherry, grows widely in Europe including Britain, where it is apparently native.

Other ornamental trees (or near trees) include *Amelanchier*, which the Americans call the shadbush; the flowering quince (*Chaenomeles*); *Cotoneaster*, always known as cotoneaster; the hawthorn, *Crataegus*, otherwise known as the quickthorn or may tree; the firethorn (*Pyracantha*); the roses of course, in the genus *Rosa* – many thousands of cultivars hybridized from about nine wild ancestors; *Sorbus*, which includes the mountain ash, or rowan, and also the whitebeam; and the florists' favourite, *Spiraea*. Some have other uses, too. Hawthorn in particular is Britain's favourite hedging plant, layered – the branches half cut across then laid sideways – to form an impenetrable thorny barricade; and sometimes left to grow into a big mature tree in the hedge, as elms often were.

Five closely related genera within the Rosaceae (though none of the important ones mentioned above) have developed symbiotic relationships with nitrogen-fixing bacteria that live in nodules in their roots. The bacteria are not *Rhizobium*, as in the Fabaceae; but *Frankia*. In all, plants from about ten families harbour nitrogen-fixing *Frankia* in nodules in their roots (one of the chief of which is the alder, *Alnus*). It would be tempting to suggest that the Fabaceae also started out with *Frankia* in their roots, but that these were later displaced by

Rhizobium. But if this were so, we would expect the most primitive Fabaceae to harbour one or other of the two bacteria. In fact the most primitive Fabaceae do not have nitrogen-fixing bacteria at all. Thus it seems that nodules to harbour *Frankia*, and nodules to harbour *Rhizobia*, are independent, parallel inventions – yet another, stunning case of convergent evolution. We also see, yet again, the propensity of organisms – one might almost say, their eagerness – to cooperate.

Then there are the Rhamnaceae, the family of the buckthorns: 850 species in forty-five genera: often thorny; some trees, some shrubs, some climbers – and again, sometimes, with nitrogen-fixing bacteria in their roots. We are familiar with a few Rhamnaceae in the temperate north: buckthorn is *Rhamnus; Ceanothus* is a highly fashionable ornamental. But the Rhamnaceae come mainly from the tropics, where many are useful. *Hovenia dulcis* is the raisin tree. *Ziziphus jujube* is the Indian jujube, alias flame-of-the-forest. The jujube grows fast on dry, poor land to form red-flowered trees that are often scrubby but can reach 24 metres. Its timber is good, and it burns well; its prickly branches make serviceable fences; its leaves and twigs are fodder for camels and goats; its wild green fruits make sherbet, sold in the markets and (it's said) much loved by students; and it is cultivated for its fruit, used for seasoning, cooked with sugar, or stored in oil or sugar syrup. Perhaps most of all, though, the jujube is a fine host for the lac insect, which sucks its sap and exudes a reddish resin over the whole surface of the twigs – which yields a dye and also becomes shellac, once used for gramophone records and still favoured for polishes and for lacquer. The jujube illustrates a general principle: how much use is made of plants that outsiders would scarcely notice, by people who know about them; how entire economies and cultures can flourish under our noses without us noticing, and how easily and often those ways of life are swept aside – for what developer would care about the wild jujube trees? The lac insect also feeds on the peepul, the rain tree and the mango.

Both the Rosaceae and Rhamnaceae families are at the edge of the Rosales order, however. The remaining families all seem to group roughly together in one great clade. And what families they are.

First come the Ulmaceae, the family of the elms (*Ulmus*) and the favoured park tree *Zelkova*. There are six genera, and about forty species, all trees, mostly in the temperate north. Elms until recent times were so common in England they largely defined the lowland landscape: they dominate John Constable's Suffolk landscapes in the east, and in the west were known as the 'Wiltshire weed'. They commonly grew in hedges and formed fine trees whose timber was often used in great slabs, for example to make the buttock-moulded seats of rural wooden armchairs, and the sides of wheelbarrows. Then in the 1970s Britain's elms were struck down by Dutch elm disease: caused by the fungus *Ophiostoma novo-ulmi*, and carried and introduced beneath the bark by bark beetles of the genus *Scolytus*. Within a few years, despite the best efforts of foresters and biologists, mature elms had all but disappeared. The original types hang on as hedgerow bushes – but as soon as they reach a critical height, within the beetle's flying zone, they are attacked again, and die off. New resistant strains are being developed, but England's lowlands will never be the same again (although, of course, the transformations wrought by urbanization and agribusiness are far more dramatic). Dutch elm disease occurs on mainland Europe and in America as well, where the fungal pathogen is carried both by *Scolytus* and *Hylurgopinus*.

Then there are the Celtidaceae, which include the hackberry or sugarberry (*Celtis*), whose colourful fruits are for the birds that disperse them, and are rarely eaten by humans. But hackberry is used for timber and grown as an ornamental.

Then comes a huge and supremely important family – the Moraceae. Its fifty-three genera (1,500 species) include shrubs, climbers and herbs – but also some intriguing and supremely important trees that grow throughout the tropics as key players in rainforest. The jackfruit or breadfruit (*Artocarpus*) has a pale grey-green warty-skinned fruit which is really a fused mass of fruits (an 'infructescence') and may be huge: as big as a sack of coal and up to 40 kilograms in weight. *Brosimum* is the breadnut. There are some fine timber trees too: the iroko (*Chlorophora excelsa*) from Central and West Africa is often used as a substitute for teak. The snakewood (*Piratinera guianensis*) from tropical America is extremely heavy (much heavier than water when dried) with a black-brown tortoiseshell pattern favoured for everything fancy, from the backs of brushes and umbrella handles to violin bows – and for native bows for shooting arrows. More temperate is the mulberry (*Morus*): the white kind grown to raise silk moths (4,000 kilograms of leaves for one silk blouse); the black kind favoured for its glorious edible blackberry-like fruits, popular from the seventeenth century and now featuring, gnarled, in many an old walled garden.

Then there is *Ficus*, the genus of the figs. *Ficus* seems to go out of its way to be extraordinary. First, it is enormously various, with about 750 known species: a huge presence throughout the tropics, primarily in tropical rainforests, but reaching too into the subtropics and Mediterranean. Then there is the way it grows. About half the species simply take root in the ground like most other trees. The other half begin their lives as epiphytes – plants that grow in other plants. The seed lodges in the fork of a branch of some forest tree; or, often, in the severed leaf-base of some palm; or sometimes in a crevice in a wall. Then as it grows, as an epiphytic cactus or orchid might, it sends down roots towards the ground – which eventually take root on their own account. Then the once-dangling roots function as stems: where previously they had carried nutrient and water downwards, from the epiphytic roots above, now they carry minerals and water upwards, from the new roots in the ground below. Generally the dangling roots/stems fuse with each other to form what looks like an exercise in macramé: you finish up with what may be a massive trunk that you can see right through. In the epiphytic figs known as 'stranglers' the probing roots twine around the host trunk. In others, like the Indian banyan, *F. benghalensis*, there is less obvious entwining. The roots merely drop to the ground eventually to form multiple trunks. The throttling of the stranglers, the monstrous weight of the epiphytic fig as a whole as it grows (whether it strangles or not), and the blotting out of light, eventually kills the host – although the obliteration may sometimes take a century or more, until which they live in grisly union.

Many trees are sacred to the Hindus and Buddhists of India and two of the epiphytic figs are among the most sacred of all. The British in India were often cavalier, but on the whole I am told (by Indians), they respected the sacred trees and many ancient ones remain even in the heart of British enclaves – including one magnificent specimen (albeit split into two by a storm in 1947) in the Forestry Research Institute at Dehra Dun, at the foot of the Himalayas. The FRI was established at the turn of the nineteenth century on the site of what had been twenty-two villages (although many of the village people stayed to work at the institute, and many of their descendants still do). The indigenous vegetation was largely cleared – but this great

Banyans form new trunks by sending down roots from above

banyan, dating at least from the eighteenth century and probably before, was spared. The greatest of all, in Calcutta, is reputed to be a quarter of a mile in circumference, with hundreds of trunks connected overhead to form a colonnade, able to provide shelter for 20,000 people. The roots are guided to the ground along sticks.

Most sacred of all the figs, however – most sacred of all trees indeed – is the peepul, *F. religiosa*, also known as the bo. For it was under a bo tree that the young Prince Siddhartha sat and meditated, some time in the sixth century BC, and received enlightenment, and thenceforth was known as the Buddha. Peepul trees are unmistakable. The leaves of many tropical trees have 'drip tips': extensions at the end that act like gargoyles, quickly getting shot of surplus rain. The bo leaf is heart-shaped, typically about the size of a man's hand, and its drip tip is enormous, about a third of the total length. This form, so characteristic, has become a Buddhist icon; bo leaves often provide the background to pictures of the Buddha. Yet they grow like the humblest nursery clone by the side of Indian roads – or sometimes right in the road, with ox carts and bright-painted, black-smoking lorries milling in the dust around them. In India, sacred means sacred. The traffic gives way.

Most extraordinary of all, however, is the manner of the figs' reproduction. The flower, or rather inflorescence, is a fleshy cup formed from the flower stem; and within that cup hundreds of flowers open *inwards*. The whole apparatus is pollinated by minute wasps: extreme specialists, for there is one dedicated species of wasp for each of the 750 species of fig. At least, one fig: one wasp is the general rule. The reality is turning out to be more complicated and even more fantastical, and is discussed at length in Chapter 13.

Finally, comes a pair of closely related families – the Urticaceae and the Cecropiaceae. The Urticaceae is the family of the nettles – which northerners know as herbs with a serious line in stinging (and which are also fine sources of fibre) – but it also includes some tropical forest trees, some of which also sting. In Queensland I was told the tale of well-known British botanist much given to the waving of arms who waved them a little too much, right through the branches of a nettle tree, and finished up in hospital.

Cecropiaceae is a most intriguing family, which includes the tropical genus *Cecropia*. Cecropias are pioneer trees par excellence, with hollow stems like bamboo for extra rapid growth, branching at the top to produce an umbrella of silvery-grey compound leaves roughly like a horse chestnut's. In the hollow stems dwell ants which, as in acacias, are the tree's housekeepers. In newly exposed land the cecropia provides almost instant shade, while its leaves, roots and latex are pharmacologically potent and are used to treat a range of conditions from hypertension and depression to gastric ulcers. In tropical America, three-toed sloths are fond of cecropias: sloths seem remarkably common, and all the ones I have seen (all as it happens in Panama) were in cecropias. But cecropias can bring bad news, too. Their silver foliage stands out in the forest, and since they grow fast in open space they show where gaps have recently formed. Some of those gaps are legitimate: old trees die naturally, and in some areas at least, favoured forest trees are selectively and carefully harvested. But often you see hillsides where no logging is allowed awash with cecropias. Then they reveal where the illegal loggers have been at work. In Brazil and elsewhere, such sights are all too frequent.

The Cecropiaceae family, as now defined by Judd, also includes *Cannabis*, provider both of hemp fibres and of marijuana, and *Humulus* (hops). This is how the molecular studies suggest they should be classified. In the past, however, *Cannabis* and *Humulus* have commonly been placed together in a separate family, Cannabinaceae, which in turn has been grouped together with the Urticaceae within their own order, the Urticales. Urticales in turn has sometimes been associated with the witch-hazels in the hamamelid group (now disbanded), and sometimes placed in the Malvales order, of which more later. *Cecropia* has been shuffled uncertainly between the Urticaceae and the Moraceae. For the time being things are as described here, with hemp and hops grouped with *Cecropia* in their own family, Cecropiaceae, among the Rosales. Let us hope it stays that way. Cannabis and nettles both produce fine fibres, good for ropes, though nettle ropes are not common these days.

Oaks, Beeches, Birches, Hazelnuts and Walnuts:
ORDER FAGALES

Within the eight families of the Fagales order are some of the most beautiful, most iconic, most treasured, and most ecologically and economically significant of temperate broadleaved trees: the oaks, beeches, chestnuts, southern beeches, birches, alders, hornbeams, hazels, she-oaks, bayberries, walnuts, pecans, and hickories. Estimates of the total species vary wildly: Judd suggests 1,115, but there could be many more, not least because many of the Chinese kinds, where the order abounds, are as yet largely unstudied. All Fagales are trees or shrubs: there isn't a herb in the entire order. The oldest fossils (of pollen and other parts) date from around 100 million years ago, when the dinosaurs were still in full pomp. *Where* they arose is a bit of a mystery. The oldest family of all seems to be that of the southern beeches (Nothofagaceae) – but they are the only Fagales family that lives in the southern hemisphere. All the rest are based in the northern hemisphere, with just a few venturing south of the equator here and there. But then, the evolutionary history of all groups is full of loose ends.

The Fagaceae family includes the best known trees of the Fagales order, and the most important economically and ecologically. The family probably first arose around 90 million years ago in tropical mountains – near the equator but nonetheless a relatively cool habitat – and it extends through the northern hemisphere from temperate lands to the tropics. All are trees or shrubs, rich in tannins. All have fruits in the form of a nut, with a spiny or scaly capsule round the outside – sometimes fully enclosing the nut, and sometimes holding it decorously like an egg in an egg cup.

Botanists tend to define and include around nine genera within the Fagaceae. The oaks, *Quercus*, are the biggest genus with 300 to 600 species, depending on who's counting (Judd opts for 450). It seems, though, that many of the Asian oaks should be placed in the tanoak genus, *Lithocarpus*, which at present contains around 100 to 200 known species from North America, with only one officially recognized from Asia.[4] *Fagus* is the genus of the beeches, with about ten species. *Castanea*, the chestnuts, also have about ten species. *Castanopsis*, which in some classifications embraces *Chrysolepis*, includes the 150 or so species of chinkapin or chinquapin, from North America, China, India and the Malay archipelago. There is just one species of *Colombobalanus*, from Columbia in South America; and two of *Trigonobalanus*, from China and Malaysia. About one in eight of all the species of Fagaceae – 12 per cent – are included in the 2003 *Red List of Threatened Plants*.

North Europeans and Americans think of the genus *Quercus* as mighty trees of temperate lands that shed their leaves in winter. This is true of Britain's two native oaks, the English or common oak, *Q. robur*, and the sessile oak, *Q. petraea*: and those two, plus ash and Scots pine, would now extend from the tundra of northern Scotland to the tip of mild and rainy Cornwall were it not for our forebears, who stripped the post-Ice Age forest to make farmland and took the oaks in particular to build the medieval and Tudor cities, and the navy that defeated the Armada. ('Heart of oak are our ships' we used to sing at school in moments of patriotism. But later, the British navy made much use of teak, from India.)

The oaks are the most widely distributed of forest trees – and not particularly northern. Indeed, common and sessile oak are the only species that venture more than 50° north of the equator (and those two extend up to 60° N). Evolutionary studies suggest that *Quercus* first appeared in South-East Asia, around 60 million years ago[5] and most still live between 15° N and 30° N, especially in Mexico and Central America, and in the Yunnan province of China. In the south the genus peters out in Colombia (where oaks live in the highlands) and Indonesia. North America has the most species; Europe and Asia have many; and a few live in North Africa. Between them the different kinds are adapted to every habitat from desert to swamp and from sea level to the highlands – up to 4,000 metres in Yunnan: as high as the highest Rockies. In the wild, the most widespread of all the individual species are red oak (*Q. rubra*) and white oak (*Q. alba*) in North America; *Q. acutissima* and *Q. mongolica* in Asia; and common and sessile oak in Europe. Many are shrubs. Many – like live oak (*Q. agrifolia*) and *Q. wislizenii* of California; or *Q. coccifera*, holm oak (*Q. ilex*) and cork oak (*Q. suber*) of southern Europe – are evergreen.

Oaks are so tremendously successful partly because, like many other successful trees, they can reproduce in so many ways. Unlike most Fagaceae, but like beeches and southern beeches, oaks are pollinated by wind, and generally bear both male and female single-sex flowers on the same tree (that is, they are monoecious). In good years the acorn crop is prodigious; they are distributed largely by mammals (especially rodents). But oaks also sprout from damaged trunks, to form natural coppices; and coppicing once supported entire industries of foresters (broadly defined) throughout Europe. Oaks also send forth shoots after forest fires, again producing entire stands that are clones of the parent individual.

Nowadays, of course, many have been transplanted from their native lands and have become ubiquitous: red oaks, from the eastern United States, are planted throughout Europe; common oak, from Europe, is now seen just about everywhere; and so on. The timber of many species is of legendary strength and beauty, and is used for everything from pit props to the finest veneers. Half of all hardwood in the United States comes from oaks. Their wood burns well: oak chips are essential for smoking kippers. Oaks are burned for charcoal, too. They make some of the finest barrels, most favoured for sherry and whisky. Their tannins are used for tanning leather. Cork oak has a thick fire-proof layer of cork beneath its bark, a key player in the traditional economies of Portugal and Andalucia – the plantations not only providing cork but also a home for the black pigs that produce such magnificent hams: a fine exercise in agroforestry. Entire streets in Andalucia are lined with shops that seem to sell hams exclusively. The young piglets are ginger-brown, and trot about in tightly coordinated squads as baby warthogs do. You meet them on Iberian hikes – torn, as wild animals so often are, between curiosity and nervousness: charming, fleeting companions. The cork forests, too, are home to some of the last of the Spanish lynx. Natural cork is now threatened by plastic: yet another degradation of the world's ecological and cultural diversity.

The tanoaks, *Lithocarpus*, have fruits like acorns but leaves like a chestnut's. Tanoaks are less legendary than oaks; but they, too, produce hard, strong timber – though mostly used for pulp and firewood – and their tannins again are used to tan leather.

Seven of the ten or so species of beech in the genus *Fagus* live in temperate east Asia. There is also one in North America, one in Europe, and one in the Caucasus. Normally they are found with other deciduous species in mixed forests, in temperate climes with soils that hold moisture well but are not waterlogged, for they like neither flooding nor drought. But they don't mind shade and will grow for decades in the shadow of other trees. So it is that the European beech, *F. sylvatica*, grows happily alongside sessile oak, common oak and the European hornbeam, *Carpinus betulus*. In plantations, hornbeams are grown with beech to 'train' them – encourage them to grow straight and tall, and to shed their lower branches as they reach up. The American beech, *F. grandiflora*, features in about twenty different

One of the temperate world's most valued hardwoods: the beech

forest types in the eastern United States, sometimes dominant but by no means always. Its companions typically include hickories and oaks, yellow birch, sugar maple, red maple, American basswood, black cherry, plus eastern white pine and red spruce; and in the south, it may grow alongside *Magnolia grandiflora* (another combination I would very much like to see in the wild, and haven't yet). All in all, beeches seem to be sociable trees. They are also long-lived. Among American broadleaves only the white oak and the sugar maple are said to live longer.

Beech, like oak, is highly prized, its silky, pale timber excellent for floors, furniture, veneers, turning and steam-bending. Beech that has been infected with fungi may be particularly valued for turnery: wiggly black lines etched by the fungus in the creamy wood. There is a parallel here with the 'noble rots', the yeasts and fungi that produce the world's great wines and blue cheeses: one of the great leitmotifs of Western art is that beauty and decay are never far apart. People as well as wildlife eat the nuts or 'mast' of beech. I have waited in traffic in Holland while pigs snuffled beechmast off the road (and rightly so). Beeches are beautiful, too. Their trunks are smooth grey columns. Their diaphanous leaves filter the light like pale-green glass. Some garden varieties are stunning, like the copper beech (in truth deep red) and the weeping kinds that grow like spheres, their branches sweeping the ground if browsing animals are kept at bay. They also make fine hedges – their leaves turning red-brown in winter but nonetheless remaining on the hedge (although beeches that are allowed to grow into forest trees shed their leaves). Altogether, *Fagus* is a fabulous genus.

The ten or so species of the genus *Castanea* are the sweet chestnuts. They are also known in some parts as chinkapins (but so are *Castanopsis* and *Chrysolepis*). Sweet chestnuts hail naturally from southern Europe, North Africa, south-west and east Asia, and the eastern USA. They are prized for their nuts, of course: traditionally roasted, and also the source of a fine stuffing, excellent with wintry goose. Before the 1930s, too, the American chestnut, *C. dentata*, was prized for its timber. But American chestnut is one of all too many trees that have been devastated by disease: in this case the chestnut blight, *Endothia parasitica*. As with Dutch elm disease, the fungus attacks larger trees, so American chestnuts are at least able to grow to the size of the shrub before the stems are killed. The roots survive, the stems regrow until they are big enough to be zapped again, and so on ad infinitum. It is all immensely sad.

The southern beech, the genus *Nothofagus*, first attracted the attention of Joseph Banks in the late eighteenth century, during his trip south with James Cook. Then in the 1830s the young Joseph Hooker was struck by the similarity between different southern beeches on different continents. In particular, New Zealand's tawhai or silver beech, N. *menziesii*, much favoured for its timber, is remarkably similar to Australia's N. *cunninghamii*, another fine timber tree, confusingly if typically known as the Tasmanian myrtle; and also to N. *betuloides* of South America. Such observations fed the growing suspicion that different species in different places must have evolved from some common ancestor. (Hooker later became great friends with Charles Darwin, another pioneer naturalist in southern climes, and succeeded his father William Hooker as director of Kew.) Botanists for many years placed southern beech among the Fagaceae, alongside beech. But the cups that enclose the nuts of beeches, oaks and chestnuts are formed from the flower stalk, while those of southern beech are compacted from bracts. So the relationship is not as close as it may seem, and southern beech now has its own family, Nothofagaceae, although still within the order Fagales. Nothofagaceae is the only Fagales family that belongs to the southern hemisphere.

There are thirty-five or so known species of southern beech, although in reality there are probably many more. Most are evergreen, although a couple are deciduous. Nine live in South America, not least in the bands of forest along either side of the Andes in Patagonia, which are dominated largely by the lenga (N. *dombeyi*) and the nire (N. *procera*), and are home to pumas, guanacos (small relatives of the ilama), southern river otters, geese, Andean condors and deer. Here is another grand place to see and wander through, now sadly threatened by too much logging, although in part protected now by Patagonia's Perito Moreno National Park. Three more southern beech live in Australia. They were far more widespread there when Australia was wetter, but in these dry and fire-prone times they are largely supplanted by eucalyptus. They flourish still in New Zealand (four species); and there are eighteen more in New Guinea, New Britain and New Caledonia (inevitably!), and a few more on other islands. There are none in Africa, but before Antarctica drifted to the Pole and was buried in ice it clearly had great forests of southern beech. Often, within their range, they are the dominant broadleaved trees, but in

New Zealand and South America in particular they tend to share their forests with the great southern conifers, the podocarps and the various araucarias. Though we can reasonably suggest that southern beech is the southern equivalent of the oaks, beeches and chestnuts of the north, in truth the southern temperate forests tend to be very different in character: typically damper and, in New Zealand, having a wondrous understorey of giant ferns, each worthy to stand in some stately conservatory in a grand ceramic pot.

Challenging the Fagaceae family in diversity and ecological range (although nothing can quite challenge the Fagaceae) is the family of the birches, alders, hazels and hornbeams, the Betulaceae. Again, all of them are trees and shrubs, widespread in the northern hemisphere – both in temperate regions, and in the most extreme north. Just a few, notably some alders (*Alnus*), drift into the southern hemisphere. In general trees of the Betulaceae are early on the scene when there is new ground to be colonized – and so they rapidly spread north after the last Ice Age, in the wake of the retreating glaciers. In the case of alders, they are helped in this by nitrogen-fixing *Frankia* bacteria in nodules in their roots. Alders in particular, too, largely because of their nitrogen-fixing bacteria, do well in water-logged soils, in the same way as mangrove. Sometimes, however – as with the endless forests of birch in Siberia and Canada – Betulaceae are the dominant forest trees. Their flowers are catkins, pollinated by wind; and their seeds are mostly distributed by wind (or water, in the case of alders and hop hornbeams); but the seeds of hazels are spread by rodents, with squirrels as key players, although they exact a huge fee for their services.

Judd recognizes six genera, with a total of 157 species: sixty birches (*Betula*); thirty-five alders (*Alnus*); thirty-five hornbeams (*Carpinus*); fifteen hazels (*Corylus*); ten hop-hornbeams (*Ostrya*), also sometimes called ironwood; and a couple of *Ostryopsis*, which resemble hop-hornbeams. Early botanists placed alders, birches and willows together in a sub-family which they called 'Amentiferae': all, after all, are trees, and have catkins. Walnuts, figs and elms were bundled in too, for good measure. But the catkins that they all shared evolved independently, not from a single common ancestor. Alders, birches and willows do share a propensity as pioneer species, however, and are used for similar things – not least to provide the charcoal needed to make gunpowder.

Helped by their nitrogen-fixing bacteria, alders grow rapidly, sometimes reaching 30 metres in a decade – and so aggressively that they are often rated as weeds. They definitely have their upside, however. Because they fix nitrogen, they are able to improve the soil significantly, and so benefit the whole forest. They are excellent pioneers; in general they are widely planted, and in particular are pressed into service for soil reclamation.

Though Judd recognizes thirty-five species of alder, this should not be taken as gospel. Stephen Harris, curator of the herbarium in Oxford University's Department of Plant Sciences, puts the figure at nearer twenty-five. But as outlined in Chapter 1, *Alnus* is one of those genera in which the concept of 'species' is hard to pin down. They have a great tendency to form polyploid hybrids, and given the enormous geographical range of *Alnus*, and the remoteness and hostile nature of many of the places it tends to live, we can also be fairly sure that many species remain to be identified.

For alders are indeed extremely widespread. A. *acuminata* lives in Central America and extends into the highlands of South America, and is planted extensively for timber and fuel. The black alder, *A. glutinosa*, is widely distributed in Europe. In workaday mode, it serves to stabilize riverbanks and roadsides, and is grown for fuel. More grandiosely, it once supplied much-valued timber for violins. In recent years, however, at least in Britain, black alder has suffered hugely from attack by the fungus-like *Phytophthora cambivora*, related to potato blight. The Nepal alder of the Himalayas, *A. nepalensis*, is also planted widely for timber and firewood, and as forage for cattle and sheep. The red alder of the American north-western floodplain, *A. rubra*, is a huge tree, up to 40 metres, favoured for building and furniture, as well as for fuel and pulp (though used mainly in mixtures). Red alder can also be a significant weed in plantations of pines. Alders do well in waterlogged places. Alders are among the many trees that tend to concentrate minerals within their cells: they pick up gold, for instance. Whether it is worth trying to extract gold from them, I do not know. Alder bark is also astringent and is used traditionally to treat burns and infections.

The oldest known fossils of alder date from the Miocene, around 18 million years ago. The genus *Betula* is much older: birch fossils date from the Upper Cretaceous, still in dinosaur times, and perhaps birches were most diverse in the Eocene, around 45 million years ago.

Fey, melancholic and wonderfully hardy: the birch

The sixty or so living shrubs and trees also live in even more diverse habitats than alders do, from temperate lands to the extreme northern limit of trees. They are present and may be dominant in peat lands: along the banks of streams and the shores of lakes; in damp woods; on the margins of roads and railways; in alpine settings; and in tundra.

Birches are pollinated by wind – and produce a great deal of pollen. Like alder pollen, this is bad for hay fever, though speaking as a sufferer, I find it a small price to pay for all that beauty. Again, birches tend to be polyploid, and prone to hybridize.

Birches are good biochemists – and so are used for many things besides their attractive white timber. Their leaves are often rich in resins and their bark (particularly from the white-barked kinds) are rich in phenolics. Some species produce betulin in their bark, which makes them waterproof. (These agents are also said to be effective 'antifeedants', repelling hungry browsing animals in winter; yet many insects feed on birch, and fungi may rot the heartwood.) Their twigs were traditionally used for punishing: the generic name *Betula* derives from the Latin for 'beat'. Birch bark (for example of the paper-bark birch, *B. papyrifera*) is used for roofs and canoes, and is the stuff of the oldest known Hindu manuscripts, dating from around 1800 BC. It is also rich in oil and starch and serves as food in times of famine. Like alder and willow, birch wood burns to make good charcoal, excellent for gunpowder (how many did the Russians fell in seeing off Napoleon?). The sap of the Appalachian *B. lenta* is tapped in the spring and fermented to make birch beer. Oil wintergreen, containing methyl salicylate (related to aspirin), can be obtained from *B. lenta* and the yellow birch, *B. alleghaniensis*. The leaves of the European species downy birch (*B. pubescens*) and silver birch (*B. pendula*) produce a green dye. Birch, like alder, accumulates heavy metals in its leaves, and can be used to reveal their presence in the soil beneath. Birches in general yield valuable timber, and also pulp. Many birches are grown as ornamentals, and quite right too.

The hornbeam, *Carpinus*, is also known as ironwood; so hard that it was the traditional stuff of axles and cartwheel spokes before iron became cheap enough to take over. A brewery in my village that has somehow escaped the corporate ravages is powered by a nineteenth-century steam engine, and the moving parts that extend from it have hornbeam cogs. They are better than iron, I'm told, because they don't shear. Hornbeam is often overlooked. It looks somewhat like beech, but its leaves are generally smaller and more deeply furrowed, its trunk is fluted, and its nuts are winged as a beech's are not. Like beech it is good for hedging, and for pleaching – when branches from adjacent trees closely planted in a row are run together and trained to form what looks like a hedge on stilts. I first met the hop-hornbeam (*Ostrya*), from Asia, Europe and America, in the Botanic Garden at Cambridge one July: its bark shaggy, its cinnamon-coloured dangly male catkins and its pale-green female catkins hanging side by side like courting couples at the tips of the twigs. A delightful tree – again with very hard wood. Most hazels and filberts (*Corylus*) are more shrub than tree, but *C. colurna* from Turkey grows to nearly 25 metres.

The Casuarinaceae family are the she-oaks. There is only one genus, *Casuarina*, which is basically Australian but also grows widely through Asia and the Pacific islands, including Fiji and New Caledonia. Like alder, it has nitrogen-fixing *Frankia* in its roots; but unlike alder, it generally favours drylands and is grown for timber, fuel, and to provide shelter belts in China and shade on tropical and subtropical beaches in Africa and America. Three species are naturalized in Florida, where they flourish as significant weeds. *Casuarina* have grooved, green, jointed twigs; and some are grown as ornamentals.

The Juglandaceae family are the walnuts, hickories and wingnuts, which have either big fleshy fruits containing nutritious, aromatic nuts that are dispersed by animals (mainly rodents), or winged seeds that are dispersed by wind. In all of the Juglandaceae the pollen is wind-dispersed, and most have catkins. Most are bona fide trees, though a few are shrubs; mostly resinous; mostly aromatic; mostly rich in tannins.

The Juglandaceae family as a whole seems, like the genus *Quercus*, to have arisen in warm latitudes at a time when the world as a whole was warm – around 40 to 50 million years ago, when there were palm trees in the Dakotas and temperate forest in Siberia; and in the present, cooler world the family spreads from tropical to temperate lands. Now there are species in North and Central America, in South America along the Andes, in Europe and in Asia – India and South-East Asia.

Most widespread (in the Americas, Europe and Asia) are the twenty or so species of walnuts (*Juglans*). They hate shade: so when they are in forests, they need to be the dominant species, or at least co-dominant, able to shade out the rest. They are helped in this it seems because they produce juglone in their leaves, bark, husks and roots; and this is said to be noxious to other trees, which give them a wide berth. Paper birch, apple and various pines are said to be particularly sensitive. Fishermen have also used bruised walnut branches, leaves and fruit to stun fish: unscrupulous if practised by sportsmen, but perhaps more excusable among people whose lives may depend on the catch.

The eastern black walnut of eastern North America (*Juglans nigra*) is famed for the fine furniture made from its timber. *J. regia* is sometimes known as the Persian walnut and sometimes as the English walnut. But although it grows naturally in a wide variety of habitats from cool steppe to moist subtropical forest, it does not grow wild in England at all. It's just that the British in general, and the English in particular, have an unsurpassed talent for expropriation. Thus by the same token, the Scots pine flourishes throughout Europe and into Asia Minor and could just as well be called the Russian pine (though it does at least grow naturally in Scotland as well). But walnuts are widely cultivated. The ancient Greeks and Romans had walnut orchards. Northern Europeans began to cultivate the walnut in the 1500s and its timber became the favourite for high-grade furniture until, from the 1600s onwards, it was ousted by mahogany from the Americas – although walnut is still used for gunstocks and the insides of prestige cars. The English have long cultivated walnuts with optimism if not always with success, and breeders at Oxford University are now seeking to create varieties that really can tolerate our decreasingly harsh but increasingly fickle climate. Already there are at least 400 varieties of cultivated walnut. Turkey is a great walnut producer, but the largest producer of all is now California.

Judd claims sixteen species of hickory (*Carya*) – thirteen in North America (one restricted to Mexico) and three in Asia. Hickories too are known both for their nuts (including pecan) and for their timber. Hickory is wonderfully shock-resistant: much favoured for the handles of hammers and axes and, in the good old days, certainly well into the twentieth century, for the shafts of golf clubs. Once, too, it was *the* thing for barrel hoops. The intricate knowledge that our forebears had of each kind of plant and its caprices and possibilities never ceases to astonish me: knowledge now largely lost, or at least confined to academic tracts or whimsical accounts like this one. Maybe when the fossil fuels run out and heavy industry has run its course, such wonders may be rediscovered. The tropical *Engelhardtia* and the wingnuts (*Pterocarya*) also provide fine timber.

Walnuts and hickories in various forms are also valued as ornamentals. So too are wingnuts, which, like walnuts, have big, sweeping, feather-like leaves. One wingnut tree that I know, in the Botanic Garden at Cambridge, has about a dozen trunks that clearly arose as suckers around a central trunk in the manner of giant sequoias; but as often happens with sequoias, the central trunk has gone, leaving the outriders in a circle, like standing stones in some ancient place of worship. Some of the world's most startling plants are in gardens. Protected over decades or centuries from predators and competitors, they can grow more extravagantly than ever they are allowed to do in the wild.

Finally, the Fagales order contains the Myricaceae family, aromatic trees or shrubs, rich in tannins and aromatic essential oils, widespread through the tropics and temperate countries. The genus *Myrica* includes the bayberry, wax myrtle and candleberry, which provide aromatic waxes. Some *Myrica* also have edible fruits; some are ornamental shrubs. They have small flowers pollinated by wind and fruits which are mostly dispersed by birds, although the small fruits of *Myrica gale* are fitted with bracts that act as floats, and are dispersed by water. Myricaceae in general are water lovers. Like alders – also in general water-lovers – they have nodules of nitrogen-fixing *Frankia* in their roots.

So that completes the Fagales. To be sure, no order compares with the Fabales. But in the second league, to which all others belong, the Fagales are certainly among the greatest.

Terminalia, Myrtles and Eucalypts:
ORDER MYRTALES

The Myrtales order is huge: 9,000 species in fourteen families. Some of those families have no significant trees. Among those that have a few intriguing kinds are the Lythraceae, which includes purple loosestrife

(*Lythrum salicaria*), which is a pleasant wayside flower in Britain but, introduced into North America, is a major weed of wetlands; and also *Punica*, shrubs and small trees which include the pomegranate (*P. granatum*). The Onagraceae family contains the evening primrose, the willowherbs and the fuchsias, which include the kotukutuku, *Fuchsia excorticata*, the unique fuschia tree which grows throughout New Zealand to a height of around 14 metres. The Vochysiaceae family of Central and South America (with a small presence in West Africa) includes several trees that I have encountered in the dry forest of Brazil, the Cerrado, some of which are used for boats and furniture.

Outstanding, however, is the Combretaceae family. It includes some important trees of the mangroves: the Pacific *Lumnitzera* of Asia, Australia and East Africa; and *Languncularia*, straddling the Atlantic in West Africa and America. The family also includes *Terminalia*, with many big tropical trees valued both for their physical beauty (they are much favoured in the grounds of gracious houses for their vast festoons of red and yellowish flowers) and for their timber. Several species of *Terminalia* are called 'Indian laurel'. *T. bialata* is known in the trade as Indian silver-grey wood; West Africa's idigbo is *T. ivorensis*; and afara, or limba, also from West Africa, is *T. superba*.

Closely related to the Combretaceae, and even more outstanding, is the Myrtaceae family. Its members are wonderfully aromatic, stuffed with essential oils. The family is named for the myrtle, *Myrtus*, which seems to be the only European genus. It also includes the clove tree, *Syzygium aromaticum*. *Pimenta* is the genus of allspice (*P. dioica* or *P. officinalis*) and of bay rum (*P. racemosa*). *Melaleuca* with its bottle brush flowers is a favourite ornamental shrub, while *M. leucadendron* provides medicinal cajuput oil. The Myrtaceae family provides some fine fruits: notably the guava, *Psidium guajava*, from tropical America and the West Indies. Most of all, though, the Myrtaceae include the extraordinary genus *Eucalyptus*, also known as gum trees. As now defined, *Eucalyptus* is a huge genus with around 700 species – so big and various that it should probably be split into several smaller genera (as with *Acacia*).

Eucalyptus is one of those fortunate organisms – you meet them in all walks of life – that found itself in the right place at the right time, ready-equipped with a bag of adaptive tricks that helped it to survive and flourish where others languished. To judge from present distribution, the genus seems to have arisen in Australia some time in the early Tertiary, around 60 to 50 million years ago. Australia was a lot wetter then than it is now, and at first the eucalypts had to compete with conifers such as cypress pine (*Callitris*) and the araucarias, and with broadleaves such as she-oak (*Casuarina*) and southern beech (*Nothofagus*). But the continent soon became a lot drier; and with aridity, comes fire. Furthermore, the land is exceedingly ancient, heavily eroded, with little volcanic action to stir the geological pot, and it has lost much of the fertility it might once have had – notably the rock-bound nitrogen and phosphorus, the principal nutrients of plants; but also potassium, sulphur and some essential trace minerals.

Present-day eucalypts live in a wide variety of habitats but overall they specialize in dryness, sucking water from great depths; and although they may burn spectacularly when things get out of hand – the crowns exploding as their essential oils are vapourized – they are in many ways fire-adapted. Like the *Banksias* (of the Proteaceae) they typically hold their seeds in little wooden capsules, and they are able to germinate only after they've been cooked. After fires the seeds are shed in vast numbers, on to ground burned bare of litter and competitors, and transiently rich in nutrients from all the ash. Like the redwoods (another striking piece of convergence), eucalypts often have buds just beneath their bark which spring to life when the bark is burnt off, so the charred eucalypt re-emerges as a coppice. Many, too, have 'lignotubers' – tuber-like swellings of the lower trunk which generally become buried, and are packed with buds and regenerative tissue, which spring up like a phoenix when the main stem is destroyed. Eucalypts cope with low fertility through close associations with mycorrhizal fungi – often more than one kind of fungus at once – which greatly extend the range and efficacy of their roots. As we saw in Chapter 5, pines in northern continents also succeed in poor soils with the help of mycorrhizae. Many trees have mycorrhizae, but pines and eucalypts seem particularly adept. Yet another example of nature's reinvention.

Thus in present-day Australia eucalypts are absent only from extreme desert, the highest mountains (not that Australia has very high mountains) and rainforest. There are vast areas that don't suit them – including the tropical and subtropical rainforests of Queensland and northern New South Wales – but there are even vaster areas that suit them very well; much better than they suit the *Callitris*, araucarians, she-oaks and southern beeches, which over the past few tens of millions of years have been largely sidelined. Over most of the continent eucalypts reign supreme, typically forming open, evergreen woodland. They have diversified wonderfully – with more than 700 species at the latest official count, arranged in thirteen different 'series' (sometimes aggrandized as 'subgenera'). All but a handful are exclusive to Australia. There are just five in some islands to the north, and a few in New Guinea.

Eucalypts vary enormously in form: some like the yellow gum, *E. vernicosa*, of mountainsides are shrubs less than 1 metre tall; some, known as mallees, have many stems springing from the ground, like bamboo or seriously coppiced chestnut; many are woodland trees, 10 to 25 metres tall; and some are huge forest trees – notably the mountain ash, *E. regnans* ('ruling eucalypt'), which can grow to nearly 100 metres. Mountain ash is the tallest of all flowering plants, almost matching California's coastal redwoods. (You will be pleased to know that I have seen the biggest of the ones in New South Wales, and been properly awed.) The variety of eucalypts is further increased by a strong tendency to hybridize; and the non-botanist may think there are even more than there really are because the young plants typically have 'sessile' leaves (without stalks) and hold them horizontally, while leaves on older plants do have stalks and are held with the blade vertical. The plant's chemistry may change at the same time, and so too its susceptibility to pests. It is a veritable metamorphosis, like caterpillar into moth: one group of genes going offline, another group coming into play. Sometimes when a eucalypt regenerates from buds and lignotubers, the new stems again produce sessile juvenile leaves.

Nowadays, of course, eucalypts grow throughout the world: Californians, Indians, Africans, even Mediterraneans may be surprised to learn that they are, in origin, so emphatically Australian. The first specimen (in fact of *E. obliqua*) was sent to Europe in 1777. Charles-Louis L'Heritier de Brutelle coined the name *Eucalyptus* in 1788. But already by about 1790 eucalypts had been taken to India. By 1804 they were growing in France. Then to South America: Chile in 1823; Brazil in 1825. South Africa acquired them in 1828; Portugal in 1829.

At first they were grown in botanic gardens, and their transition into commerce wasn't always smooth. The first imports had little genetic variety, and did not necessarily lend themselves to much improvement. Some of those transferred from botanic gardens to plantations turned out to be hybrids. The first generation of hybrids – 'F1' – often grow very well. But in subsequent generations the genes are mixed up ('recombined') and the F2s, F3s, and so on are completely inconsistent. It took time, therefore, to create colonies of trees that performed well, bred true, and yet were not too inbred.

Yet it was clear from the early nineteenth century that the right eucalypts in the right places could grow at an astonishing rate. Plantations were soon established in Brazil and South Africa, to help build railways and as fuel for locomotives; in Brazil they were also used for charcoal, for smelting iron and making steel; and in Chile and South Africa for pit props. Eucalypts were also deployed widely as windbreaks, and for land reclamation. The oil from their leaves was a bonus. The aroma was thought to discourage the mosquitoes that carry malaria and they became known as 'fever gums', and were planted even more enthusiastically than ever. Their flowers also provide bees with honey.

But in the twentieth century, eucalypts were cultivated more and more for the great mass of short and uniform fibres in their wood, valued for paper for all purposes, from decrees nisi to disposable nappies. But still the wagon rolls on and eucalyptus is being used more and more in construction – for sawn wood, MDF (medium-density fibreboard), even veneers, and in plastics. More humbly, eucalypts have become hugely important for fuel in rural communities in India, China, Vietnam, Peru and Ethiopia. Indeed, eucalypts are now grown in more than ninety countries. The total area of plantation seems hard to estimate: some say around 9.5 million hectares in 1999, rising to 11.6 million by 2010; some say 16 million in 1999, rising to 20 million by 2010. In any case, it's a lot – though much less than the 130 or so million hectares reckoned to grow wild in one form or another in Australia (although Australia also has plantations). The growth rate of eucalyptus becomes more and more fabulous. The average worldwide is probably about 20 cubic metres of timber per hectare per year but some plantations claim to achieve 60 cubic metres. The big and popular *Eucalyptus grandis* typically achieves 40 cubic metres per hectare per year and is harvested, as sizeable trees, in six to eight years.

Beyond doubt, eucalypts are immensely valuable. Beyond doubt, too, they are a bandwagon: the more that is invested in them, the more their genetics and biology are studied, the more they are 'improved', and the more the gap grows between them and other species that should be given more of a chance. In much of the world they have become weeds, ousting the native flora. In some places, because of their great thirst and their supreme ability to drag water from the depths, they dry the land too much for native species to cope – and this among other things encourages fire that tips the ecological balance in their favour even more. So eucalypts overall are a mixed blessing. Their safe and advantageous deployment requires aesthetics, restraint and good husbandry, and not just an eye for the expedient. That of course is true of life as a whole.

Lindens, Cocoa and Baobabs: ORDER MALVALES

The Malvales are named for the Malvaceae family – which has always been a most intriguing family and, as modern taxonomists get to work, is becoming more so. Traditionally the Malvaceae included the homely mallows, *Malva*, and the English garden hollyhock, *Althaea*, but also the more exotic *Hibiscus*, including *H. esculentus* whose fruits manifest as okra, alias lady's fingers, alias bindhi; and the cotton plant, *Gossypium*, whose hairy seeds, even more than the latex of the rubber tree, have changed the world. Yet modern DNA studies suggest that Malvaceae should be defined even more broadly. According to Judd, several families that have traditionally enjoyed independence – including several of very significant and sometimes extraordinary trees – should now be subsumed within the Malvaceae. These include the Tiliaceae, Sterculiaceae and Bombacaceae. Thus the newly-styled Malvaceae is rich and various indeed. For the purposes of this book, Malvaceae in the old sense would hardly deserve a mention at all, but in its new form it certainly does. In the following account, however, both for general ease and to facilitate cross-reference to traditional texts, I shall stick to the old-style family names.

Tiliaceae is the family of the limes, alias lindens, and the American basswood (where 'bass' is pronounced with a short 'a' as in the fish, not as in the large guitar). Limes put up with heavy pruning and thus are often butchered as street trees (though they get their own back by attracting aphids, which secrete gum, which in turn attracts fungi, and so coat the cars parked beneath in what seems like tacky soot), but when allowed to grow to full magnificence they are unsurpassed in avenues, seen wondrously in English estates and in Berlin's Unter den Linden, the lovely road that once ran from the Brandenburg Gate to the palace of the Kaiser Wilhelm. Lime timber is excellent. Europe's *Tilia cordata* makes musical instruments and fine furniture (including the carved fronts of many a pulpit). America's basswood, *T. americana*, is used for turnery (and even before America had lathes it was used for bowls, if the wedding sequence from Henry Wadsworth Longfellow's *Hiawatha* is to be believed: 'Sumptuous was the feast Nokomis/ Made at Hiawatha's wedding./All the bowls were made of basswood/ White and polished very smoothly'). Limes are pollinated by bees, and provide excellent honey. Several other members of the Tiliaceae family provide fibres for ropes, notably of the genus *Corchorus*, grown in India and to some extent in Africa for jute. The West African danta, *Nesogordonia papaverifera*, provides fine flexible timber used for everything from telegraph poles to gunstocks, carriages, boats and veneers.

The name of the Sterculiaceae family comes from the Latin *stercus* meaning 'dung'. The huge 'wild almond' (up to 36 metres) of India is called *Sterculia foetida* ('stinking'), as if to emphasize the point. As D. V. Cowen writes in her classic *Flowering Trees and Shrubs in India* (Thacker and Co Ltd, Bombay. 6th revision 1984): 'Coming across a wild almond in bloom one's first thoughts would be that one was near an open sewer and many parts of the tree when bruised or cut emit this rank, unpleasant odour. It is unfortunate as the tree is extremely handsome: tall and straight, its well shaped crown swathed in deep coral, often without a single touch of green, it stands out among the surrounding verdure in great beauty and dignity.' Flowers of ordurous odour (there are many such) are invariably pollinated by flies. As always in the genus *Sterculia*, the flowers come out before the leaves. The fruits develop in April soon after the leaves appear and, says Mrs Cowen, they are 'large as a man's fist, woody and purse-shaped ... like odd, dark objects casually thrown into the tree'. Yet the wild almond has many good points. The leaves and bark are medicinal, a useful gum comes from its trunk and branches, the bark yields fibres for cord, and the seeds can be eaten when cooked. For good measure the wood does not split and is used for spars. (D. V. Cowen was a true memsahib: a fine hostess who also painted and wrote beautifully about plants, was a competent birder, and, for good measure, a champion golfer. Her book is a grand piece of publishing. Everyone should own a copy. I got mine in Delhi.)

There are many other fine genera of trees in the Sterculiaceae, too, as well as some shrubs and climbers. *Pterygota* has wooden fruits almost as big as a croquet ball, packed with winged seeds. It includes some fine, tall avenue trees and others that yield valuable dark-flecked yellowy timber. *Heritiera* from South-East Asia yields the dark, much-valued hardwood mengkulang. *Theobroma cacao* is cocoa: *Theobroma* means 'food of the gods'. *Cola nidia* and *C. acuminata* are the trees that provide cola nuts, the original ingredient of the soft drinks, the political and economic influence of which this past hundred years can hardly be overestimated.

Perhaps most remarkable of all the extraordinary trees in the Bombacaceae family is the baobab tree, *Adansonia digitata*, also known as the monkey-bread tree. Of the eight remaining species, seven are indigenous to Madagascar; one also grows naturally in Africa; and one more is indigenous to northern Australia. It used to be thought that the baobab was another ancient Gondwanan genus, with just a few species surviving on various Gondwanan landmasses. But DNA studies suggest the genus arose in Madagascar, long after the break-up of Gondwana – and that a few seeds managed to float across the Indian and Pacific oceans and took hold in Africa and Australia. These things happen. Several species have also been distributed by human beings and, for example, there are now plenty of baobabs in India.

Their appearance is extraordinary. They are not on the whole outstandingly tall (up to about 20 metres or so) but the trunk is swollen with water, filled to bursting like an over-stuffed sausage, and can be huge: up to 10 metres in diameter, or 33 metres in girth. The mop of twisted (but often vast) branches at the top look more like roots. Thus have arisen various myths. One has it that all the animals were given trees of their own. The baobab went to the hyena – who was so disgusted by it that he turned it upside down. Another version has it that the first ever baobab was extremely beautiful – and far too proud of its beauty. So the gods, to punish its conceit, stuck it back in the ground the wrong way up. The story echoes that of Arachne in Ancient Greece – far too proud of her sewing for the gods' liking, and turned into the world's first spider.

In truth, the trunks serve as giant water-butts, so the baobab is marvellously resistant to drought. The spongy timber is of little use, but the rest of the tree is very valuable indeed. The enormous flower buds – 'like balls of pale-green suede', says Mrs Cowen – open to form big, creamy-white flowers which appear at midnight one day in July and are wilted by morning. (Many plants flower remarkably briefly: clearly they have enormous confidence in the animals that pollinate them.) After the flowers come the fruits, white and gourd-like. The woody shell is rich in protein and in Africa is used to feed livestock, while in India, Gujurati fishermen tie them to their nets as floats, and monks employ them as water pots. The seeds provide valuable oil. Each seed is surrounded by pulp (which goes powdery), and is extremely rich in vitamin C; in both Africa and India it makes a cooling drink which protects against scurvy, and is otherwise med-icinal. The pulp is held in position around the seeds by small fibres that are used for stuffing cushions. The leaves are eaten too, and strong rope is made from the bark. In Gujurati the baobab is known as *gorak chinch* after a monk, Gorak, who taught his disciples under the shade of one. In Zimbabwe Phytotrade Africa is marketing baobab products – part of its broad initiative to gain economic value from native plants for the benefit of local people.

Nobody knows how long a baobab may live. They are so vast, so monumental, the biggest seem to be thousands of years old. But they grow very quickly, and perhaps 500 or so is the limit. Age is particu-larly hard to judge because, with age, they grow hollow – and then, perhaps with further hollowing, they are put to all kinds of uses. Often they are filled with water, as village reservoirs. The hollow spaces may be as big as a fairly sizeable suburban dining room – and some in their time have served as pubs and post offices. Spookily, Africans sometimes inter their dead in hollow baobabs whereupon, in the dry heat, the bodies mummify without further treatment. It is wise to take care when entering a hollow baobab.

There are still more outstanding Bombacaceae. *Bombax* is the genus of the various kapok trees: their seeds are hairy (like the cotton plant, of which they are of course distant relatives) and the hairs are used for stuffing mattresses, sleeping bags, Mao Tse Tung-style quilted jackets, and what you will. Kapok trees are widely grown. I have enjoyed the shade of some truly magnificent *Bombax ceiba* growing as specimen trees in Belém in Brazil, defining the boundaries of the cathedral garden there. India's red silk cotton, *Bombax malabarica*, is found throughout India and into Malaysia and Myanmar, and is now cultivated widely in the tropics, including Africa. Several other trees are also called 'silk cotton', with various coloured flowers (white, yellow), some of which belong to the Bombacaceae and some of which do not. The form of the cottony seed has been widely adopted, and clearly has evolved more than once.

Like *Sterculia*, *Bombax* produce their flowers when the tree is bare of leaves. Those of the red silk cotton are bright red through pink to orange, and fall to the ground to be eaten by deer – or by the villagers, who put them in curries. The finger-like fruits that follow the flowers harden and split to release the cottony seeds which, before the wind does its work, seem to smother the tree in cotton wool. Crows, bulbuls, mynahs, rosy pastors, sunbirds and flower-peckers flock to feast on the oily seeds, though monkeys seem to be deterred by the vicious spikes on the smooth grey trunk and branches. The wood, whitish, soft and light, is known as 'simul' and is used for dugout canoes, floats, matches and coffins.

The tree with one of the lightest timbers of all (though not quite the lightest) is also from the Bombacaceae family: balsa, *Ochroma pyramidale*, native to tropical America from Mexico through to Brazil (and also Cuba), and now planted in India and Indonesia. Every child knows balsa as the stuff of model-making. Grown-ups use it for rafts, aircraft, insulation, equipment for water sports, and of course for theatre and film props, to reduce damage to actors. The main commercial source nowadays is Ecuador.

Finally we might mention the durian, *Durio zibethinus*, whose fruits are huge and spiky. Some find them delectable: another food for the gods. But they stink (shades of *Sterculia* again) and are expressly forbidden on aeroplanes.

In its new form, then – subsuming the Tiliaceae, Sterculiaceae and Bombacaceae – Malvaceae emerges as a truly remarkable family of truly remarkable trees. But the Malvales order also includes another arboreal family that is clearly discrete from the Malvaceae: the Dipterocarpaceae.

The dipterocarps are hugely various: 680 or so species in sixteen or seventeen genera (different taxonomists split them slightly differently). The principal genera are *Shorea* (by far the most important), *Dipterocarpus* and *Dryobalanops*. Mostly the dipterocarps come from south or South-East Asia – Malaysia is the focus, with 465 species in fourteen genera – but there are also forty-nine species in three genera in Africa, and one each in South America and the Seychelles. Between them the different types grow from coast to uplands in many (tropical) climates and in soil both fertile or – often – extremely infertile. Some grow in dry land but on the whole they prefer the wet: there are many kinds in swamps and the biggest grow where there is year-round moisture. I am told there is an old plantation of dipterocarps outside Kuala Lumpur where many individuals exceed 60 metres.

Dipterocarp means 'two-winged fruit' – and indeed they have fruits roughly like sycamore keys, though often much larger, and sometimes brightly coloured. People make wide use of dipterocarps. The fruits of many *Shorea* and some *Dryobalanops* are boiled as vegetables. The seeds of both *Shorea* and *Pachycarpae* are extremely rich in fat – up to 70 per cent – which is similar to cocoa butter though harder, and is much favoured for chocolate and cosmetics. (And note again the loose phylogenetic relationship between the dipterocarps and the cocoa tree, in the Sterculiaceae. Biochemistry runs in dynasties.) Most dipterocarps too produce useful resins; and a form of camphor comes from *Dryobalanops*, used as incense.

But above all, the dipterocarps dominate the international market in tropical timber. *Shorea* species from South-East Asia are marketed as 'meranti'. Various kinds are sold as light-red or dark-red meranti – red indeed and finely figured, and used for all purposes. Another group of *Shorea* feature as white or yellow meranti, also with many uses, from floors to ships, plywood and veneers. Mersawa and krabak are two more, similar, South-East Asian timbers from the genus *Anisoptera*. Kapur from Malaysia and Indonesia is the genus *Dryobalanops*. Altogether, the Dipterocarpaceae are a formidable family – comparable in their part of the world with the oaks and beeches from further north.

Frankincense and Myrrh, Oranges and Lemons, Maples, Mahogany and Neem:
ORDER SAPINDALES

The Sapindales are closely related to the Malvales. As in the Malvales, reclassification is in train, so the traditional list of eleven Sapindales families is now reduced to eight. Six of those families – one of them now much expanded – contain trees that are at least intriguing, if minor (ecologically and economically), and others that are of supreme importance both in the wild and to humanity.

Minor but intriguing is the Simaroubaceae family, a hundred species of trees and shrubs in twenty-one genera from throughout the tropics and subtropics. They are biochemically potent and widely deployed in medicine, especially *Quassia* from Africa; while *Picramnia* from the Americas was once exported to Europe to treat erysipelas and venereal disease. The white syringa of Africa, *Kirkia acuminata*, grows to 18 metres, provides useful timber, and also has swollen roots that store water which knowledgeable locals tap in times of drought. Best known in the gardens and streets of the West is the tree of heaven, *Ailanthus altissima*. It was first imported to England from its native China in the mid eighteenth century and thence to the rest of the West. It resists pollution, can grow to 30 metres sometimes in less than twenty years, and has compound feathery leaves like an ash, although each may be nearly a metre long.

Minor too by world standards is the Burseraceae family – but extremely intriguing, as the family of both frankincense and myrrh. In all there are about 500 species, in seventeen genera, throughout the tropics but mainly in Malaysia, tropical America and Africa, and many provide resins and aromatic oils for perfumes, soaps, paints, varnish and incense. Frankincense comes from *Boswellia carteri*, of Somaliland; myrrh is from various *Commiphora* species, notably *C. abyssynica*, which grow and are now cultivated in Arabia and Ethiopia. The gifts of the Magi to the infant Christ in Bethlehem were exotic indeed. The first known government-sponsored plant-collecting expedition was in search of myrrh: reliefs on the Temple of Deir el Bahair at Karnak show myrrh trees being transported from the Land of Punt, around 1495 BC. The Burseraceae family also provides some useful timbers, widely used in Malaysia and Africa. Outstanding is gaboon of Africa, *Aucoumea klaineana*, used for everything from cigar boxes to sports gear and high-class furniture. Sometimes gaboon is mottled and striped, and then it is highly valued for veneers.

The Rutaceae family is named after the rue, *Ruta graveolens*, a small aromatic shrub, toxic but also medicinal, and grown in herbal gardens for centuries. It pops us here and there in the plays of Shakespeare, who surely was a competent naturalist, while his son-in-law was an outstanding apothecary. But the Rutaceae overall include about 900 species in 150 genera that grow throughout all the warmer reaches of all continents, including Australasia. By far the best known and important is the genus *Citrus*: *C. limon*, the lemon; *C. medica*, the citron; *C. aurantium*, the sour Seville orange – a variety of which is also the source of bergamot, the stuff of Earl Grey tea; *C. sinensis*, the ordinary, sweet orange; *C. reticulata*, different varieties of which are mandarins, satsumas and tangerines; *C. aurantifolia*, the lime (not of course to be confused with *Tilia*); and *C. paradisa*, the grapefruit. Closely related to *Citrus* is *Fortunella*, the genus of the kumquat, which turns up sometimes pickled on smart dinner tables, though not in my opinion to very obvious advantage.

The Rutaceae also provide some valuable timbers. Various species of *Flindersia* feature as the cinnamon-coloured 'Queensland maple' – again for prestige furniture, gunstocks, oars, and what you will. Southern silver ash from Eastern Australia, *F. schottiana*, is just as versatile, but pale yellow. From South America comes pau marfin,

Balfourodendron riedelianum: tough, pale, flexible; wonderful for oars, tool handles, shoe lasts, furniture, marquetry. Ceylon satinwood, *Chloroxylon swietenia*, from the southern Indian subcontinent is called satinwood because it seems to shimmer like folded silk, and is highly prized for panels and veneers. Pale-golden West Indian satinwood, *Fagara flava*, was used widely in the eighteenth century by the great English makers of fireplaces and furniture: Adam, Sheraton and Hepplewhite.

But satinwoods have largely been ousted from favour by members of the Meliaceae family. The fifty-one genera (with 550 species) include several outstanding timber trees – indeed their present and historical importance can hardly be over-emphasized. World star of the families for almost four centuries is the genus *Swietenia*: the American or 'true' mahoganies. There are three species – closely related, and often hybridizing in the wild where their ranges overlap. *S. humilis*, with small leaves, prefers drier country, and spreads north into Mexico. *S. mahogoni*, the first of the trio to be named (by the Austrian botanist Gerard von Swieten) is from the Caribbean and Southern Florida. *S. macrophylla*, the big-leaf mahogany from Brazil and Honduras is the giant: an emergent tree, towering above the seasonally dry rainforest canopy at about 70 metres, with huge buttress roots to reinforce a trunk that can be 3.5 metres in diameter.

Or this, at least, was the pristine state. There is little or no forest left where *S. mahogoni* and *S. humilis* once lived. The big-leaf mahogany is still to be found in the forests of southern Amazonia – anywhere between one big tree in every 10 hectares to three trees per hectare – and might still, some say, be harvested sustainably and profitably from the wild. It is difficult to increase the proportion in the wild because mahoganies are light-lovers, and need open ground or clearings to get going. If they are simply planted among other trees they fail. But big-leaf mahogany and *S. mahogoni* are now widely grown in plantations, particularly in tropical Asia and Oceania, and increasingly in their homelands in the American tropics. Yet they have suffered enormously in plantations from an insect pest, the mahogany shoot-borer, a species of *Hypsipyla*, which bores into the shoots and turns what should be a straight proud tree into a mean shrub. The pest can be controlled; but largely because it is perceived as a problem, the world's plantations of mahogany are only one twentieth those of teak. Wild mahogany is now listed by CITES, and trade in general is more and more regulated. But it will be interesting to see what happens to mahogany, wild and cultivated, over the next few centuries.

Several close relatives of true mahogany are also fine timber trees: pride of India, *Melia azedarach*; African walnut, *Lovoa trichilioides*; South American cedar, species of *Cedrela*; Asian-Pacific red cedars, species of *Toona*; the sapele, *Endophragma cylindricum*; and *Endophragma* and *Khaya*, which are sometimes called 'African mahogany'. Several trees that have nothing to do with Meliaceae at all are also sold as 'mahogany', including several dipterocarps (of the genus *Shorea*) and the occasional eucalypt.

Less closely related to *Swietenia*, yet still within the Meliaceae, is the wondrous neem tree, *Azadirachta indica*, which featured earlier as the antidote to noxious tamarind. The neem grows to 20 metres or more, with evergreen, roughly ash-like leaves, much valued for their year-round shade. It is native to south Asia, but its deep roots enable it to thrive on dry, poor soil and it has been planted and become naturalized throughout all tropical Asia, while the British took it to Africa in the early twentieth century to slow the spread of the Sahara. It has also been taken to Fiji, Mauritius, Saudi Arabia, and all tropical and subtropical America, including Florida, Arizona and California. In the USA there are neem plantations.

There are many outstanding chemists among plants, but the neem is among the greatest of all. For centuries, indeed for thousands of years, the Indians have treasured the neem for medicines: it features in some of the most ancient Hindu texts. Many Hindus begin the new year by chewing neem leaves. Many clean their teeth with neem twigs. They treat skin disorders with its juice, and drink infusions as a tonic. Gum from the bark is used for dye. Neem has also proved active against more than 200 species of insects, preventing them from feeding, inhibiting their reproduction – discouraging egg-laying and disrupting the development of any eggs that are laid. As if to make the point a plague of locusts in India in 1959 destroyed just about everything that grew, except the neems. The timber is termite-proof, and perfect therefore for hot climates, for everything from furniture to tool handles. Indians put the leaves in cupboards, to safeguard the contents. The leaves also make good fodder, while the seeds are 45 per cent oil and provide excellent seed cake for livestock, or oil for lamps. Very properly, the neem is venerated. Many Indian place names incorporate its name: Neemuch, Neemrana, Nemawar, and hundreds more. The neem is said to have been blessed with nectar, sent from heaven.

Science has reinforced the folk law. Different parts of the neem, but particularly the seeds, contain a host of potent organic compounds shown to be active against just about everything pestilential: bacteria, fungi, viruses, nematodes and mites, as well as insects. A powerful spermicide is in there too, raising hopes in some circles of an effective male contraceptive. Yet the extracts do not seem to harm mammals (including people) or birds. Outstanding among the compounds studied so far is azadirachtin. It is now being incorporated into commercial pesticides not only because it seems innately effective but also as part of a general swing away from industrial chemistry to biotech, based on natural processes and materials.

At this point, the story becomes less pleasant. In the 1990s US-based companies began patenting various components of the neem which the many excellent scientists of India had not bothered to do because, under Indian law, such medicinal, natural materials are not subject to patent. Thus the beneficent, sacred neem, rooted deep in Indian soil and Indian culture, has become the subject of squalid legalistic wrangling.

The Anacardiaceae are a wonderfully distinguished family. Its seventy genera (with 600 species) include some of the finest ornamentals like sumac, *Rhus*, and the smoke bush, *Cotinus*, with its round leaves and wispy grey inflorescences. But these tend to be shrubs rather than trees. The pistachio, *Pistacea vera*, is a bona fide tree, however, up to about 10 metres. It is native to the Near East and Central Asia but has long been cultivated in the Mediterranean and the southern United States for the delectable green kernels of its nuts, eaten as savouries and marvellous in ice cream and its Indian equivalent, kulfi. The cashew tree, *Anacardium occidentale*, is native to tropical America but is now grown widely in India and East Africa – a very useful crop on land somewhat drier than most fruit or nut trees prefer. Cashews are of course delectable and also nutritionally potent: 45 per cent fat and 20 per cent protein. They are commonly served as nibbles before a meal but in truth, like peanuts, are as rich as any dinner is liable to be. Trees that disperse their seeds with the help of animals must in general attract the animals' attention and none does so more promiscuously than the cashew. For the nut itself is presented at the tip of a cashew 'apple', which is more like a yellowish-reddish pear. The effect is of a sculpture on a plinth, though held upside down on the twig. Some people ferment the 'apples' to make kaju, a strong liquor. The shells of the cashew contain an oil that irritates the hands but is also used in industry. I encountered my first cashew tree in Pakistan and although the trip as a whole was memorable, connoisseurs of trees will understand when I say that this was among its highlights.

But the Anacardiaceae offers yet more tropical delights: the mango, *Mangifera indica*. The mango originated somewhere in India or Burma but it thrives in all kinds of soil and is now grown throughout the tropics and into Egypt and Florida, often as a street tree (as in large areas of Belém, although you might think Brazil had enough trees of its own), its dark-green, shiny, willow-like leaves providing admirable shade. Often indeed mangoes grow virtually as weeds: I have stood beneath a bus shelter in Panama in a storm while mangoes thudded on the roof, and very acceptable they proved to be. They also provide energy (10 to 20 per cent sugar), and are particularly rich in vitamin A. Nowadays deficiency of vitamin A is a huge issue. Forty million children are believed to be affected, and many are blinded by it (it leads to the drying of the cornea known as 'xerophthalmia'). Modern biotech companies are using genetic engineering to produce 'golden' rice that contains some vitamin A, and are claiming thereby to be socially responsible, not to say heroic. But all dark-green leaves are rich in vitamin A, and all that's really needed is horticulture – which was always a part of traditional farming; and with a few mangoes or papayas around, the problem is solved. If high-tech vitamin-A-rich rice is ever of help at all, it is only in regions where traditional agriculture has been shoved aside by high-tech, industrialized, monocultural farming.

Finally, completing this lightning sketch of the Sapindales order and also the Rosid eudicots as a whole, comes the extremely distinguished Sapindaceae family. Like the Malvaceae discussed above (and the

Cupressaceae, in Chapter 3) the Sapindaceae family is interesting in its traditional form but has become even more interesting of late as other plants including many fine trees have been included within it. Thus the Sapindaceae family as now described by Judd embraces the old-style Aceraceae (maples), and the erstwhile Hippocastanaceae (horse chestnuts). In the following, in the interests of clarity and of continuity with most extant texts, I will discuss the three traditional families separately. But in the decades to come, I imagine they will all be discussed as Sapindaceae.

The traditional Sapindaceae family is distinguished enough. It includes about 2,000 species in 150 genera – mostly trees or shrubs but also about 300 climbers (which are of the kind that climb by tendrils, and are often of huge ecological significance). The family is tropical and subtropical, occurring through the tropical Americas, sub-Saharan Africa, India and east Asia, Malaysia and Indonesia, and down into Australia. The family takes its name from *Sapindus*, which provides oil for soap. It includes some fine fruit trees. The sweet-acid lychee, *Litchi chinensis*, is from southern China. Botanically the fruit is an aril, like that of yew. A close relative is *Melicocca bijuga*, grown in America. The rambutan, *Nephelium lappaceum*, is like a lychee with mad hair. *Blighia sapida*, named after the unfortunate Captain Bligh of the ill-fated *Bounty* (who in truth was a distinguished amateur botanist and no mean artist) is known in Africa as the akye, and in the West Indies as the akee. It is the national fruit of Jamaica. Akee is again an aril, which tastes like scrambled eggs when cooked, but is poisonous if plucked at the wrong stage. The Sapindaceae family provides some fine ornamentals, too, like *Koelreuteria* (a street tree) and *Xanthoceras* (grown for its flowers). The taun tree, *Pometia pinnata*, from the South Pacific is magnificent at up to 45 metres, with a trunk up to a metre in diameter and a lovely, smooth, reddish fine-grained timber valued for everything from joists and rafters to pianos and the bars of ships' capstans.

The Aceraceae, the old-style maple family, includes at least a hundred species in the genus *Acer*, plus two in *Dipteronia*. Maples live all over the northern continents, with a huge representation in China, and one in North Africa. This is the field maple, *Acer campestre*, which is also native to the British Isles and is our only maple. China has scores of native species. Japan has nineteen. North America has about a dozen. Most maples are small to medium-sized but some are large and evergreen, including a few through Malaysia and into Java. As with oaks, we see that in any one genus some species may be evergreen, and some not. Maples are easily recognized for their fruits: paired keys, which spin helicopter-style in the wind, elegantly known in botanical circles as 'samaras'. In *Dipteronia*, by contrast, the seeds sit in the centre of a round wing like a yolk in a fried egg. Many maple species produce fine timber, including *A. pseudoplanatus*, which the British call sycamore and the Americans call 'plane'. Sycamores in Britain are often magnificent but are commonly perceived as invasive weeds. They are despised in part because they do not support such a wide variety of native insects and other animals as oaks do, for example. But they do support an enormous biomass of insects, which in turn support birds. The British are almost certainly right to despise feral rhododendrons, which really do belong elsewhere, although buzzards like it for nesting. But we should perhaps ease up on the sycamore. Maples are good for other things, too. Sugar maple, *A. saccharum*, and others are the source of maple syrup, to which it is easy to become addicted. Of the many routes to obesity on offer in the USA, that of maple syrup, crisp streaky bacon and a 'long stack' of buckwheat pancakes is perhaps the most seductive of all.

The Hippocastanaceae is the family of *Aesculus*, thirteen species of horse chestnuts and buckeyes, widespread in North America, southern Europe and temperate South-East Asia; and of *Billia*, just two evergreens from Mexico and tropical South America. *A. hippocastanum* is 'the' horse chestnut: *hippo* is Greek for horse and *castanum* means chestnut. In 'nature study', children sketch the horse chestnut's huge palmate leaves, the intriguing horseshoe-shaped scars they leave on the stems, and the big resinous scaly buds (or they did when I was at school), and prize their big brown seeds as 'conkers'. Grown-ups value them primarily as ornamentals, in many a lovely avenue. Some *Aesculus* too are in various ways medicinal, and extracts have been used in North America to stun fish (which seems to emerge as a national hobby); and the wood, light and not durable, is used for boxes and charcoal.

10

From Handkerchief Trees to Teak: The Daisy-like Eudicots

The handkerchief tree – named by the children of the British Raj

The second great group of eudicots are the asterids – named after the daisy family, Asteraceae. To be sure, they don't all look like daisies, any more than the rosids all look like roses. But again, details of their DNA and of their microstructure, particularly the ovules, suggest that the asterids do all derive from a common ancestor – that they form a true clade. The modern taxonomy favoured by Judd divides the asterids into ten orders. Three of them contain little in the way of trees, although they do include some shrubs. Thus the order Garryales has a few shrubs from Central America, some of which are grown as ornamentals, and include the bayberry. The Apiales is named after the family Apiaceae, formerly known as the Umbelliferae, which is best known for carrots, celery and coriander – but does include the umbrella tree, *Schefflera*. The Dipsicales includes teasel and honey-suckle, and its greatest claims to arborescence lies with the snowberry, *Symphoricarpos*, and the elder, *Sambucus*. But the remaining seven orders between them contain some of the most magnificent and valued trees of all.

Dogwood, Tupelo and the Handkerchief Tree:
ORDER CORNALES

Judd recognizes 650 species of Cornales, in three families, but only the Cornaceae include any significant trees. The family is mostly (though not exclusively) north temperate. *Cornus*, for which it is named, includes the forty-five species of dogwood, mostly shrubby. *Davidia* is the lovely handkerchief tree. It was first reported from the mountains of western China in 1869 by the Jesuit naturalist Father David, who also made the giant panda known to the West, and gave his name to Père David's deer. Another missionary priest sent seeds of *Davidia* to Europe thirty years after David found it. Just one seed germinated of the thirty-seven that were planted. It grew into a smallish tree with leaves somewhat like a lime tree – pleasant, yet nothing special. But when it flowered in 1906 Europeans saw for the first time the wonder that had first enchanted Father David: for each flower is flanked by two petals as big as your hand that look for all the world like rich creamy-white leaves. The tree in full bloom is festooned. The children of colonial civil servants in India are said to have given *Davidia* its common English name, comparing its flowers to the handkerchiefs of the crowds who waved them off from the quay. It is also called the ghost tree, or dove tree. All the names suit. I met one in full bloom early one sunny July morning in the University Botanic Garden at Cambridge. Stunning.

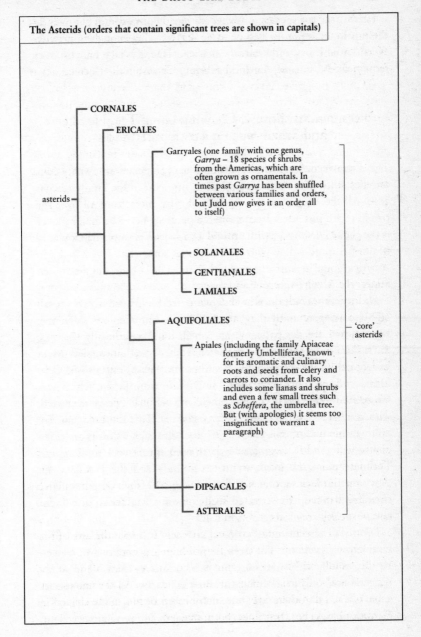

The Asterids (orders that contain significant trees are shown in capitals)

CORNALES

ERICALES

asterids

Garryales (one family with one genus, *Garrya* – 18 species of shrubs from the Americas, which are often grown as ornamentals. In times past *Garrya* has been shuffled between various families and orders, but Judd now gives it an order all to itself)

SOLANALES

GENTIANALES

LAMIALES

AQUIFOLIALES

Apiales (including the family Apiaceae formerly Umbelliferae, known for its aromatic and culinary roots and seeds from celery and carrots to coriander. It also includes some lianas and shrubs and even a few small trees such as *Scheffera*, the umbrella tree. But (with apologies) it seems too insignificant to warrant a paragraph)

DIPSACALES

ASTERALES

'core' asterids

Finally, the ten species of *Nyssa* are mostly ornamental but include the tupelo, *N. aquatica* and *N. sylvatica*, trees big enough to provide North Americans with railway sleepers. (Once *Nyssa* had its own family, the Nyssaceae, but Judd subsumes it within the Cornaceae.)

Persimmons, Ebony, Chewing Gum, Tea, Heather and Brazil Nuts: ORDER ERICALES

The Ericales order includes around 9,450 species in twenty-four families, with many outstanding trees, not least in the family Ebenaceae. All the Ebenaceae are trees or shrubs: most from all over the tropics, with just a few from more temperate zones. The major player is the genus *Diospyros*, with around 450 species: 200 or so in lowland Malaysia; quite a few in tropical Africa; somewhat fewer in Latin America; some in Australia and India; and a few outliers in the United States, the Mediterranean and Japan.

Diospyros includes a whole range of edible fruits, which are all highly astringent until they are fully ripe, but then are delicious. Best known are the persimmons, which look superficially like big, thick-skinned tomatoes, and are eaten fresh, cooked or candied. Most widely cultivated – especially in China and Japan, but also in California and the south of France – is *D. kaki* from Japan, which was introduced to the US in the late nineteenth century. Native to the US, with smaller, dark red fruits is *D. virginiana*. This is a tall, thin tree up to 30 metres and not much cultivated, although its fruits are often picked wild (and *D. virginiana* is often used as rootstock for *D. kaki*). The date plum, *D. lotus*, is grown in Italy and the Far East. *D. digyna* is the black sapote. Not all the *Diospyros* fruits are particularly friendly, however: the crushed seeds of some Malaysian and Indonesian species are used to poison fish.

Diospyros also includes various extremely valuable timbers of the kind known as ebony. The trees are not huge – generally around 15–18 metres tall, with trunks around 60 centimetres thick – and so the timber is sold only in short lengths. But the heartwood of some species is jet black, and others are deep rich brown or alluringly striped in brown or black. For their fine colour, strength and prodigious weight,

– far heavier than water – the ebonies have been valued since ancient times. The pharaohs had their glossy black furniture made from ebony. It is excellent both for sculpture and for turning – door knobs, the butts of billiard cues, chess pieces – and marquetry, piano and organ keys, clarinets, and the chanters of bagpipes.

Various species are harvested in Africa, where ebonies can be important forest trees, including the very dark *D. crassiflora*. *D. reticulata* from Mauritius is highly prized. *D. ebenum* from Sri Lanka is known as Ceylon ebony and is often called 'true ebony' because its timber is a uniform jet black. *D. marmorata* from the Andaman islands is a small tree (only about 6 metres) but it yields a fabulous brown-black mottled timber known as Andaman marblewood. In sharpest contrast, America's native persimmon yields a straw-coloured sapwood that is marketed as 'white ebony' (and is also known as bara bara, boa wood, butter wood, possum wood or Virginia date palm). It is used for tool handles, and textile shuttles made from it are said to last 1,000 hours before they wear out. The pale sapwood and the thin dark heartwood are sometimes used together to make a fine veneer.

The trees and shrubs in the great family of the Sapotaceae bestride the wet, lowland tropics: 1,100 species in fifty-three genera. Many have edible fruits. The huge genus of *Pouteria*, with 325 species, includes the mamey sapote, *P. mammosa*, and the eggfruit *P. campechiana*. The seventy species of *Chrysophyllum* include the star apple, *C. cainito*, which is also grown ornamentally as the satinleaf. All these sapotaceous fruits are said to be delicious; though the only one I can vouch for personally is the sapodilla, *Achras sapota*, which has beautiful, barley-sugar flesh. Many tropical fruits are disappointing. Many are fine, but should probably be left for local people to enjoy. But you could munch out on sapodilla; it could well join the banana and the mango as a world favourite. The shea butter tree, is *Butyrospermum paradoxum*, the source of an edible oil.

There is much more to the Sapotaceae. *Manilkara zapota* is chicle: its latex is the stuff of chewing gum. Several of the 110 known species of *Palaquium*, especially *P. gutta* from Sumatra, Java and Borneo, yield latex that hardens to form gutta-percha, which is chemically related to rubber (another polymer of isoprene). Nineteenth-century industrial chemists found that they could mould it every which way when it was warm and soft, and it held its shape when cool and hard. Soon it was used to fill teeth, make golf balls, and insulate electric and undersea telephone cables. The latex was tapped by cuts in a herringbone pattern, as with rubber, which damaged the wild trees, so plantations were established in Java and Singapore. Now, alas, only dentists still make use of gutta-percha, for temporary fillings.

Many Sapotaceae yield fine timber, too. Some species are big – up to 30 metres tall, with trunks 2 metres thick. Some have heavy timber, spiked with silica. Other have lighter timber, often free of silica. Among the seventy-five or so species of *Sideroxylon* is the timber known as buckthorn, ironwood or mastic. *Minusops* is marketed as cherry mahogany. The makore (*Tieghemella heckelii*) from West Africa is a huge tree up to 45 metres tall and 1.2 metres thick. Its heartwood is a pale blood-red to reddish brown, its sapwood is slightly lighter, and some logs have the mottled, lustrous look of watered silk. It is said by aficionados to have a much finer texture than mahogany, and is much favoured for furniture, veneers and turning; and for laboratory benches, parts of carriages and boats, and for marine-quality plywood. Various species of *Palaquium* (the gutta-percha genus) and of *Payena* from Malaysia and Indonesia are bundled together under the trade names of nyotah or padang. The timber is deep-pink to red-brown, often with dark streaks: again, excellent for furniture and doors, and also outdoors, for shingles to roof or clad buildings like the scales of a fish.

Finally, the mournful tambalacoque tree of Mauritius, *Calvaria major*, is from the Sapotaceae. Exactly why it is mournful will be revealed in Chapter 13 (though it apparently has less cause to be downcast than was once thought).

There is much less to report from the Myrsinaceae family though it is large enough, with 1,000 species in thirty-two genera, spread over warm temperate lands and the tropics. The *Ardisia* genus imposes itself most on human consciousness, with several ornamentals, a few fruits grown in gardens and greenhouses, such as *A. crispa*; *A. squamulosa*, used to flavour fish in the Philippines; *A. colorata*, whose leaves Malaysians take to settle the stomach; and *A. fulginosa*, which the Javanese boil in coconut oil and use to treat scurvy.

The Theaceae family, however, is full of interest. The 300 or so species in twenty genera include the showy flowers of the Franklin tree, *Franklinia*; and among the eighty-odd species of the genus *Camellia* is *C. sinensis* – tea. It was first grown in China, probably as a tonic, rich in caffeine and essential oils Now it is drunk by half the people on earth, and grown on the largest scale in India and Sri Lanka, but also in East Africa, Indonesia and Russia: another of those plants that have transformed the economy and politics of the entire world. *C. sinensis* left to itself grows into a respectable tree, as tea plantations demonstrate when abandoned – including one I have been told about in Uganda, which was abandoned in the time of Idi Amin, and quickly grew into a veritable forest. But in active plantations its tips are picked every fifteen days or so, depending on variety, place and weather, and in effect it is bonsai'd into a hedge, around waist high: a meticulous exercise in topiary. Typically the bushes are grown on steep hillsides (and you really have to go to Asia to see how steep a hillside can be; those classical Chinese painters do not exaggerate). The bushes are brightest green and cover the hills as far as the eye can see, zig-zagged with dark narrow gulleys for the pickers. Tea grows best in shade, and among the shade trees grown in Kerala, where I once stayed on a tea plantation, is the Australian silk oak: pruned to filter just the right amount of light, and to provide fodder for buffalo. The landscape is as magical as Alice's wonderland – bonsai'd, topiarized, tessellated, and dotted with trees like feather dusters. Traditional farms, orchards, and plantations worldwide show that beauty and productivity can go hand in hand, and in a crowded world, so they must.

The Ericaceae family, for whom the whole order is named, is also full of good things. Defined broadly (as Judd does) it now includes 2,700 species in 130 genera of climbers and shrubs as well as trees. They grow almost worldwide (although they never made it to Australia – at least as wild plants), especially on uplands and typically on acid soils, relying very heavily on the mycorrhizal fungi in their roots. A few are epiphytes. Some have evolved into parasites, and have abandoned chlorophyll. The Scots at the edge of Europe's tundra know the family mainly for *Calluna*, the heather, the stuff of purple hillsides. But *Calluna* is but a windswept northern outlier. The related heaths in the genus *Erica* are particularly various in South Africa (which has 450 species), and grow to at least head height.

Where exactly the Ericaceae first arose is not clear, but the Himalayas is a good bet. At least 700 of the 1,200 species of *Rhododendron* and related *Pieris* grow where some of the world's mightiest rivers begin – the Brahmaputra, the Mekong, the Yangtse – and many achieve tree-like dimensions, though since they have many stems they are generally rated as shrubs. There are another 300 species in New Guinea, which are apparently an offshoot of the original Himalayans. Rhododendrons are considered bad news in Britain, where they grow as wild and rampant exotics, though they provide excellent nesting grounds for buzzards.

Arbutus is the genus of the strawberry tree and of the lovely madrones of North America – with smooth red bark that peels away to reveal yellow-grey beneath. I have walked among madrones in the hills of California, north of San Francisco: they are among the glories of a glorious landscape. Deer and quail enjoy their orange-red fruits. So, too, did the Native Americans. The wood is used locally and makes fine charcoal. Madrones do not grow big, but they can live for at least 200 years.

In the Lecythiidaceae family are some of the world's greatest, most intriguing, and most economically important trees. The family includes 400 or so species in thirty genera – of shrubs and climbers as well as trees; centred in South America but also found in Africa, Madagascar and tropical Asia. Smallest of all is the eccentric *Eschweilera nana*, which grows out in the Cerrado and often has an underground trunk like the *Attalea* palm, a device that protects against fires. Many, however, are emergent trees that grow through the canopy to tower above the rest. Tallest of all are *Cariniana micrantha* and *Couratari stellata*, both up to 60 metres. *Cariniana* and *Couratari* have the sky to themselves and are pollinated by wind, which is unusual among tropical trees (and among Lecythiidaceae). The long straight boles of *Cariniana* make excellent timber, and forest hunters tip their arrows with a poison from the bark of *Cariniana domestica*. *Cariniana* is the longest-lived of the Lecythiidaceae – or indeed of all the neotropical trees. Some have been dated at 1,400 years.

Most striking, however, are the fruits of many of the Lecythiidaceae: big wooden globes and cylinders, generally born directly from the trunk, and packed with seeds. The wooden armour has evolved, presumably, to deter predators, although the capuchin monkeys of Amazonia sometimes get the better of it. Sometimes the seeds are big and fleshy and eminently edible, and are dispersed by animals – sometimes by fish. But in some genera (like *Carinaria*) the seeds are winged like those of an ash and, when the casing breaks, are dispersed by wind. These seeds are too light to carry much nutrient and to make up for this the seed leaves are green and begin to photosynthesize the moment the seeds germinate.

Most important by far to human beings is the Brazil nut, *Bertholettia excelsa* – up to two dozen triangular desperately hard nuts packed like the segments of an orange within their desperately hard casing. The Brazil nut tree is almost as tall as *Carinaria*. It can provide good timber but few would cut it for such a purpose. It's the nuts that matter: they are 66 per cent fat and 14 per cent protein – and, more to the point, their flavour is sublime. Europeans and North Americans import about 50,000 tons of them a year from Brazil and Venezuela, some from wild trees but also from plantations, not least around Belém at the mouth of the Amazon. The wild trees are protected, and often these days you see them gaunt and abandoned in the middle of nowhere: the rest of the forest, which they have evolved to look down upon, now cleared to make way for sun-stressed cattle and soya. Brazil nut trees are very susceptible to fire, and although they are huge they are not long-lived (in contrast to *Cariniana*). The biggest of them are less than three centuries old.

The wooden orb that encloses the Brazil nuts has a neat cap at the top, which comes off when the fruit falls to the ground. But the nuts remain trapped: the Brazil nut tree at this point relies upon the good offices of the agoutis, which are long-legged relatives of the guinea pig, to make the hole bigger and carry the seeds away. The agouti eats some of the nuts, and buries others for later just as a squirrel buries acorns (such stashing is a very rodent trick). It doesn't recover all of them, however, and the seeds that it forgets grow into new trees – though they take twelve to eighteen months to germinate. Thus the Brazil nut tree relies not simply upon the agouti, but upon the agouti's amnesia. The empty Brazil nut cases fill with rainwater and then form nurseries for insects and frogs.

The *Lecythis* genus produces sapucaia or paradise nuts, which are said to taste even better than Brazils. I cannot vouch for their excellence; I wish I could. Clearly there is another market here – another economic reason for conserving the Brazilian forest, rather than cutting it down, to add to its aesthetic and ecological advantages. The

Cannonball fruits grow straight from the trunk

wooden spheres that enclose the nuts are known as 'monkey pots'. Apparently you can catch monkeys by putting a sweetmeat inside an empty pot. The monkey comes along and grabs it, cannot withdraw its closed fist, and refuses to release its booty even when the hunter comes along to blip it on the head. Monkeys are standard fare for the people of tropical forests. *Lecythis pisonis* of Amazonia and the Brazilian Atlantic forest has the largest fruits of all the family: it takes them about a year to grow as big as your head. The fruit opens while still on the tree to reveal seeds that are each coated in a bright fleshy aril. On the same night that the fruits open the seeds are dispersed by bats. The efficiency both of animal pollination and seed dispersal can be staggering: co-evolution really works. The flowers of *L. pisonis* are pollinated by carpenter bees, which are among the biggest bees of all (even bigger than bumble bees). *L. grandiflora* yields a much-valued timber marketed as wadadura.

The wooden fruits of *Couroupita guianensis* grow straight out from the trunk and may be 20 centimetres across, giving it the name of 'cannonball tree'. The fruits break as they hit the ground to expose a bluey-green pulp, packed with hairy seeds. To people the pulp smells foul but peccaries (American wild pigs) love it. The seeds, or some of them at least, pass right through the peccary, apparently protected from its digestive juices by their hairiness, and may germinate in the dung. The pulp of cannonball trees is also fed to domestic pigs and poultry. The cannonball tree is often grown as an ornamental: its intriguing fruits follow a show of large, waxy, sweet-smelling flowers.

Also grown as ornamentals are species of *Barringtonia*, the most important genus from the Old World. *Careya*, of Malaysia and India, gives useful timber known as tummy wood.

Nightshade, Borage – and Some Very Fine Trees:
ORDER SOLANALES

It isn't a hundred per cent clear that the Solanales do form a coherent group, a true clade. For the time being, however, the order is taken to include around 7,400 species of herbs, shrubs, vines and trees in six families, of which three are of particular interest.

The Solanaceae are the family of potatoes, tomatoes, capsicums and aubergines. All are pharmacologically potent, at least in the wild form: wild potatoes and tomatoes are generally poisonous. Also in the family, as if to rub the point home, are tobacco, mandrake and the nightshades. The shrub jimsonweed (*Datura*) provides steroids that are used to prepare oral contraceptives. Only a few Solanaceae are trees, however. I have seen some of them out in the Cerrado. Attractive, but not large.

The family Boraginaceae may or may not belong among the Solanales. Its position, says Judd, is 'somewhat problematic'. In any case it includes 2,650 species in 134 genera of herbs (including borage and comfrey), climbers and shrubs both tropical and temperate – and some very convincing and commercially significant trees. The 320 or so known species of the genus *Cordia* include *C. abyssynica* and *C. millenii* from East Africa, and *C. platythyrsa* from West Africa, all marketed as 'African cordia', though all have euphonious local names as well. They grow up to thirty metres, and up to a metre thick. The bole has an irregular shape but the sapwood is a pleasant cream and the heartwood is golden brown, and although it is weak and soft it is good for decorative trimmings, like the edges of shelves in libraries, and for veneers. More to the point it resonates well, and is favoured for drums and sounding boards; and the people of West Africa make boats from cordia. Brazil has its own *Cordia: C. goeldiana*, known as freijo (or in Brazil as frei Jorge) and also, in the US, as cordia wood or jenny wood. Some species of *Cordia* live around the coast, and their corky seeds are dispersed by water.

Coffee, Quinine and Some Excellent Timber:
ORDER GENTIANALES

Within the Gentianales order are four families containing about 14,200 species, with many excellent and important trees. The very large family of the Rubiaceae contains around 9,000 species in 550 genera. Most live in the tropics and subtropics, and most of the tropical species are trees or shrubs. They include many fine ornamentals, notably the much-valued *Gardenia*, some of which are trees in the wild. Most important by far of the many Rubiaceae that are pharmacologically significant is the genus of coffee, *Coffea. C. arabica* produces the best quality coffee, and is traditionally grown in tropical America, notably Brazil; while *C. canophora* yields 'robusta' coffees which are generally less fine, but higher yielding, and have been largely grown in Africa and Asia. Coffee is grown in various ways but the quality is finest when the bushes are grown in shade, because then the fruits mature more slowly; the shade trees can be of many kinds but often leguminous types are grown which yield excellent fodder as a bonus. Everyone should do well out of coffee, but thanks to modern globalized trading laws all farmers in the tropics just about everywhere are now being exhorted to grow more and more of it – in places like East Timor and Vietnam as well as the traditional countries – so now there is world glut: prices fell by nearly 70 per cent in the five years up to 2004. Farmers borrow from banks or sometimes from money-lenders to start their coffee plantations, then find themselves selling for less than the cost of production, and go bust (I have seen the failing plantations in eastern Brazil). Myth has it that the consumer benefits from their misery but high street prices have not dropped noticeably and in truth the only beneficiaries are the traders. 'Twas ever thus, it seems.

Several species of *Cinchona* were also of great pharmacological significance as the source of quinine, for a long time the front-line drug against malaria, much favoured by the British colonialists. Whether the British obtained the rubber tree from Brazil by underhand means is somewhat arguable, but the removal of cinchona seeds from Bolivia in the nineteenth century was explicitly against Bolivian law. This seed founded the Javanese quinine industry, which dominated the world's trade until 1939.

The Rubiaceae family also provides some fine timbers. Abura (*Mitragyna ciliata*) of the west coastal swamps of tropical West Africa grows to 30 metres and provides a pale-yellow to pinkish-brown timber, excellent for mouldings and floors, and also used to encase accumulators (batteries) because it is resistant to acid. Even taller, at around 50 metres, is the opepe, *Nauclea diderrichii*. Its creamy-pink sapwood and golden-yellow heartwood are valued both for their looks – for furniture, floors, turnery, veneers – and for rough work outside, from jetties to railway sleepers. Degame, *Calycophyllum candidissi-mum*, from Cuba, Mexico, Venezuela and Colombia is smaller than its African cousins (up to 20 metres tall) but again excellent for sculpture, turnery, furniture and floors.

The Apocynaceae family is again big and highly various, with around 3,700 species of trees, shrubs, climbers and herbs in 335 genera, again mainly tropical and subtropical with just a few in temperate lands. Most of the genera contain poisonous species which, as is the nature of things, are often converted to valuable drugs – none more so than the Madagascar periwinkle (*Catharanthus*) source of some of the most effective drugs against leukaemia, and *Rauvolfia*, which provides a drug that's used to treat hypertension. But the Apocynaceae also include some pleasant and quaint ornamentals including periwinkle (*Vinca*), oleander (*Nerium*) and the carrion flower (*Stapelia*), a peculiar worm-like succulent whose evil smelling blooms are pollinated by flies and yet is beloved of collectors (among whose ranks I once was, and perhaps will be again). The trees of the Apocynaceae, all tropical, are often huge and impressive. The genus *Dyera* includes *D. costulata* of Malaysia and Indonesia, known as jelutong. It stands up to 60 metres tall and yields creamy timber that matures to the colour of straw, excellent for sculpting and carving, not least of wooden clogs. It also yields a latex that is extracted for chewing gum – and the process of extraction commonly leads to fungus infection that colours the timber which, as in beech, or indeed in blue cheeses, can increase its value. Species of *Aspidosperma* from south-eastern Brazil are marketed as rosa peroba. The trees stand around 15 metres tall; the sapwood is creamy yellow, the heartwood rosy red, hard, heavy, very dense, and used for everything from building ships to parquet floors, turnery and very beautiful veneers.

Best known of all the trees of the Apocynaceae, however, is the frangipani, *Plumeria rubra*, alias the temple tree or pagoda tree, sometimes growing as a small tree up to 6 metres and often as a large bush. The frangipani was imported to India from its native Jamaica, Mexico and Ecuador. Mrs Cowen, in *Flowering Trees and Shrubs of India*, writes that with its 'pale swollen limbs the tree is in itself no thing of abundant beauty'. But 'It claims affection for its sweet-scented flowers which, nearly throughout the year, open, bloom, and fall to lie immaculate on the earth beneath. To both Buddhists and Mohammedans the tree is an emblem of immortality because of its extraordinary power of producing leaves and flowers after it has been lifted from the soil. For this reason it is frequently planted near temples and in graveyards, where daily the fresh, creamy blooms fall upon the tombs.

Hindus make use of the flowers in worship and they are frequently given as votive offerings to the gods.' Frangipani finds many uses in medicine and its yellow-brown and reddish timber, though it comes only in smallish pieces, is easily worked.

Among the 970 or so species (in 84 genera) in the Gentianaceae are the gentians for which the whole family and order are named. These are not trees, but ornamental herbs. Some members of the family are extreme parasites with small leaves and no chlorophyll (showing once again that flowering plants have a considerable predilection for parasitism). Some are pleasant shrubs. Many provide medicines and flavourings. But there would be precious few significant trees among the Gentianaceae were it not that the Loganiaceae, which is the last family of the Gentianales, has recently been re-examined and reorganized. So the tropical Asian genus *Fragraea* has been transferred to the Gentianaceae from the Loganiaceae, where it traditionally resided: and *Fragraea* includes some significant timber trees, one of which (*F. fragrans*) has particularly huge flowers and is also planted as an ornamental. The plants that remain in Loganiaceae include nothing that need occupy a book on trees.

Olive, Ash, Jacaranda and Teak:
ORDER LAMIALES

Taxonomists have got seriously stuck into the Lamiales this past few years. The order has grown: it now embraces the traditional order Scrophulariales, which included the antirrhinums and foxglove; and subsumes the old Bignoniales, which contains some magnificent trees as will shortly be described. In addition, there has been much chopping and changing of genera between the different families that made up the original three orders that are now included within the Lamiales. I will not go into details because most of the changes do not involve trees – and I am mentioning all this only to provide continuity with traditional texts. (If you read Heywood, for example, as everyone should, you will find the order Scrophulariales, and might be wondering what happened to it if it wasn't mentioned here.)

The new-style, expanded Lamiales now includes 17,800 species in twenty-four families. Nineteen of those families contain mainly herbs, some of which are parasitic, like the serious tropical weed *Striga* and the temperate wayside broomrapes, *Orobanche*. There are many epiphytes (notably in the African violet family, Gesneriaceae) and also some vines and shrubs (the garden favourite *Buddleia* has been imported into Lamiales from its former position among the Gentianales), but at best these families include only a few small trees. Among five of the Lamiales families, though, there are some very significant trees indeed.

The Avicenniaceae family includes *Avicennia*, which is described in the discussion of mangrove swamps in Chapter 11. Until recently, *Avicennia* was sometimes placed within the Rhizophoraceae, alongside what is perhaps the best-known of the mangrove genera, *Rhizophora*. Now it is clear that its resemblance to *Rhizophora* is an example of convergent evolution – two plants with a similar way of life, adopting similar solutions to it.

The Myoporaceae family – about 150 species in four genera from Australia and the south Pacific, with a few in South Africa, south Asia and Mauritius – are mainly trees and shrubs. Best known are the highly drought-resistant genus of *Eremophila*, which includes the colourful emu bushes and some useful timber trees; and *Myoporum*, also with some pleasant, aromatic shrubs and some timber trees.

Ecologically and economically, though, the Avicenniaceae and Myoporaceae families are minor. Of supreme importance in all ways is the Oleaceae family. It includes 600 or so species in twenty-nine genera, grows in tropical and temperate climates, and includes climbers, shrubs and trees. The shrubs include privet, lilac, forsythia and jasmine. The trees include the sixty or so species of ash (*Fraxinus*); and the twenty or so species of olive (*Olea*).

The European ash, *Fraxinus excelsior*, is one of the great trees. If it weren't for human beings changing the landscape so radically, present-day, post-Ice Age Britain would be covered virtually from end to end with forests of Scots pine, oak, or mixtures of oak and ash. The timber is creamy white to light brown, sometimes with dark heartwood that is marketed as 'olive ash'. *F. americana* is the American or white ash. The timber of both species is wonderfully springy – excellent for oars, tool handles and baseball bats – and also lends itself to steaming and bending: and so it turns up too in bentwood furniture, umbrella handles, and so on. *F. ornus* is cultivated in Sicily for the sweet gum that it exudes, known as 'manna'.

The olive of the Mediterranean, *Olea europea*, is the species grown for its delectable, oily fruits – not just a treat in those parts but traditionally a staple. The trees are generally small and misshapen but other olive species from Kenya, Tanzania and Uganda such as *O. hochstetteri* and *O. welwitschii* grow to 25 metres and provide valuable 'olivewood': pale brown with attractive dark curly streaks, highly resistant to abrasion, and valued for floors, furniture, sculpture, turning and veneers.

The family Bignoniaceae includes some very important trees – although most of the family are climbers and many more are mere shrubs. The 800 species (in 113 genera) are mostly native to South America but others come from throughout the tropics and subtropics, and a few are temperate. The forty-odd species of *Jacaranda* from South American include *J. mimosaefolia*, a native of Brazil but grown throughout the tropics and subtropics for its beautiful blue flowers. Jacaranda has fine timber too: pleasantly scented, often patterned with dull purple streaks, and favoured for pianos in Egypt for some reason. *Catalpa* is one of the Bignoniaceae that grows in temperate climes and in Britain is sometimes called the 'Indian bean' because its fruits are long pods, although it is neither a bean nor (I believe) Indian. The extraordinary sausage tree, *Kigelia*, has fruits that are indeed like big fat salamis hanging from the rafters of some Italian kitchen, though biscuit-coloured. I have seen it in India and my wish list includes a possible meeting in its native Africa. The tulip tree, *Spathodea campanulata* (not to be confused with *Liriodendron* from the Magnoliaceae), comes from tropical Africa but was introduced to India in the late nineteenth century both for its shade and for its marvellous masses of scarlet flowers, which earned it the soubriquet of 'flame-of-the-forest'. Some Bignoniaceae are useful timber trees. *Paracetoma peroba* from the coastal forest of Brazil grows to 40 metres and its pale, golden-olive heartwood is known as white peroba and is highly favoured for furniture, boats, floors and veneers. Even catalpa provides fence posts.

Now comes scope for possible confusion, which I will do my best to ameliorate. In traditional texts you will find two families that have long been thought to be closely related: the Verbeneacae and the Lamiaceae. (Confusion is compounded because the Lamiaceae was formerly known as the Labiatae, and members of the Lamiaceae are still commonly called 'labiates', at least by people like me, in whom old habits die hard.) The Verbenaceae family was named after the perennial *Verbena* and also included *Lippia*, the lemon verbena, and *Lantana*, the South American ornamental that is now the mother of all pests, taking over tropical waysides and forests just about everywhere it has been introduced, and spurned by elephants, which gives it even more scope to grow. The traditional Verbenaceae was best known, however, for *Tectona grandis*, alias teak: which, when both quantity and quality are taken into account, is by far the most valuable of tropical hardwoods. The Lamiaceae (formerly Labiatae) included a few big woody plants including the fifty-seven known species of the genus *Gmelina* from India, Malaysia and Australasia, such as the all-purpose timber tree *Gmelina arborea*. But the Lamiaceae family was best known for its range of culinary herbs: mint, oregano, basil, rosemary, sage, lavender, thyme.

Of late, however, taxonomists have found that the old-style Verbenaceae was not a clade; more of a mixed bag. Most of its members were more closely related to other families, or needed to be placed in their own families. Much more to the point, about two-thirds of the traditional Verbenaceae now seem to belong in the Lamiaceae. So now Verbenaceae is a much less interesting family. It still includes lemon verbena, verbena and lantana. But the jewel of the family, *Tectona*, now finds itself grouped with mint, thyme, oregano and the rest in the Lamiaceae. Modern taxonomy makes some strange bedfellows. The Lamiaceae now emerge as a huge and cosmopolitan family with nearly 7,000 species in more than 250 genera.

Teak grows naturally in mixed tropical forests in a range of conditions, although it is generally supposed to prefer a climate with a dry season, as opposed to uninterrupted rain. It grows naturally alongside many other species in forests in India, Myanmar, Thailand and Laos – although it may not be native to India: Hindu monks may have brought it in from Indonesia in the fourteenth century. There are huge plantations too throughout the tropics, not least in Brazil.

Teak grows slowly – traditionally harvested in India at eighty-year intervals

Some Indian foresters feel that teak, for all its magnificence, is somewhat overplayed: at the start of 2004, to help restore the balance, the director of the Forestry Research Institute in Dehra Dun, Dr Padam Bhojvaid, organized a conference to re-focus attention on some of the other 400 or so Indian species (out of more than 4,000 Indian natives) that have been put to use this past few thousand years. But the attractions of teak are clear. Its timber is strong and durable, enduring the great outdoors without treatment, and weathering to a characteristic, sober grey. As a bonus, it is available in long lengths. The Mesopotamians recognized its worth in the third millennium BC. The Europeans came to it somewhat later, in the sixteenth century, and from then on used it to build much of their navies. It is now grown in plantations throughout the tropics, not least in Brazil, and is the subject of huge research endeavours in genetics, tissue culture and the control of pests and diseases. A particular target is a moth that regularly defoliates the trees – which is disfiguring and also reduces growth rate. By 1998 the total area of teak worldwide was estimated at 28 million hectares – although the lion's share of this was still in natural forest. *Tectona grandis* is the valuable teak. Four other known species (*T. australis*, *T. hamiltoniana*, *T. philippinensis*, *T. ternifolia*) can be ecologically but not economically important locally.

Holly: ORDER AQUIFOLIALES

The only family in the Aquifoliales order is the Aquifoliaceae. It contains about 400 species in three genera – and about 97 per cent of them are in the genus *Ilex*: the trees and shrubs known as hollies. Most of them live in tropical mountains but they are very widely spread and include only one of two (at most) species of evergreen broadleaved trees in Britain. Many are cultivated for their glossy and often prickly foliage and for their bright red or yellow fruits (distributed by birds), but they are rich in caffeine and some are cultivated for medicine. The leaves of *I. paraguariensis* are brewed to make maté, high in caffeine, while the native people of the south-eastern United States make 'black drink' from the leaves of *I. vomitoria*. Holly timber is hard, white, smooth and much prized.

Daisies and Just a Few Trees: ORDER ASTERALES

Twelve families and nearly 30,000 species make up the Asterales. The Asteraceae family – formerly known as the Compositae – contains most of them, and may be the biggest plant family of all (although some say the orchids have more species). Among the Asteraceae are many ornamentals such as daisies, chrysanthemums and marigolds, and edible and medicinal herbs including endive, artichoke, sunflower, Jerusalem artichoke, dandelions and lettuce. There are few convincing trees, although a few (including some from the Brazilian Cerrado)

Handsome in winter and with fine white timber: the holly

do provide timber that is locally useful. The muhuhu, *Brachylaena hutchinsii*, is one of the few bona fide trees, and an impressive one. It comes from the coastal belt of East Africa and the highlands of Tanzania and Kenya, grows to 25 metres, and provides short lengths of very heavy timber with grey-white sapwood and orange-brown heartwood, which looks good, wears well and is favoured for everything from floors to animal carvings. The muhuhu also smells pleasantly and its oil is distilled as a substitute for sandalwood – and it is even exported to India, as an aromatic fuel for cremations.

III

The Life of Trees

How Trees Live

Mangroves: how can trees grow with their feet in the sea?

A century or so before Aristotle the Presocratic philosophers of Greece proposed that all material things, including those that are alive, are composed of just four elements: earth, air, fire and water. This sounds quaint to modern ears, and is sometimes taken as evidence that the Greeks, for all their sophistication, were in truth rather primitive. Yet in this, as in so many things, they were spot on.

Nowadays, to be sure, the word 'element' does not mean what the old Greeks meant by it. Chemists now recognize about a hundred basic 'elements', of which all the tangible components of the universe are constructed. Thus water (H_2O) is compounded from two atoms of the element hydrogen and one of oxygen. Carbon dioxide (CO_2) is one carbon with two oxygen. Ammonia (NH_3) is one nitrogen with three hydrogen. And so on.

Early in the nineteenth century came the revelation that flesh, too, is chemistry: that it is not extra-special 'vital' stuff, but is made from the ordinary elements of the universe. Thus the science of biochemistry was born. Carbohydrates such as sugars, starch and cellulose, and fats including fish, vegetable oils and waxes, are all compounded exclusively from carbon, hydrogen and oxygen. Proteins are made from carbon, hydrogen, oxygen and nitrogen, with a touch of sulphur. The stuff of which genes are made, DNA, and its companion nucleic acid, RNA, is compounded of carbon, hydrogen, oxygen, nitrogen and phosphorus.

In practice life is not so simple, and virtually all organisms also need a fairly long catalogue of additional minerals to fill out the details, including metals such as calcium, sodium, potassium, magnesium, iron, zinc, and manganese; and non-metals or quasi-metals such as molybdenum, boron, chlorine and (in animals) iodine. But the bulk of all flesh is compounded from the big six – carbon, hydrogen, oxygen, nitrogen, phosphorus and sulphur. Carbon is the key player, forming the core structure of all life's most characteristic molecules. The grand word 'organic', at its most basic, simply means 'containing carbon'. I have known non-scientists to object to such discussion: to 'reduce' life to mere chemistry, they argue, is to demean it. But this is to misconstrue the nature of chemistry. That the simple elements, suitably arranged, can give rise to living things, shows how wonderful they really are. Life, in all its extraordinariness, is implicit in the fabric of the universe. We can only guess what else the universe might be capable of.

All this seems to make the Presocratic Greeks look a little silly. The tangible universe and all living things within it are constructed from a hundred elements, in a billion billion combinations, each interacting with all the rest. What sense can it make to suggest that everything is made from air, fire, earth and water?

All the sense in the world, is the answer – at least when we are talking of trees.

EARTH, WATER, AIR AND FIRE

Living tissue is complicated: it is built from many different components. More to the point, it is 'alive'. It is constantly replacing itself, even when it seems to stay the same. It is not a thing, but a performance. The physical components from which living tissue is built are acquired in the form of nutrients; and the incessant self-renewal – metabolism – requires a constant input of energy.

For animals, nutrients and energy seem to amount to the same thing. Both must be supplied in their food. Animals get most of their energy by breaking down carbohydrates and fat. They acquire this energy-rich provender by either eating other animals, or by eating plants – or both, as human beings do. Organisms like us, which need their food ready-made, are called 'heterotrophs'. But the buck has to stop somewhere – and in most earthly ecosystems, it stops with plants. Plants make their own carbohydrates, fats, proteins, and everything else they need from raw materials – simple chemical elements, and the simplest possible chemical compounds. They obtain the energy to do this from the sun. They are 'autotrophs': self-feeders.

The key to autotrophy is photosynthesis. Within their leaves plants harbour the wondrous green pigment known as chlorophyll. Chlorophyll traps units of energy – photons – from sunlight. Then, acting as a catalyst, it uses the photon energy to split molecules of water. Where there was H_2O, now there is H plus O. The O – oxygen – floats away into the atmosphere as oxygen gas. If it weren't for photosynthesis, there would be no oxygen gas at all in the atmosphere, and creatures like us could never have evolved at all.[1] The interesting bit in this context is the hydrogen, which is then combined, within the leaf, with carbon dioxide gas from the atmosphere. Thus simple organic acids are created, compounded from carbon, hydrogen and oxygen. These simple compounds, with a little more manoeuvring, are transformed into sugars (the simplest carbohydrates). When the sugars are modified a little more, they become fats. Add nitrogen, and they can be made into proteins. Incorporate a few other chemical elements, and all the components of living tissue can be made. Chlorophyll itself is basically a protein, with some magnesium at its centre.

Green plants are engines of photosynthesis. It is what they do, their *raison d'être*, and we should be properly grateful that it is, for without their ingenuity and labour, insouciant heterotrophs like us could not exist. Trees are the greatest of nature's engines of photosynthesis. Their need to photosynthesize explains the whole, vast, elaborate architecture of the tree. Leaves are the meeting place of carbon dioxide (wafting in from the air), water (drawn up from the ground) and sunlight. All are brought together in the presence of chlorophyll, which acts as host and mediator. Leaves archetypically are flat and thin, to expose the chlorophyll within them to as much sunlight as possible. The chlorophyll is held in loosely-bound cells in the middle layers of the leaf – a spongy arrangement, so air can circulate freely. The air enters through perforations underneath the leaf, known as 'stomata', which open and close according to conditions (generally closing when it is too dry, and the leaf is in danger of wilting, and also, typically, when it is dark). All green plants do all this – but trees, the greatest of plants, hold their leaves as high in the sky as possible, for maximum exposure to air and sun. The water (and minerals) come mainly from the ground – sometimes from deep below the ground – and so must be carried upwards through all the length of the roots and trunk and branches to the leaves aloft.

Yet the trunk of the tallest trees – redwoods and Douglas firs and some eucalypts – may be 100 metres tall. The roots may add a great deal more to their length: those of trees such as eucalypts that may live in semi-desert, and the native trees of Brazil's dry Cerrado, may reach down for tens of metres. The longest known roots of all belong to a South African fig: 120 metres. The whole vast and intricate structure is evolved to bring air and water together in the presence of sunlight; and the water and attendant minerals come mainly from the earth.

So the old Greeks were absolutely right. Trees at least are compounded from earth, water and air, and the sun that powers the whole enterprise is the greatest fire of all, at least in our corner of the universe. Other ancient myth-makers conceived of trees as the link than this. The tension within them is enormous: the threads are taut as piano wires. Yet, except under conditions of severest stress, they do not break. Water molecules cling tightly together. Their cohesive strength is prodigious. Were it not so, trees could not pull water from below, and could not grow so tall; but in practice the forces are such that a tree could grow to a height of three kilometres if the tensile strength of water was the only constraint on its growth. Even as things are, the threads of water may sometimes break – an accident known as 'cavitation' – leaving a space in the vessel that a plumber would call an airlock, and a surgeon would call an embolus. Given time and favourable conditions, plants can eventually fill this space again, and normal service is resumed. Otherwise, if cavitation is too great, the tissues that depend on the vessels may die. Parasites such as the mistletoes increase the tendency to cavitation because they take in the conducting vessels of their hosts and extract water from them by transpiring more quickly than the host, thus creating even greater osmotic tension than the host itself. Sometimes in these conditions the water supply gives out. Mistletoes are wonderful and have launched a thousand myths, but they may kill their hosts, not least by desiccation.

The final evaporation of water from the leaves, out through the stomata, should perhaps be seen as a side effect of the whole mechanism. The evaporation may bring benefit because it cools, as sweating does, and it's hot out in the sun. Some drought-tolerant plants practise a refined form of photosynthesis known as crassulacean (because it was first discovered in succulent plants of the genus *Crassula*, much beloved of gardeners) that is designed (or evolved) to reduce the loss of water. In such plants, the stomata open only at night. The carbon dioxide that is taken on board during the night is then put into temporary chemical store, and is released again next day when the sun comes out and photosynthesis can resume. No tree that I know practises crassulacean photosynthesis, so it is not directly relevant here – except to say that these plants at least demonstrate that it is possible to live out in the sun without overheating, even when water is not lost by evaporation. So it seems that most plants (including all trees) lose water through the stomata simply because this is very difficult to avoid; or at least, the loss is a price worth paying to maximize the efficiency of photosynthesis. The *point* of the plant's architecture – all those conducting vessels, all those perforated leaves – is to bring the Greek elements together: to present water to the sun, in the presence of air. But it is hard to bring them together without losing water, and sometimes losing more than the plant would like.

The overall effect is a flow of water from the roots, through the vessels to the leaves and out to the atmosphere: trees act like giant wicks. The final loss of water by evaporation is called 'transpiration'; and the total flow of water from soil to atmosphere is the 'transpiration stream'. The overall magnitude of this stream, especially when several trees are gathered together, can be prodigious; and its effect on soil and climate, and thus on surrounding vegetation and landscape, is critical to all life on earth, including ours. (I discuss this further in Chapter 14.)

So to the earth, the fourth of the four Greek elements. In this context, earth means soil.

THE SOIL

Air and water provide the carbon, oxygen and water that are the most basic materials of plants. The soil provides all the rest: nitrogen, phosphorus, sulphur – which (apart from carbon, hydrogen and oxygen) are the materials required in the greatest amounts – and a host of metals. All these extra elements from the soil are collectively known as 'minerals'. If any of these essential minerals is lacking (or indeed if carbon or water are lacking) then growth is restrained, if not imposs-ible: whichever ingredient it is that is deficient, and holding up the rest, is called the 'limiting factor'. In most soils, the most likely limiting factors are nitrogen, phosphorus and potassium: and these are the three standard components of the artificial fertilizers used in agricul-ture, which accordingly are packed in bags marked 'NPK' (K being the chemical symbol of potassium).

In truth, such fertilizers should probably contain sulphur as well: but the fields and forests at least of industrialized countries have been well if dubiously served this past 200 years by the sulphur-rich smoke from coal-burning factories, which has kept crops and trees alike well supplied. Now that factories are burning cleaner fuels, crops could well become short of sulphur and we are likely to see fertilizer bags marked 'NPKS'. Nitrogen, too, has rained on plants from on high, mainly in the form of the ammonia and nitrate from car exhausts; indeed there has been enough nitrogen from such pollution to keep the forests of Europe ticking over, even when the timber and fruits are harvested. It is indeed an ill wind that blows no good. On the other hand, sulphur and nitrogen in the form of sulphuric and nitric acids fall as acid rain, and then have often proved extremely harmful. To enrich the soil by polluting it is a precarious way to proceed.

Nitrogen is the mineral needed in greatest amounts. As a chemical element, it is tremendously common: it accounts for nearly 80 per cent of all the gas in the atmosphere. But in gaseous, elemental form, nitrogen is of no use to plant or beast. For plants to absorb it, it must first be converted into some soluble form – of which the commonest by far are nitrate and ammonia. Of course, organic gardeners contrive to supply their plants with nitrogen in organic form – which, broadly speaking, means nitrogen in the form of protein (or the broken-down products of protein) in manure and rotting vegetation, and so on. Fair enough. But the plants cannot absorb the nitrogen in the organic material, and hence make use of it, until it too is broken down (by soil bacteria) to ammonia or nitrate. Those two simple compounds are the ultimate currency of nitrogen.

In nature (unassisted by nitrogenous car exhausts), soluble nitrogen comes from four sources. Some may come from ground rock. Organic material – the rotting corpses of animals, plants, fungi and bacteria, and the faeces of animals – is also important. (But the dead leaves that form most of the leaf litter on the forest floor are typically low in nitrogen, for the trees withdraw the nitrogen from them before shedding them.) Then there is 'nitrogen fixation', in which nitrogen gas is chemically combined with hydrogen (derived from water) to form ammonia, which may then be oxidized in the soil to form nitrate. This happens in two ways. First, lightning fixes a surprising amount of nitrogen: the necessary chemistry is brought about by the electric flash, and the ammonia thus formed is carried into the soil by rain. But a wide range of bacteria can also pull off this trick, albeit with somewhat less drama.

These nitrogen-fixing bacteria live in a variety of ways. Many cyanobacteria are nitrogen fixers. You often see them on the boughs of trees as a dark bluish slime (hence the misleading soubriquet of 'blue-green algae'); but you won't see the ammonia (converted to nitrate) that they produce, which is carried down the trunk in solution when it rains, runs into the soil, and so nourishes the tree. Many nitrogen-fixing bacteria live free in the soil, and to a large extent (it seems) they are nourished by carbohydrates that the tree 'deliberately' exudes to keep them happy. This is symbiosis of the mutually beneficial kind, known as 'mutualism': the tree provides the bacteria with sugars, which they absorb like any other heterotroph; and the bacteria in turn provide soluble nitrogen, which the tree would otherwise lack.

But about 700 species of tree are known to form much more intimate, mutualistic relationships with nitrogen-fixing bacteria (and another 3,000 tree species are suspected of doing so). In these, the bacteria lodge in custom-built nodules on the roots.

Most of the plants that have such nodules on their roots are in the Fabaceae family, the 'legumes' – like acacia, mimosa, *Robinia* and the tropical American angelim. It comes as no surprise to any gardener that these leguminous trees are nitrogen-fixers – for so too are peas and beans, from the same plant family. In all of the legumes, the nitrogen-fixing bacteria in the roots are from the genus *Rhizobium* (though there are many different species of rhizobia). Most gardeners would be surprised to learn, however, that various species from ten other families of flowering plants are also known to fix nitrogen. Like the Fabaceae, all of those families come from rosid orders. Among the nitrogen-fixing, non-leguminous trees are the she-oak, *Casuarina*, and the alder, *Alnus*. In the non-legumes, the nitrogen-fixing bacteria are not rhizobium but from a quite different genus, *Frankia*.

Whatever the details, nitrogen-fixing trees in general can grow on particularly infertile soil, since they make their own fertilizer: and thus we find alders on dank and impoverished river banks. Nitrogen-fixing trees also tend to provide particularly nutritious leaves, for fodder. The leguminous trees especially are the arborescent equivalents of clover, alfalfa and vetch, which enrich the world's grasslands and are much favoured by livestock farmers. Since the nitrogen-fixing nodules are leaky, they release surplus nitrogen – and so they serve to enrich the soil at large. For all of these reasons, nitrogen-fixing trees are often of particular use to foresters – and especially to agroforesters, who seek to raise other crops, or livestock, among the trees. Thus acacias and *Robinia* are highly favoured the world over not simply on their own account but also to help all else that grows.

Clearly, close cooperation (via root nodules) between plants and nitrogen-fixing bacteria has evolved more than once: once in the Fabaceae with *Rhizobium*, and also in other rosid groups with *Frankia*. We have already seen many times how nature has constantly reinvented the same kinds of structure or modus operandi so this need not surprise us. Indeed, such associations seem such a good wheeze – the plant apparently gets free nitrogen fertilizer – we may wonder why all plants don't do it. But nothing is for nothing. The nitrogen-fixing bacteria are not altruists. They want something in return – that something being sugars. Hence legumes and alders and the rest must use some of the organic molecules that they create by photosynthesis to feed their nitrogen-fixing lodgers rather than directly for their own growth. Clearly, it's often worth it. Worldwide, the Fabaceae are a particularly successful family. Leguminous trees are a huge presence throughout the tropics, where soils are often low in nitrogen. There are many places, not least the cold, dank woods of Latvia, where alders flourish. She-oaks too find their special niches. Equally clearly, it is sometimes just as easy to do without bacterial residents, and get your fertility from elsewhere (not least from neighbouring legumes).

Far more common and widespread than such arrangements with nitrogen-fixing bacteria, are the associations between trees and fungi that invade their roots – not as parasites, but as useful and in some cases essential helpmates. These associations are called 'mycorrhiza', which means 'fungus-root'. Most forest trees and many other plants too make use of mycorrizae: some, like oaks and pines, seem particularly reliant on them.

Fungi in general consist of a great mass of threads (known as 'hyphae', which collectively form a 'mycelium') and a fruiting body that typically appears only transiently, and often manifests as a mushroom or toadstool. Many of the toadstools that are such a delight in autumn, and are avidly collected by gourmet peasants in France and Italy and elsewhere, are the fruiting bodies of fungi which, below ground, are locked into mycorrhizal associations with the roots of trees, and help them to grow. Thus the fungi are even more valuable than they seem. The wild mushrooms and toadstools are often only a tiny part of the whole fungus. The whole subterranean mycelium, including the mycorrhizae, sometimes covers many acres and weighs many tons. Forest fungi, mostly hidden from view, include some of the largest organisms on earth.

Mycorrhizae take various forms. Some simply seem to ensheath the fine roots of the trees. Sometimes the hyphae penetrate between the cells of the root, and often, in various structural arrangements, they invade the cells themselves. The relationship, in short, can be extremely intimate. Often, a tree will form mycorrhizal associations with more than one fungus at once, each with a different invasive strategy. Leguminous trees such as acacias, which harbour bacteria in root nodules, commonly have various mycorrhizae as well. Their roots are a veritable zoo.

The arrangement between trees and fungi, like those between trees and nitrogen-fixing bacteria, is extremely advantageous for both participants. The fungi gain because they take sugars from the tree, the products of photosynthesis. The fungal hyphae in turn are functional (and indeed anatomical) extensions of the roots, and hugely increase their efficiency. The hyphae commonly spread far beyond the normal limits of the roots, and vastly increase their effective absorptive area. They also function in the way that fungi always do – by producing enzymes that digest surrounding materials, and then absorbing them. Thus they make direct use of organic materials in the soil and may also break down phosphorus-containing rock – and lack of phosphorus (in the form of phosphate) is often a huge problem for growing plants. Then again, fungi are heterotrophs – they live by breaking down organic material; and so an oak or a pine or an acacia whose roots are fitted with mycorrhizae has the advantage both of autotrophy (through photosynthesis) and of heterotrophy (via its fungal helpmeets). Furthermore, a single fungal mycelium, sometimes covering several hectares, may interact with many different trees. Thus all the trees in a wood, even of different species, may be linked to others; and each may therefore share to some extent in the benison of all the others. Trees collaborate one with another in several ways, as we will explore in the next two chapters. Here is one: cooperative feeding.

Many trees including pines are as successful as they are largely because they have evolved particularly advantageous relationships with mycorrhizal fungi. Astute foresters commonly supply young trees with cultures of mycorrhizae to set them off. Many tropical trees prefer to grow as far away as possible from others of their own species (for reasons discussed in the next chapter) but young temperate oaks are said to grow best when close to others of their kind. Close together, they gain from each others' mycorrhizae.

Indeed, although we often think of fungi as pests of plants (and they often are: mildews, rusts, wood-rotters and the rest) they often emerge as key allies. Lichens are associations of fungi with algae: and lichens are found everywhere, on rocks and as epiphytes, in thousands of forms. Indeed, many botanists now feel that the association of plants with fungi is intrinsic to the success of both. Both groups evolved initially in water. It seems at least possible that neither could have invaded the land except by cooperating with the other. There is indeed some fossil evidence that the very first algae that came on land had fungal companions. Since then, fungi have evolved in all directions, not least to produce the magnificent creatures that we now know as toadstools; and plants have evolved this way and that, and now include the world's wonderful inventory of trees. But the old habits persist. Plants and fungi still stick together to their mutual advantage as, apparently, they always did.

All soils are different, but this, in broad-brush detail, is how all trees cope with all of them: the ground rules. Some soils, however, are more different than others. Some are positively weird. But there are trees to cope with some of the weirdest.

STRANGE SOILS: MANGROVES AND OUTRIGHT TOXICITY

Around the shallower shores of the tropics and subtropics, in 114 countries and territories, are the forests known as mangroves – a word that is also applied to the individual trees and shrubs that live within them. The mangrove forest is typically low, but some trees within it sometimes grow to a height of 50 or 60 metres. The mangroves generally extend only a few kilometres inland and cover only 181,000 square kilometres worldwide, and yet are supremely important. Like any forest, they are habitats for a huge array of land creatures – insects, spiders, frogs, snakes, birds, squirrels, monkeys – plus a host of epiphytes; and they provide local people with food, fuel, and timber for shelter. Like any forest, they lock up a significant amount of carbon and so help to protect the world's climate (of which more later). Unlike most forests, they lack an understorey of specialist shade-loving plants: at ground level, there are just roots, water and mud. Unlike other forests, too, they serve as the breeding ground for a long inventory of marine creatures, including fish, crustaceans and molluscs, and around their roots are trails of marine algae. Thus the mangroves link the food webs of land and sea. For good measure (in a natural state) they filter the silt that may flow from the land and so they protect the beds of seagrass that generally lie further out, permanently submerged, and the coral reefs that often lie beyond the seagrass. They also protect the land from excessive seas – the tsunami that struck so devastatingly at the end of 2004 might well have been less devastating if some of the shores had retained more of their mangroves. If we take away the mangroves, then all the creatures that they are home to, and all the seagrass beds and reefs and coastal lands that they protect, are liable to disappear as well.

Most plants, like most creatures of any kind, are killed by too much salt: but mangroves spend much of their time with their roots in pure seawater. This is sometimes diluted by rain, but at other times it evaporates to become as saturated with salt as water can be, until the salt begins to crystallize out; and for good measure, much of the tree roots are intermittently exposed as the tides rise and fall. In addition, the mud in which the trees are rooted is often shallow and is invariably starved of oxygen except for the top few millimetres – yet roots need oxygen. In temperate latitudes, willows, poplars and alders are among the trees that cope with waterlogged soils that may similarly be deprived of oxygen – but at least, in their cases, the water is fresh. Salt water raises a whole new raft of problems.

To be sure, tropical seashores are in many ways favourable for growth (nutrients, water, sunshine), but conditions overall are as tricky in their way as in any desert, or on any frozen tundra. So you might expect that only a very few, extremely specialist plants could live there. Yet mangrove forests worldwide contain many species of trees and shrubs and there is a core group of thirty or forty that occur in most of them, and these come from several plant families. The core group includes various 'white mangroves' from the Combretaceae family; the red mangrove, *Rhizophora*, and others from the Rhizophoraceae family; *Xylocarpus* from the mahogany family, the Meliaceae; *Avicennia* from the somewhat recondite Avicenniaceae family in the order Lamiales (the order of teak and mint); and palms of the genus *Nypa*.

Independently of each other, the different tree families of the mangroves have evolved a series of physiological tricks to cope with the otherwise disastrous conditions in which they find themselves. Thus, the tissues that form a sheath around the xylem of the roots provide 'ultrafiltration', preventing the salt from entering the conducting vessels and thus polluting the rest of the plant. All mangrove species can do this but some are particularly good at it, including the red mangrove. Some that are less efficient at filtering out the salt, like *Avicennia*, do absorb at least some salt: but then they excrete it actively (by processes that use energy) from special glands in their leaves, where it may dry in the sun to form palpable crystals. Many seabirds do the same thing through their beaks.

Mangrove trees in general combat the airlessness of the soil by ingenious anatomy. Most have at least some aerial roots, directly exposed to the air. Their surface is perforated with 'lenticels', apertures that enable air to enter, and inside the root the tissue is spongy, with huge air-spaces between the cells that may account for 40 per cent of the total volume. Many have stilt roots, mostly out of the water. In *Rhizophora* these stilt roots arch away from the trunk, enter the mud, but then may re-emerge and form another loop, snaking along half in, half out. In some species, including *Avicennia*, the stilt roots thicken to form buttresses. Many species from unrelated groups have independently evolved 'pneumatophores', which grow vertically into the air to act as snorkels. In *Avicennia* the pneumatophores are thin like pencils. In other species they may be secondarily thickened, and develop into tall, substantial cones. As a final refinement, it seems that the air that does get into the roots is not left simply to find its own way around. The rising tide pushes the old air out; and when it recedes, fresh air flows in again through the lenticels and pneumatophores. Thus the roots of the mangrove trees effectively *breathe*. They use no muscle power to do this, as an animal must. The sea is their diaphragm. The tide to aerate their roots; wind and fleets of obliging animals to spread their pollen and seeds. Trees just don't need the elaborations of muscle and blood and nerves on which animals expend so much.

All in all, the main problems for the trees of the mangroves are chemical – all that fierce salt; too little oxygen. Yet, chemically speaking, the seashore is by no means the most hostile environment that the earth provides.

In particular, land that has been polluted by volcanoes – naturally polluted, that is – may contain an array of metals in concentrations that would be lethal to most plants, and indeed to most life. Some of them are simply innately toxic. Others are present in most soils and may be essential in very small amounts but are lethal in high concentration. Yet again, many plants, from many different families, have evolved tolerance. So it is that an array of plants from various families has been definitely shown to be highly tolerant to nickel, zinc, cadmium or arsenic. Others have been reported (though not confirmed) to withstand high doses of cobalt, copper, lead or manganese.

Most outstanding is a tree from New Caledonia, the island that is so extraordinary in so many ways. *Sebertia acuminata*, of the Sapotaceae family, grows in soils rich in nickel. It does not exclude the metal from its conducting vessels, as mangrove trees generally contrive to exclude salt. Instead *Sebertia* accumulates the metal. Indeed it accumulates so much that if the trunk or branches are damaged then the rubbery sap within (the latex) runs bright blue. Analysis reveals that the latex contains 11 per cent of nickel by weight – and it accounts for an extraordinary 25 per cent of the dry weight.

Exactly why some trees accumulate such metals (although few match *Sebertia*) is unknown. Clearly it's a way of coping with metal-rich soils. But some other plants may grow in the same soil without accumulating the metals – simply excluding them, as mangroves exclude salt. Perhaps the metal- or arsenic-rich stems and leaves are natural pesticides. Some experiments suggest that this is so; others give less clear results. In any case, such accumulators seem to offer a means of replanting and indeed reforesting land that has been polluted unnaturally, for example by mining. There are many examples world-wide of such reclamation. It has even been mooted that soils might be freed of toxic metals by growing hyper-accumulators and then harvesting and removing them. But this could be too slow to be worthwhile and also raises the problem of what to do with the har-vested, metal-rich plants. Perhaps the metals might be recovered from them, and pressed back into industrial service. But calculations suggest that this would rarely be worthwhile economically: it might be for nickel and cadmium, but not for zinc. Meantime, the hyper-accumulation of metals remains a botanical oddity.

So trees make use of what the soil and the atmosphere has to offer them; and so too they have evolved to endure the extremes of both. But they do not merely endure. Trees are not passive players. They are much more subtle than that.

HOW TREES KNOW WHAT TO DO (AND WHAT TO DO NEXT)

Trees live simple lives – or so it may seem to us: nothing to do all day but stand in the sun with their feet in damp and nutritious earth. But there's a lot more to their lives than meets the eye. Trees, like all of us, have to do many different things, and they have to do the right things at the right times. Taken in the round, their lives are as intricate as those of Hamlet or of Cleopatra, albeit without such conspicuous drama. Living is innately complicated.

All living things must respond to their surroundings, and trees respond in many ways. Many trees, like all plants, can move bits of themselves as the world changes around them, opening and closing their flowers or the stomata of their leaves, and these movements are known as 'nastic'. More broadly, all plants – including all trees – shape themselves according to circumstance, their stems growing away from gravity and towards the light, their roots generally doing the same in reverse. Trees do not simply grow: they grow *directionally*, most economically to fill the space available, and this directed growth is called 'tropism'. Growth towards the light (as in stems) is called 'positive phototropism'; and growth away from the light (as in roots) is 'negative phototropism'.

More cleverly yet, trees do not respond simply to the here and now. They anticipate what is to happen next. The deciduous types of the north, like oaks and birches and limes, have to shed their leaves in autumn and send out their hopeful flowers in the spring – and both of these procedures, the shedding and the blossoming, must be prepared for weeks in advance: the shedding in the height of summer, the flowering at a time when all thought of tender growth seems ludicrous.

Finally, trees like all of us, have to find mates, and interact with them; and all trees of the same type must be sexually active at the same time, so each must know what the others are up to – or at least, each must respond to the same clues of climate, or length of day, or whatever, so that all are coordinated. Many – particularly though not exclusively in the tropics – rely on insects, too, or birds or bats, to spread their pollen, and on yet more animals to disperse their seeds; and so they must attract their collaborators – and again, must make sure that they do all that is necessary in due season. But trees too, like all of us, are besieged from conception to the grave by potential parasites and must find ways of warding them off.

Trees have no brains or nerves and instead run their entire lives with the aid of a remarkably short shortlist of chemical agents: just five basic hormones, plus a handful of pigments, and a miscellany of other materials through which they convey information to others of their own species, or to other organisms including those that would attack them. The hormones control their growth and hence their overall body form, the emergence of buds and the shedding of leaves. Of the essential pigments, the interplay of just two in particular enables them to keep track of the seasons, and to anticipate what is to come. The various agents by which they communicate with other trees and with animals seem diverse but even so belong to only three classes of chemical compounds. These are of the kind known as 'secondary metabolites' – virtual by-products of metabolism. All green plants produce the basic five hormones and the principal pigments, but only some plants produce just some of the secondary metabolites that help plants to communicate: they are a moveable feast. The chemistry of animals, by which they coordinate their lives and communicate with others, is at least as complicated – yet they have nerves and brains as well. But then, a tree might ask, why bother with brains and all the expense and angst that goes with them, when you can run your life just as well without?

HOW TREES SHAPE THEMSELVES

The five hormones by which plants run most of their lives are auxin, the gibberellins, abscicic acid (ABA), the cytokinins and ethylene – a gas. Hormones in general (whether in plants or animals) affect body cells by interacting with receptors on the surface of the cell. The receptors in turn link up to secondary messengers within the cell, which transmit the information of the hormone to the parts of the cell that are supposed to respond. Immediately there is scope for further subtlety, because different cells have different receptors, linked to different secondary messengers. On cell A, a hormone may make contact with receptor X, and have one effect; and on cell B, the same hormone may be picked up by receptor Y, and have a different effect. In short, each cell extracts from each particular hormone the information it wants to extract – just as any of us, reading a text, focuses on certain aspects of it. The message is partly in the words, and partly in the particular interest of the reader.

In addition, the different hormones work together in pairs, or groups. Sometimes two acting together will enhance each other. Sometimes a particular cell will not respond to a particular hormone unless some other hormone has first primed it to do so. Sometimes two hormones oppose each other's action. And so on. Five hormones many indeed seem ridiculously few. But when each can have different effects (depending on the receptors), and when they act in permutations, then the total amount of information that the simple few might convey becomes effectively infinite. Even so, the underlying simplicity of the system is wondrously elegant.

Among the first to study plant responses seriously were Charles Darwin and his son Francis. In particular, as they described in 1881 in *The Power of Movement in Plants*, they studied the way plants modify their growth according to the light – phototropism. The Darwins studied not trees in this context but oats, which are easier to work with – but what works for oats in the first few days of their life also works for oaks and redwoods and all the other forest giants through century after century.

Oats (like oaks and redwoods) grow up towards the light; and when the light shines from the side, they bend towards it. The Darwins showed that it was the region just below the growing tip that changes direction: bending occurs because, when the light shines from the side, the tissue on the shady side grows faster than the tissue on the side that's lit. They found, too, that if they put little opaque caps over the growing tips, the bending stopped.

About forty years later (in the 1920s), other biologists revealed the mechanism. The growing tip of the oat (or the leading apical bud in a twig) produces a chemical that flows down to the tissue below, and prompts it to grow. But this chemical, it turns out, migrates *away* from light. So if the light shines from the side, it finishes up on the shady side of the stem – and so stimulates the growth on the shady side, and not on the illuminated side. Nothing could be simpler. An engineer who came up with such a scheme would warrant a Nobel Prize. The chemical involved is the first of the five major plant hormones, and is known as auxin.

But auxin does not always act as a growth promoter. Auxin flowing down from the apical bud of a twig or from the lead shoot of a tree *suppresses* the development of lateral buds lower down. If the apical bud is damaged, the flow of auxin ceases, and then the subsidiary buds burst into life. So in northern woods we may often see a conifer with a kink in the trunk: some time in the past the growing tip must have been damaged and the topmost lateral bud has taken over the job as lead shoot.

Often, too, the trunk of a tree seems simply to stop, and out of the top grows a mass of branches like a bush. Again, the terminal bud at some time in the past has been removed, and not one, but in some cases dozens, of lateral buds beneath have been liberated as the flow of suppressive auxin stops. Sometimes this happens in nature – I remember a huge *Terminalia* tree at the Indian Forestry Research

Institute that took this form; in the 1940s, so the FRI's Dr Sas Biswas told me, the young tree had been cut off by lightning. Disease, too (like the shoot-boring caterpillars that infest mahoganies), may produce this effect. But in addition, foresters and horticulturalists may cut the tops of trees deliberately to provide an instant source of sticks and staves from the lateral buds below. This is called 'pollarding'. Pollarded willows, hornbeams, oaks, hazels and chestnuts have for many centuries been major industries in Europe. Some of England's best-loved woods are former pollard plantations. Topiary too makes use of this effect. Yew, privet, beech and box all make respectable trees if left alone – but if they are clipped the surface is crammed with subsidiary twigs that otherwise would remain repressed. City trees, such as planes or limes, are often lopped to produce neat tops like mops or lollipops: a straight stem, and then a crown of more or less equal branches.

Auxin also prompts cut stems to produce roots of the kind known as 'adventitious', which are those that grow directly from stems. Many trees produce adventitious roots in various circumstances – indeed as we saw in Chapter 7, *all* the roots of a palm are adventitious. Growers make use of this propensity. They dip cuttings in auxin to help things along. In similar vein, a steadily growing catalogue of valued trees – including coconuts and teak – are now raised by 'micro-propagation': the production of whole plants from cells grown in culture. Auxin is essential in this (though it is not the only hormone involved).

Fruits won't normally develop unless the flowers are first fertilized, and so produce seeds – and it's auxin produced by the seeds that makes the fruit grow fleshy. Pick off the seeds from a strawberry, and the succulence stops. But some plants will produce seedless fruits if treated artificially with auxin, and so we are now regaled with seedless tomatoes, cucumbers, aubergines and grapes. Cultivated bananas produce seedless fruits as a matter of course. Presumably they contrive to produce auxin even in the absence of seeds.

The role of auxin in prompting trees to drop their fruits and shed their leaves in autumn ('abscission') is still uncertain. Auxin levels drop as the leaves fall, but correlation is not cause. Yet the addition of auxin can prevent leaf fall; and so it is applied commercially to stop the leaves and berries falling from decorative holly, and to keep oranges firmly on their branches until the pickers are ready. On the other hand, large amounts of auxin can *promote* fruit drop – and so it is sometimes deployed to thin crops of apples and olives, enabling the remaining fruit to grow bigger.

Finally, auxin has been modified to make weedkillers, basically by promoting over-rapid growth. This has great commercial value in agriculture and is not all bad: chemical control can be relatively benign for wildlife, if used selectively and decorously. But auxin has also been modified to make the infamous Agent Orange, which was used to defoliate trees in Vietnam and thus (or so it was intended) to reveal the Vietcong. The policy was horribly destructive of wildlife as well as of people and was of very limited military use, not least because the Vietnamese army dug themselves in, and lived largely under-ground. Agent Orange also contained dioxin as a contaminant, which causes horrible blistering of the head and body, and is probably carcinogenic. Thus the findings of science may be corrupted. Manufacture of Agent Orange is now banned in the US.

The gibberellins, also discovered in the 1920s, also promote growth as auxin does – but again, they do many other things as well. In particular, they are highly concentrated in immature seeds and help to break their dormancy. Gibberellins too, like auxins, can cause fruits to flesh out even in the absence of seeds. So they are used to produce seedless apples, mandarins, almonds and peaches, as well as currants, cucumbers, aubergines and grapes.

Soon after the gibberellins were discovered, abscisic acid or ABA came to light. In contrast to auxin and the gibberellins, its prime role is to suppress cell division and expansion – in other words, to suppress tissue growth. It also serves in seeds to suppress premature germination – germination before conditions are favourable. But despite its name, abscisic acid seems to pay very little part either in the shedding (abscission) of leaves or fruit.

The fourth of the major plant hormones, the cytokinins, have the opposite effect to ABA: they prompt cells to divide. The cyto-kinins were first discovered in 1941. They turned up first in coco-nut milk, but they are now known to occur in all plants. In some contexts cytokinins act in opposition to auxins, and override them:

so horticulturalists may use cytokinins to make side-buds sprout, even when the apical bud is still in place.

The fifth of the basic hormones is ethylene, which affects many different aspects of a plant's life, from the ripening of fruit to the falling of leaves and a great deal more besides. Ethylene is a chemically simple gas, and it may seem an odd choice for a hormone, for gases are wayward clouds of molecules that seem too unruly for precision work. But then, one of the surprises of recent years has been that the physiology of animals – not least of human beings – is profoundly influenced by nitric oxide, which is even simpler than ethylene. Like ethylene, nitric oxide seems to be involved in just about every system that has been looked at. It is the key even to Viagra – which operates by prompting release of nitric oxide, which in turn, in this context, acts as a muscle relaxant, and so allows engorgement. Ethylene does not act simply as a hormone within any one plant. It may also travel from plant to plant and so acts as a pheromone: an airborne hormone that acts on creatures other than the one that produces it. Whether nitric oxide also behaves as a pheromone, as ethylene does, is a most intriguing question.

Ethylene's role as a hormone was first revealed in the 1880s, when the trees that lined the streets in many a city began to lose their leaves. German scientists were the first to find the reason: the street lights were run on gas, and some of it escaped unburned. In 1901 one D. Neljubov showed that none of the several components of the gas had any defoliating effect, except ethylene. Ethylene was active at astonishingly low concentrations: 0.06 parts per million (or six parts per 100 million). On the other hand, the light from street lights can prompt city trees to retain their leaves, so that the same tree may lose its leaves on the shaded side, but keep them far longer on the side nearer the lamp. City trees in the days of gaslights must have been horribly confused.

Ethylene soon proved to have other effects too. It causes fruit to drop, as well as leaves: and, like auxin, ethylene is used to thin commercial crops of plums and peaches, and to loosen cherries, black-berries, grapes and blueberries in preparation for mechanical harvest-ing. Ethylene also promotes ripening, and herein lies another bizarre tale. In the early 1900s growers ripened fruit and walnuts in sheds which they warmed with kerosene stoves. But the more go-ahead growers switched to electricity. This was altogether more satisfactory: cleaner, more reliable, more *modern*. The only trouble was that the fruit no longer ripened. It wasn't the warmth that did the ripening. It was the ethylene, leaking from the smelly and despised old heaters.

So it is that plants control their form. Darwin wrote of the English wayside as a 'tangled bank', and in the jungle the tangle can be beyond unravelling. Yet each plant in the mêlée knows what it's doing. Each contrives to position its leaves in the light, and send its roots to the ground (except for a few specialist epiphyte orchids, which send some of their roots upwards), while the climbers wrap around others for support. In short, all in the apparent havoc know exactly what they are doing. Each adjusts its shape to the conditions, and to the presence of the others; and all this achieved, it seems, through astonishingly simple chemistry. The workings of trees, like plants in general, are indeed wonderfully elegant.

But trees do not dwell only in the present. They remember the past, and they anticipate the future.

THE PAST AND THE FUTURE: MEMORY AND ANTICIPATION

How trees remember, I do not know: I have not been able to find out. But they do. At least, what they do now may depend very much on what happened to them in the past. Thus if you shake a tree, it will subsequently grow thicker and sturdier. They remember that they were shaken in the past. Wind is the natural shaker, and plants grown outdoors grow thicker than those in greenhouses, even in the same amount of light; and so it is too that parkland oaks grown in splendid but breezy isolation are much more sturdy than those of the forest. Similarly, a larch tree remembers attacks by caterpillars. In the year after an assault it produces leaves that are shorter and stouter than usual. (Larches are among the few deciduous conifers.) Short, stout leaves do not photosynthesize as efficiently as the thinner, longer kind, but they are better at fending off pests. In subsequent years, however, the once-infested larch can and does revert to normal foliage. By then, the population of predatory moths would have plummeted, since the offspring of the original plague of caterpillars (which became moths) would have found nowhere to feed.

Most trees, like most plants of all kinds, are also aware of the passing seasons: what time of year it is; and – crucially – what is soon to follow. Deciduous trees lose their leaves as winter approaches (or, in the seasonal tropics, as the dry season approaches) and enter a state of dormancy. This is not a simple shutting down. Dormancy takes weeks of preparation. Before trees shed their leaves they withdraw much of the nutrient that's within them, including the protein of the chlorophyll, leaving some of the other pigments behind to provide at least some of the glorious autumn colours; and they stop up the vessel ends that service the leaves with cork, to conserve water.

How do the temperate trees of the north know that winter is approaching? How can they tell, when it is still high summer? There are many clues to season, including temperature and rainfall. But shifts in temperature and rain are capricious; they are not the kind of reliable signal to run your life by. Sometimes a winter may be warm – but frost is never far away. Some autumns and springs are freezing, some balmy. The one invariable, at any particular latitude on any particular date, is the length of the day. So at least in high latitudes, where daylength varies enormously from season to season, plants in general take this as their principal guide to action – while allowing themselves to be fine-tuned by other cues, including temperature. So temperate trees invariably produce their leaves and/or flowers in the spring, marching to the rigid drum of solar astronomy; but they adjust their exact date of blossoming to the local weather. This phenomenon – judging time of year by length of day – is called 'photoperiodism'. Most of the basic research on photoperiodism has been done on crop plants, which for the most part are herbs. But trees and herbs work in the same way. What applies to spinach and tobacco applies to trees too.

Knowledge of photoperiodism again dates from the 1920s, when agricultural scientists in America found that plants like tobacco, soya, spinach, and some wheat and potatoes would not flower if the days were shorter than a certain critical number of hours (often around twelve). But other plants would not flower if the days were too long:

strawberries and chrysanthemums were among those that remained resolutely sterile if the days were longer than sixteen hours. There were some, though, that didn't seem to mind the length of day. The three groups became known as 'long-day', 'short-day', and 'day-neutral'. Long-day plants generally flower in high summer, and short-day plants in spring or autumn. As a further refinement, plants also seem 'aware' that absolute day-length has different significance at different latitudes. At very high latitudes, the longest days are twenty-four hours – the sun never sets – and a fourteen-hour day is of modest duration. But in the subtropics, fourteen hours is a long day – as long as any day gets. Sometimes the same species may grow both at high and at low latitudes, including for example the aspens of North America. Then the northern ones will treat a fourteen-hour day as short, and the more equatorial ones will treat a fourteen-hour day as long. Adaptation is all.

In the late 1930s it became clear that plants do not measure the length of the day, but of the night. If the light is turned on even briefly during the night – a minute from a 25-watt bulb would do – then short-day plants such as strawberry will not flower. Contrariwise, a long-day plant that flowers in sixteen hours of light and eight hours of dark will also flower with eight hours of light and sixteen hours of dark – if the darkness is interrupted by a brief light. In truth, long-day plants should be called short-night plants; and short-day plants are really long-night plants.

In the next few years the underlying mechanism became clear – and again it is remarkably simple. Inevitably it depends on a pigment – for pigments by definition are chemical agents that absorb and reflect light, and so mediate a plant's (or an animal's) responses to it. In this case the pigment is phytochrome. Phytochrome exists in two forms which either suppress or promote flowering; and light flips them from one form to the other.[2] Again, these insights have been put to use. Growers of chrysanthemums used to keep the lights in the greenhouse on at night to delay flowering until Christmas – until, in the 1930s, they saw that a brief burst of light at night would produce the same effect, and much more cheaply. Contrariwise, appropriate flashes will bring long-day plants rapidly into bloom, by artificially shortening the nights.

All these mechanisms are evolved – they have been shaped by the experiences of past generations. They can succeed, and serve the plant well, only if present and future conditions are like those of the past. If conditions change slowly over time, then any lineage of creatures, animals or plants, can adapt to the change. But if conditions change rapidly, then creatures that have evolved their survival strategies in earlier and different times find themselves caught out.

Human beings are changing the world profoundly and – by biological standards – with extreme rapidity. In particular, we are altering the climate. Present-day pines and oaks and birches in northern latitudes are adapted to the idea that long days are warm and short days are cold. Everything they do – germination, dormancy, the shedding of leaves (in the deciduous types), the production of flowers and cones – is geared to this assumption. If long days turn out to be cooler than expected, or significantly hotter, drier or wetter, and if the cold days are not particularly cold, then the whole life cycle can be thrown out of kilter. The confusions of urban trees, when light and temperature are out of synch, are just a warning of what may happen to all the world's forests when the interplay of light, warmth and moisture is altered on the global scale. If plants are seriously incommoded – whether wild trees or domestic crops – then everything else must suffer too. Of all the threats to the present world, this is one that matters most. Yet, as discussed further in Chapter 14, the effects of climate change on plants is extraordinarily difficult to predict. The insights of modern science are wonderful, but absolute knowledge is a logical impossibility. In the end, we are just going to have to wait and see.

This, then, in broad-brush terms, is how plants keep themselves alive. But as living creatures they need to carry out two more tasks. They need to reproduce; and they need to get along with their fellow creatures, of their own and other species. How they do this is discussed in the next chapter.

12

Which Trees Live Where, and Why

California's coastal redwoods get much of their water from mist

Similar places all the world over pose similar kinds of problems – of light, dark, heat, cold, flood, drought, altitude, toxicity – and all the many varied trees that live and evolve in any one place tend to come up with the same kinds of solutions. Thus the Douglas firs, pines and spruces of the extreme north, and the rimus and kahikateas of New Zealand's south, are all tall and steeple-like, to catch the light that comes at them from the side; while the cedars and umbrella pines of the Middle East and Mediterranean have flat tops, aimed at the sunlight beamed from overhead. The trees of tropical rainforest grow straight up through the crowd while those of the Brazilian Cerrado, the African savannah, or the Australian bush, spread themselves like cats. So it is that all the world's forests conform to a score or so of different ecotypes – variations on a theme of boreal, temperate or tropical; wet forest (rainforest) or distinctly dry; seasonal or aseasonal – where seasonal means winter/summer, or wet season/dry season. Within this general framework are a series of specialisms. There are forests that follow rivers ('riverine', sometimes known as 'gallery' forests). Those in mountains are called 'montane'. At moderate heights they are 'alpine'; but in some wet warm places, as in much of South-East Asia, the trees become lost in mist towards the tops and so become 'cloud forest'. Some forests have their feet in water: swamp forests, with willows, alders, swamp cypresses, and the rest; and mangroves, at the edge of tropical, shallow seas.

Yet no two forests are alike. They are like art galleries: they all have pictures, but they don't have the same pictures. The forests of South-East Asia are rich in dipterocarps. Eucalypts are virtually confined to Australia – or would be, were it not for human beings, who have planted them virtually everywhere. Africa and Australia both have acacias in their wide open spaces – but they are different acacias. America, China and Europe all have plenty of oaks – but each has its own selection. Oaks and willows in general (with very few exceptions) are confined to the northern hemisphere. Southern beeches (*Nothofagus*) are indeed inveterately southern. Araucarias too, at least in these modern times, belong exclusively to the south. Some species – and indeed some genera or even families – grow only in particular islands, to which they are then said to be 'endemic'. New Caledonia has thirteen endemic species of *Araucaria* out of a world total of nineteen. Madagascar has six of the world's eight species of baobab, and is the only place with the extraordinary trees of the Didiereaceae. Britain, on the other hand, has a miserable native list of only thirty-nine species, *none* of which are endemic. All our natives occur elsewhere as well, mostly in much larger numbers than in Britain. Of course, lists of 'British' trees may contain hundreds of species, many growing wild; but the vast majority are imported. The British are supremely acquisitive.

So the first question is 'Why?' We would expect each region to contain plants that are adapted to it – for if they were not, then they would soon be ousted by those that are. But why does each region have its own characteristic suite of native species? Why are some species (or genera, or families) very widespread, while others are confined to single islands? Why are some islands rich in endemics (New Caledonia, Madagascar, Hawaii, the Canaries) while others (like Britain) have none?

There's another kind of puzzle, too. Whatever group you look at – birds, butterflies, fish – you find there are many more species in the tropics than in the north or south; indeed the further you travel from the equator, the more the variety falls off. With trees the falling off is striking. The apparently endless boreal forest of Canada is dominated by only nine native species: a few conifers and the quaking aspen. The US as a whole has around 620 native trees. India (much smaller than the US) has around 4,500. In the Manu National Park of Peru, almost on the equator, twenty-one study plots with a total area of 15 hectares have yielded no fewer than 825 species of tree – about one-fifth the total inventory of all India, and considerably more than the US and Canada combined. As we saw in Chapter 1, the Ducke Reserve of Amazonia has more than 1,000 different trees. Tropical America as a whole, from Brazil, Peru and Equador up to Mexico, has tens of thousands of species. The true number can only be guessed. Why so many?

Both kinds of questions have been exercising biologists for several centuries (at least) and are still a hot topic: I attended the latest international conference on these matters at the Royal Society in London in March 2004. Hundreds of putative explanations are out there which between them encompass every aspect of the life sciences – and of the earth sciences, too. Some have to do with plant physiology, some with genetics, some with history, some with evolutionary theory. All are pertinent; all, indeed, are interwoven. The following is a rough guide to the main threads.

WHY TREES LIVE WHERE THEY DO

Each lineage of trees began with a single tree: the first-ever oak, the first-ever kauri, and so on. So – to begin at the beginning – where did those 'founders' arise? What is the 'centre of origin' of each species (or genus or family or order)?

It's at least commonsensical (and we have to start somewhere) to guess that the founders arose in the places where their descendants now live in the greatest variety. Eucalypts are extraordinarily various in and almost exclusive to Australia and there, surely, is their most likely origin. But of course life is not so simple. Oaks, for instance, span the northern hemisphere, and are most various both in North America and China – which are divided by the Pacific if you go round one way, and by the rest of Eurasia and the Atlantic if you go round the other. Even if we assume that oaks arose in either North America or China, they must at some point have travelled to the other far-distant continent. But if they can make such a journey as that, might they not have begun in the middle, in Europe? Or could they have begun in some completely different place, where they no longer exist, such as Africa? Either way it's clear that the centre of origin, even if we can work out where that was, does not by itself explain the present distribution. Clearly some trees in the past – perhaps most of them – began in one place and then dispersed to others. If they found their new locations congenial, they could then have formed entire new suites of species – so that these outposts then become secondary centres of diversification. Sometimes, too, the secondary outpost might be the kind of place that *encourages* the formation of new species. Thus there are many different pines in Mexico, but we need not assume that this is where they first arose. They are diverse because the first to arrive there found it congenial and the mountains provide many different niches where semi-isolated populations can each evolve along their own lines. Just to confuse the picture a little more, any particular lineage of trees might well be extinct in the place where it first arose. The places where particular trees now flourish may well be secondary outposts – or indeed outposts of outposts, or outposts of outposts of outposts.

'Diversity', though (like all terms in biology), has various connotations. In this kind of context, it should not be measured purely in terms of number of species. We need to see how different the various species are, one from another – which is where molecular studies (of DNA) come into their own. Thus it may transpire that the twenty or so species of oak, or pine, or whatever in place A all have very similar DNA. Place B may have only half a dozen species, yet the difference in their DNA may be profound.

It would be reasonable to conclude that the species in place A all arose from a single ancestor, who arrived there fairly recently, found the place agreeable, and diverged rapidly (and perhaps re-hybridized, as outlined in Chapter 1). But the greater genetic diversity found in place B could be explained in two different ways. Perhaps all the trees did indeed arise in situ from a single founder, who arrived or originated in that place a great many years in the past, giving its descendants plenty of time to diverge. Or perhaps some at least of the very distinct trees originated in other places, and simply converged on the place that's being studied. But then we can ask a further question. It is possible to infer from their DNA which species in any particular family (or which family in any particular order) is the most primitive – this being the one that seems to have most in common with the original ancestor. Common sense suggests, then, that sites that have the greatest true diversity of species (the greatest variation in DNA) and/or include the species that are known to be the most primitive, are at least reasonable candidates as the true centre of origin. But of course such a site could just be an ancient secondary centre of diversity. The trees might be completely extinct in the place where the group truly arose.

This is where fossils come to our aid. In more and more sites all around the world, palaeontologists are now finding fossils of truly astounding quality, that reveal the structure of ancient plants in the most minute detail. Pollen is particularly informative. It is highly characteristic, and often allows identification at the level of the genus. It is also very enduring, often to be found in the deepest mud beneath lakes, or in rock that derived from the mud of lakes that are now long gone. Pollen is to palaeobotanists what teeth are to scholars of ancient mammals. Fossil and sub-fossil pollen sometimes provides a continuous record of ancient floras over tens of millions of years.

Fossils can be deceptive, however. Fossilization is a rare event. The oldest fossils known of any particular group do not necessarily represent the very first of that group. Indeed – given that all groups are rare in their early stages – the oldest known fossils are most unlikely to represent the first that actually existed. Neither do the latest ones known necessarily represent the most recent. The most recent fossils of *Metasequoia* and *Wollemia* are both millions of years old; yet both these trees have proved to be alive and well, in China and Australia respectively.

But fossils do give us some certainties. If a fossil of a particular tree turns up in a rock that's 100 million years old then we know that that tree did indeed live in that place, and that its species was there *at least* 100 million years ago. Thus we know for sure that the family Araucariaceae, now confined to the south, did once live in the northern hemisphere. Contrariwise, the absence of southern beech fossils in the northern hemisphere does not prove that the southern beeches never came north of the equator. As the adage has it, 'Absence of evidence is not evidence of absence.' Even so, the fact that thousands of trawls from hundreds of sites over many decades have failed to produce any southern beech pollen in the north is at least a strong suggestion that they have always been out-and-out southerners. Southern beeches, presumably, really did arise in the southern hemisphere.

But there is another huge complication. Conifers as a whole first arose several hundred million years ago, and some modern conifer families are well over 100 million years old. Flowering plants as a group are much younger, but still many families of them are tens of millions of years old. Since the time when many plant families began, much of the land on which they stand has moved dramatically.

AND YET THEY MOVED

The first suggestion that the continents are moving around the globe came from the German geologist Alfred Wegener, who in 1915 coined the expression that translates as 'continental drift'. Wegener found that if you cut up a map of the world and shuffle the bits around then the existing continents and the big islands, particularly of Australia, Africa, South America and Antarctica (and Madagascar and New Zealand), fit together like a toddler's jigsaw. In the north, the eastern coasts of North America, plus Greenland and Iceland, when shoved sideways, abut neatly with the west coast of Europe. The Atlantic is shaped like a snake. The coincidences just seemed too great. Surely, he said, the different continents and islands must once have been joined together, then split and drifted apart. At first, many scientists were thrilled with Wegener's idea. Then they decided that it was impossible (meaning that they could not think of a mechanism) and most then declared that it was obviously ludicrous. By the time of his death, in 1930, he was more or less outcast. Only a few brave hearts supported him.

But the brave hearts turned out to be right. The continents have moved, and, measurably, are still moving. The mechanism that drives them began to become apparent after the 1950s. The centre of the earth is hot, as indeed had been known for some decades. Indeed it is so hot that the entire interior swirls with convection currents, like a simmering kettle. The interior rock is the magma that flows out when volcanoes erupt. The continents are made of lighter rock, and float on the restless magma like froth on a slow-moving river.

The continents move slowly – only a few centimetres a year – but they have had plenty of time, and over the past few billions of years their peregrinations have been prodigious. Five hundred million years ago (in Cambrian times) the present land masses were scattered islands. Places now in the tropics have at times been near the poles, and vice versa. Places that are now in the heart of continents have been islands, and present-day islands have been part of mighty continents. Most of what is now North America was an island that straddled the equator. Siberia was a subtropical island in the southern hemisphere. And so on.

All this was before any plants had come on to land, and long before there were trees. But by about 265 million years ago (the mid-Permian), when already there were plenty of trees, all the islands had massed together to form one great land mass known as Pangaea. By about 200 million years ago (in the early Jurassic), when the conifers and cycads were in their pomp and the flowering plants were

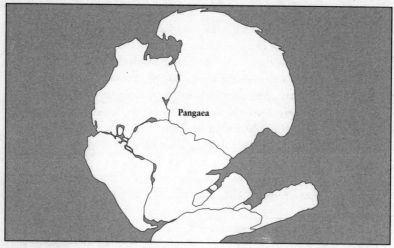

About 225 million years ago the world's continents were joined to form a supercontinent called Pangaea.

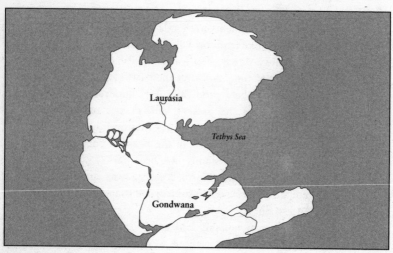

Eventually Pangaea started to break into two separate continents.

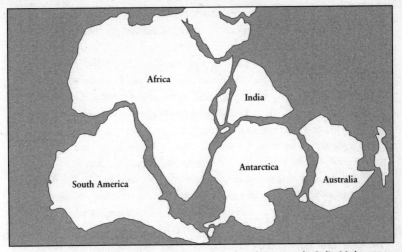

Gondwana broke up to form Antarctica, Africa, Australia and South America; plus India, Madagascar, New Zealand and New Caledonia. Many (though by no means all) of the trees that now live in those Sourthern lands were Gondwanan in origin.

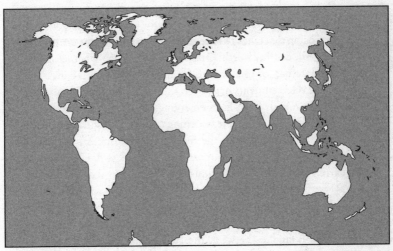

The continents are still drifting. Perhaps in a few million years Australia will collide with Southern Asia, as India once did. Perhaps it will crunch into Japan. Or perhaps it will slide past Japan and into the North Pacific. Each of the possible scenarios will be dramatic, though none is urgent.

yet to come on the scene, Pangaea began to split more or less in half to form two great 'supercontinents': Laurasia to the north, and Gondwana to the south.

Ever since – through all the time that the flowering plants have been evolving – those two great supercontinents have been breaking up. Laurasia has split to form present-day North America, Greenland, Europe and most of Asia. Gondwana has fragmented to become present-day Antarctica (a huge continent), South America, Africa, Arabia, Madagascar, India and Australia, plus a fairly long catalogue of today's islands including New Zealand and New Caledonia.

The details of this redistribution are roughly outlined in the figure on pp. 284–5. Two features in particular are outstanding. Note, first, that South America was an island for tens of millions of years, until it finally made contact with North America, via what is now called Panama, around 3 million years ago – which in geological time is very recent. India too was an island that drifted north for tens of millions of years until it finally crunched into the south of Asia. The crunch took place about 60 million years ago (not long after the dinosaurs disappeared) – and the impact, slow and inexorable as it was, caused the rise of the Himalayas (and the Himalayas would still be rising were it not for the erosion that tends to keep them at the same height). Australia is still an island, drifting north. Opinion is divided on whether it will miss Asia altogether, or crunch into China, or obliterate Japan.

All the time the continents have been shifting, so the modern groups of plants and animals have been emerging. Modern land mammals, which could not easily cross the widening oceans, still show its effects very clearly. So it was that in the nineteenth century Alfred Russel Wallace pointed out that the mammals of South-East Asia were quite distinct from those of Australia and New Guinea. In Asia there were and are cats, pigs, deer; while Australia was and is dominated by marsupials, and contained the unique egg-laying monotremes – the platypus and echidnas (though there is also an echidna in New Guinea). The boundary between the two faunas became known as 'the Wallace Line'. Although Wallace didn't know it, his line marks the border of Laurasia and Gondwana.

Continental drift must also enrich our attempts to explain why modern trees are where they are. To be sure, each kind must have arisen in one particular place, its centre of origin. Each may then have spread to other places, and then spread again, and perhaps the earliest ones in the original centre went extinct. But all the time this was happening the ground itself was shifting, as the continents wandered the globe, breaking apart and sometimes joining up with others.

Through such ideas we begin to see why different modern-day islands have such very different characters. Madagascar has been an island for a very long time but it used to be part of Gondwana. When it broke from the mother continent it contained a random collection of plants and animals – the latter (perhaps) including the ancestors of lemurs, but not of modern carnivores or hoofed animals. Whatever plants it had on board were free to evolve in all kinds of strange directions, without competition from continental types. The traveller's palm, *Ravenala*, the baobabs, and the Didiereaceae are among the results. New Caledonia too is a fair-sized, long-isolated fragment of the erstwhile Gondwana; but New Caledonia began its island existence with a different collection of Gondwanan creatures on board which now are almost as spectacularly strange as those of Madagascar. In contrast, Hawaii, the Galapagos, the Azores and the Canaries did not begin as bits of continents. They arose from the bottom of the ocean, in volcanic eruptions, and began life as bare rock. They too have their own unique collections of endemic creatures. But all their inhabitants are descended from ancestors that were brought in by wind or water, or flew or swam, or hitched a ride on the flyers or swimmers. Creatures with no such abilities stayed away. Conifers (apart from one bird-dispersed juniper on the Azores) are notably absent from mid-oceanic volcanic islands. New Caledonia is in the middle of nowhere but it is a fragment of continent, and its indigenous conifers, borne from Gondwana, are among the wonders of the world.

Britain was also part of an ancient continent, as Madagascar was, and has been linked to mainland Eurasia from time to time in the past few million years as the sea level fell during the ice ages. Yet it lacks much of the variety found in neighbouring Eurasia and has no endemics to speak of, apart from a few fish, including various subspecies of char and trout. But here we encounter yet another complication. Britain endured the rigours of the ice ages. During the ice ages the British Isles were linked to mainland Europe, which should have enriched its flora and fauna. But instead, or at least equally, the ice wiped out what was there. Madagascar has long been in tropical latitudes, and escaped any such purge. This story belongs to the next discussion, however: why the tropics are so much more various than temperate latitudes.

With these general principles in place (centre of origin, secondary centres of diversity, continental drift); and aided by modern techniques (analysis of DNA and the steadily improving fossil record), biologists are now putting flesh on the theoretical bones. The Royal Society meeting of March 2004 showed how rich – and unexpected – the realities are now proving to be.

REALITY: A FEW CASE HISTORIES

Take, for example, the American tropics (the neotropics), the most diverse ecosystem of all. South America is a fragment of Gondwana that broke away from Antarctica some time in the Cretaceous, rafted north as an island for the best part of 100 million years, and finally docked with North America around 3 million years ago. (Although 'finally' in this context simply means 'most recently'. The travels of continents, like the Flying Dutchman's, will never end.) The relationships of animals can be easier to see than those of plants, and it's clear that South America has many strange animals – notably the sloths, anteaters, armadillos and its own suite of marsupials – that are peculiar to itself, and reflect the Gondwanan origins of South America, and its long isolation as an island. But South America also contains northerners, including jaguars, pumas and various deer, which came in from North America. So we might expect to see the same pattern among trees: basically a Gondwanan island flora, with a few incursions, mainly from North America.

Yet this commonsense assumption holds up only to a very limited extent. Thus, the characteristic flora of northern South America at present is tropical rainforest. If the trees of the forest are indeed Gondwanan in origin then their ancestors ought to have been there tens of millions of years ago. But Dr Robyn Burnham of the University of Michigan has studied fossil plants in Bolivia and found very little evidence of any tropical rainforest at all before about 60 million years ago. The oldest rainforest fossils seem to date from around 50 million years ago. But she has found clear evidence of tropical rainforest in several sites in North America that are more than 60 million years old. This suggests (it does not prove, but it suggests) that the great rainforest of the neotropics, including South America, arose in the northern continent, Laurasia; and that the trees found their way to South America long before South America met up with North America around 3 million years ago, in the Pliocene. This fits in with other evidence of many kinds which suggests that South America had contact of a kind with northern continents long before the Pliocene. For example, the *small* south American cats – the ocelot and margay – seem to have been in place since well before the Pliocene. So too have South America's monkeys, like the howlers, capucins and tamarins. The ancestors of these animals also arose on other continents but they too got to South America before North and South America most recently collided. Presumably there were other land bridges in the past, now long gone – or if not literal bridges, then at least chains of islands, as are still to be seen through the Caribbean.

However, molecular evidence from Dr Toby Pennington, at the Royal Botanic Gardens, Edinburgh, strongly suggests that the present-day trees of the South and Central American forest did not come from North America either. Their closest relatives, at least of the groups he has looked at, often seem to be in Africa. This seems to take us back to the original idea – that the plants of Africa and South America are both simply Gondwanan. But close comparison of DNA suggests that many of the present-day South American plants originated *after* South America broke from Gondwana – and that by some means or other they have entered from Africa. The picture grows more and more mysterious.

As with South America, so with Australia – but with a different twist. As with South America, we would expect from first principles that its flora, including its trees, would basically be a selection of those that were originally on Gondwana. Again, up to a point, this is true. But again there are some serious complications.

Thus it is that the present-day vegetation of southern Australia is very varied. Some is alpine (low mountain). Much is shrubland and grassland. Grasses came only very recently to Australia, but much of the rest, as we would expect, is basically Gondwanan.

But Robert Hill, at the University of Adelaide, concludes from the fossils that when Australia first parted company with the rest of Gondwana around 70 million years ago, the south of Australia was largely covered in rainforest. (Strange to think of rainforest inherited from Antarctica; but then the world is strange.) But as Australia moved north it was cooled by the cold current that began to circulate around Antartica. As the island continent cooled so it dried, and by 5 million years ago much of it was arid. Australia, too, is heavily eroded and short of nutrients – so the plants had to cope both with lack of nutrient and with lack of water. In addition, Australia has been beset by ice ages.

Specifically, Mike Crisp of the Australian National University in Canberra concludes that Australia inherited southern beeches from Gondwana, and that these sat around for a bit and then diversified impressively. But then the aridity got to them, and now there are very few. So the southern beeches were largely replaced by the eucalypts, which cope with aridity very well. Professor Hill feels that the eucalypts were a late development, and indeed may have arisen within Australia itself. In contrast, Professor Crisp believes that, like southern beech and indeed like the *Banksias* (in the same family as *Grevillea*, the Proteaceae), the eucalypts arose in Gondwana and were present in Australia from the start of its career as an island. Where Australia's acacias (in those parts known as wattles) came from is not obvious, but they clearly diversified mightily as the continent became more arid.

In general, too, the picture of Australia's trees and other plants has been hugely complicated by various animals, including humans. Australia used to have some bigger mammals than it has now, including an (almost) rhino-sized wombat called *Diprotodon* and a giant kangaroo that stood around 3 metres tall. There were more huge reptiles, too – which in Australia's early days as a solo land mass included dinosaurs. Some of these may well have helped to disperse the seeds of Australia's early trees; and with them gone, the trees would suffer. The large mammals (including *Diprotodon*) evidently became extinct some time after the Aborigines arrived from South-East Asia, at least 40,000 years ago and perhaps as long as 80,000 years ago. Aborigines, too, have long made extensive use of fire to encourage grasses and so attract animals for hunting. Their fires may well have altered the vegetation of central Australia irrevocably. The Europeans, of course, who probably first set foot on Australia in the seventeenth century and finally got to grips with it through James Cook's voyages in the eighteenth, have transformed much of the country, with a huge array of imported plants and animals including quasi-wild creatures such as rabbits and foxes (which affect the local marsupials, which in turn interact with the plants) and domestics such as sheep, water buffalo and cats. Even in the deep past, however – millions of years before human beings arrived, and indeed before they arose as a species – the pristine Gondwanan flora that Australia inherited was complicated and largely transformed by climate, erosion, and the native animals.

The northern hemisphere has comparable stories to tell. Thus Michael Donoghue of Yale University has examined the relationships (as revealed by their DNA) between the plants of eastern Asia (mainly China) and those of eastern and western North America.

We know that a few tens of millions of years ago, all of the present land mass of Europe and Asia was linked at its western extreme to Greenland and Iceland, which in turn were linked to what is now North America. Eastern North America was then much nearer to east Asia than western North America was. Common sense says that the easiest way for plants to get from eastern Asia to North America in the deep past would have been via Europe and Greenland. So we would expect the trees of eastern North America to be more like those of eastern Asia than the trees of western North America are. (It's useful to have an atlas handy when reading this account.)

In fact, though, says Professor Donoghue, the trees of *western* North America are more like those of China than those of eastern North America are. It seems, indeed, that many of the trees in the west of North America arose in China – and then evidently crossed what is now the Pacific Ocean. This seems most unlikely until we consult the atlas and perceive that the gap between Siberia, to the extreme north-east of Asia, and Alaska, to the extreme north-west of

North America, is small. At present that gap is linked by a chain of islands. But when the ice ages descended in the past, the sea level fell by up to 600 metres (since so much water was trapped, as ice, on the continents both of the extreme north and the extreme south – including Antarctica of course, but also Australia). During those times, there was dry land between Siberia and Alaska – a land known as Beringia, which at times was huge: the size of present-day Poland. Many animals are known to have crossed from Eurasia to North America via that route, including lions, bison and ancient elephants; and many others crossed from the Americas to Eurasia, including dogs and rhinoceroses. Human beings also reached America via the Beringian land bridge. Professor Donoghue's studies suggest that many plants made use of this bridge too. In short, many of North America's present-day plants, including many trees, seem to have originated in eastern Asia (notably China). But they did *not* (as the history of continental drift would lead us to expect), move west to North America across Eurasia. They moved east to North America via Beringia.

More generally, we see that the broad general principles do indeed explain a great deal. Each lineage of plants did arise in one particular place; each group may then have spread to secondary centres, and diversified again; and all the time the stage has shifted, as the continents processed around the world. But if we set too much store by the broad principles, we are deceived. The actuality of each tree's history, unearthed as best we can by their fossil spoor and their relationships as reflected in their DNA, reveal layer upon layer of complexity. We could tell quite a good tale now of the origins and migrations of trees. But in twenty years it will be different and surely richer; and in a hundred years it will be different (and richer) again. Of course we can never be sure of anything, and least of all of events in the deep past. But it is tremendous fun finding out – or simply enjoying the fruits of others' scholarship.

What of the other question – why there are so many more species in the tropics than in high latitudes?

WHY SO MANY TREES IN THE TROPICS?

In print at present are approximately 120 recognizably distinct attempts to explain why the tropics are so various, and why they are so much more various than the high latitudes. Many if not all them are bona fide scientific hypotheses – not just top-of-the head speculations that may or may not be true, but ideas that give rise to predictions that can be tested. In practice, some of the predictions remain untested, and the tests that have been done sometimes seem to support the underlying hypotheses and sometimes simply raise more questions. Some of the explanations complement each other, while others are definitely at odds. It really isn't easy to convert the observations of natural history (in this case, that there are huge numbers of species in the tropics and many fewer in temperate lands) into hard science.

In a nutshell, the accounts are of three kinds. Some ascribe the diversity of the tropics to physical factors, notably the abundance of heat and light. Some home in on logistics – the notion of complexity, the outworking of natural selection, and so on. Some cite history, suggesting that the diversity of tropical forests and the relative impoverishment of temperate ones depend on what has happened in and to tropical countries over the past few decades or millennia or millions or even hundreds of millions of years. In truth, of course, *all* phenomena of all kinds should be discussed from these three angles: the physical facts of the case; logistics; and history. In the following I will discuss the first two kinds of ideas together, and treat history – always the joker in the pack – separately.

Heat, Light and Logistics

'Energy is one of the best predictors of diversity': so says Douglas Schemske of Michigan State University records in 'Ecological and evolutionary perspectives on the origins of tropical diversity' (*Foundations of Tropical Forest Biology*, 163–73). Energy, in this context, means warmth and light, including ultra-violet. This seems to be true everywhere. Even within Britain, places that have more sunshine tend to have more species than places with less. Why should this be?

For starters, and most obviously, more warmth should mean that more is happening. Plants certainly grow quicker in warm places. With such thoughts in mind, some biologists have suggested that if creatures grow quicker, then they can reach maturity more quickly. This means they can fit in more generations in a given time – and so, we might expect, they can evolve more quickly.

The first bit of this argument (that organisms can grow quicker when warmer) stands up to an extent, but is not simple. For example, mammals and birds are 'warm-blooded', meaning they achieve some independence from background temperature by creating their own body heat. By the same token, they may grow very quickly even in the cold. Nothing grows quicker than a baby blue whale out in the chilly ocean, while the growth rate of Arctic goslings or of baby seals on ice floes is prodigious – it has to be, because they have only a few brief weeks to grow before they must take to the air or put to sea. Plants, however, clearly can and do grow much quicker when it's warm (but not too warm), and in general, as we might expect, the tropics do produce much more biomass per unit time and space.

But the idea that creatures that can grow fast are likely to have shorter generation times and so may evolve more quickly, is far more equivocal. Many tropical forest trees take an unconscionably long time before they set their first seeds (bamboos take several decades) and many animals that live in the tropics, from scorpions to elephants, can be slow to mature and even then produce very few offspring, and only at long intervals. Human beings first evolved in the tropics, and we too take our time to reproduce. In short, although common sense suggests that tropical creatures *might* mature earlier and reproduce more quickly, in practice nature is far more subtle than that. There is no simple correlation.

Still, though, where there is more energy there is in general more *life*. Greater biomass is liable to be generated in a given space and time in the tropics than in temperate climates. The biomass is divided among many different individuals, which suggests there will be many more individuals. More individuals means more competition, and competition is the stuff of evolutionary change by means of natural selection; so we might expect more and more new types to be intro-duced after all, as the plethora of individuals battle it out.

But why should that produce more *species*? After all, we could perfectly well envisage that the individuals of one species would out-compete those of other species, so that the less-adapted species would disappear altogether. Then we would have a lot of individuals right enough – but they might all be of one species: the one that has adapted most adeptly, and succeeded at the others' expense. So greater biomass in a given time and space, and more individuals, doesn't necessarily lead to more species. Again, at high latitudes, we find forests with huge biomass but very few species; and in the cold oceans, too, we find some of the greatest concentrations of biomass in all the world, in the form of the planktonic, shrimp-like crustaceans known as krill. But the krill is all of one species.

A second kind of idea – of a logistic kind – was first proposed formally by Alfred Russell Wallace and was then taken forward by two great evolutionary biologists of the twentieth century, the Ukranian-American Theodosius Dobzhansky and Britain's R.A. (Sir Ronald) Fisher. In essence, it says that complexity builds on com-plexity. Every individual in a community (a community is a collection of creatures in one place, that may or may not be of the same species) obviously limits the space and resources available to other individ-uals. But at the same time, each individual may provide new niches for other individuals. Trees provide the supreme example. They have leaves, buds, flowers, fruits, twigs, a trunk with wood and bark, plus roots which create a special environment around themselves (the 'rhizosphere'). Trees provide heavy or partial or intermittent shade – a variety of light regimes. In rainforest, it will typically be wet around the roots of trees while their tops, perhaps 30 metres up, will be in burning sun, potentially as strapped for water as any desert. Thus any one tree provides a host of micro-worlds, and a host of feeding opportunities: on the leaves, in the leaves, under the leaves, in the wood, on the fungi and protozoa that grow on bark or leaf, and so on. Each creature that exploits any of those potential niches in turn provides opportunities – and raises problems – for others. Thus every beetle on every leaf has its retinue of parasites and predators. Any one niche may be exploited in a host of different ways (some insects bite, some suck, and so on). Each method of exploitation provides new niches for other creatures. For instance, suckers of plant sap may suck up parasites at the same time (as aphids suck up viruses) and then pass them on to new hosts. Thus we have a positive feedback loop: as diversity increases, so it encourages further diversity.

But here we get into some more twists, this time of a genetic nature.

First, if many different species are crammed in to any one place, then the population of each species is bound to be small. But when populations are small, they start to lose genetic variation as the generations pass. Each parent in each generation passes on only half of his or her genes to each offspring. If the total number of offspring produced is low (as it will be if the population of parents is small), then it becomes very possible that some of the parents' genes will not be passed on at all. Thus, as the generations pass, small populations tend to become more and more genetically uniform, as the rarer types of genes within the population fail to get passed on. This loss is called 'genetic drift'.

Genetic drift generally has negative effects. For as a population becomes more genetically uniform, so the different individuals within it become more and more genetically similar. When individuals that are too genetically similar mate together, the offspring are liable to suffer from inbreeding; the same phenomenon that has wiped out many an aristocratic family who loftily refused to conjugate with commoners, and bedevilled many an isolated village (in turn inspiring many a gothic novel). Genetic drift, in short, often leads to extinction.

But in 1966 A.A. Federov put a more positive spin on genetic drift. He pointed out that the loss of genes with each generation produces *qualitative* shifts in the succeeding generations. For instance, a parent generation of a plant with predominantly white flowers may contain rare individuals that have genes for red flowers, and so produce some offspring with red flowers or indeed (depending on the degree of genetic dominance) in varying shades of pink. But loss by drift is a random process, and, quite randomly, the rarer red gene will be lost. Then *all* individuals in later generations will have white flowers. A shift in flower colour may not be important – but we can easily envisage other changes that could be. For example, a shift in the assortment of genes could result in a change in mating pattern – so that later generations flowered at a different time from the parents. If this happened, then the later types would no longer be able to mate with any of the parent types that might still happen to be around. Once two groups are reproductively isolated then they will each evolve along different lines – and effectively form new species. Putting the whole thing together: having a lot of species in one place means that the populations of each species are smaller; the different populations lose genetic variation by drift, and so (if they don't go extinct!) they quickly become qualitatively different as the generations pass; and these qualitative changes can lead to the emergence of new species. Here we have a genetic reason why diversity could lead to still greater diversity.

In 1967 another American biologist, Daniel Janzen, proposed yet another twist, both ecological and genetic in nature. He suggested that since tropical species live in very favourable conditions (of heat, light and moisture), they probably could not tolerate a wide range of different conditions. (This idea was widely held in former times: it was assumed that tropical trees in general must be sensitive plants.) Janzen then proposed that if two populations of trees from the same species were separated by a mountain that was even of modest height, they would probably be completely isolated since they would not be able to live even at the modest altitude of the land that divided them. By contrast, he suggested, temperate plants are much tougher, and although they would be kept apart by bona fide ranges like the Rockies or the Alps they would not let a mere hill come between them. For this reason, he said, there is liable to be *more* isolation of different populations in the tropics than in temperate lands. Cogently and poetically, Janzen called his paper, 'Why mountain passes are higher in the tropics' (*American Naturalist*, vol. 101, 1967, 233–49).

Intuition suggests that all these forces of change and diversification could be operating, and it's easy to see how they might all act together: isolation caused by inability to cross apparently innocuous boundaries could lead to small populations and so cause further changes by genetic drift, and so on. Such ideas are very difficult to test, however. It is very hard to find out anything for certain in tropical forests, and to test subtle hypotheses that have to do with the rates of gene loss in small populations is difficult indeed. (Although, as described later, this is precisely what scientists in the Dendrogene Project in Brazil *are* now doing. But they are doing it for different reasons; not directly to test ideas such as this.)

Yet there are some relevant observations, and they don't all provide support for ideas like Federov's and Janzen's. For instance, it is easy to see how a tree provides niches for hundreds or thousands of species of other kinds of organism – fungi, epiphytic ferns, insects, mites, and so on. But the question here is not why there are so many species of ferns and insects and so on in a tropical forest. We are asking why there should be so many kinds of *tree*. Why should the presence of any one tree provide more niches for other trees? This clearly is not the case, for instance, in the coastal redwood forests of northern California.

One possible answer is as Janzen suggested: that tropical forest trees are highly specialized. For example, it is clear that the soil in Amazonia, say, may differ markedly from place to place, for example in its mineral content. If the trees were really highly specialized, then we would expect different types to grow in different places. Then again, some trees are shade-lovers, while others like bright sun and are inhibited by shade, and many others prefer shade when young but come into their own when it's sunny. Pioneer trees are generally sun-lovers, and so they quickly occupy any space that appears when some forest giant collapses or is felled, leaving a clearing. The coastal redwoods of California grow very slowly when shaded but then zoom up as soon as their neighbours fall and so let in the light. Thus Emanuel Fritz observed a coastal redwood 160 years old that was a mere 100 feet high, which meant it had been growing by only 1 per cent per year. But then a gap appeared in the canopy and for the following decade it grew at an astonishing 20 per cent per year – to take its place almost instantly as a respectable, 300-foot redwood.[1] By such means, we can see how the comings and goings of the different trees would create opportunities for others.

In truth, though, trees in tropical forests seem to be far less specialist than might be supposed. Individuals of any one species are to be found growing on a wide variety of soils. Nick Brown from Oxford's Institute of Forestry, too, has found that whereas mahoganies in Amazonia are adapted to grow at forest margins – that is, when they are young they flourish in the light – they may soon be overtaken by other trees spreading from the forest behind them, so that by the time they are mature they are in the middle of dense forest, in shade. In other words, most mahoganies, most of the time, are growing in conditions that for them are sub-optimal. But they grow just the same. But if it's the case (as it seems to be) that tropical forest trees are really quite versatile, then they would not obviously benefit from special niches created by the presence of other trees, or necessarily be separated by modest hills that provided only a slightly different environment. But if the individual trees are versatile, then there seems no obvious reason why any one species, or just a few, that happened to be more robust than the rest, should not take over the whole region – just as seems to happen in northern forests. So the observation that tropical forest trees are more flexible in their tastes than might be supposed, throws doubt on all hypotheses which suppose that tropical forests are varied because the different kinds of tree require extra-special niches.

But there is one idea at least that really does seem to hold water. There is indeed a very great deal of life in general in the tropics. There are millions of creatures milling about, with billions of possible interactions between them. Not all of those interactions are antagonistic – cooperation is an important fact of life – but some definitely are. Predator–prey relationships are antagonistic: creatures that kill, versus those that are killed. So, too, are parasite–host relationships. All trees suffer both kinds of attack in abundance: battalions of sapsuckers, leaf-eaters, bark-borers, root-miners, fruit-stealers and seed nibblers, from viruses through bacteria and fungi to weevils, toucans and orang-utans.

Of particular interest in this context are the many disease-causing parasites of trees, including viruses, bacteria and fungi. In general, these small parasites like and need large and dense populations to work on. Unlike big tree predators such as orang-utans and toucans, they cannot travel easily from host to host. They prefer close contact. But most such parasites are highly specialist. They do not leap easily from species to species. So they thrive best in large, close-set populations of the *same* kind of tree.

So a tree that seeks to avoid parasites is advised to stay as far away as possible from others of its own kind. Thus it is commonly observed (though not always!) that when seedling trees grow up close to their parent tree they die more quickly than those that are set further away. Young trees in general are more vulnerable than older ones, and the ones that stay close to home are killed off by their parents' parasites. If trees are killed off when they grow too close to own kin, then those of the same kind will finish up growing a long way apart. The gaps in between will be filled by trees of different species – each of which is anxious to put as much distance as possible between itself and others of its own kind. Thus we finish up with enormous diversity and enormous distance – half a kilometre is commonplace – between any two of the same species.

So the secret of tropical diversity – or least, of the diversity of the trees themselves – may lie with parasites. This may seem humbling: that such magnificence, sheer variety, should have such a squalid cause. But then, the great English biologist W.D. (Bill) Hamilton proposed that it was the need to avoid parasites that prompted the evolution of sex – for sex produces the generation-by-generation variation that makes life difficult for parasites, which tend to be highly specialized, to get a hold.

This simple if unsavoury idea seems to stand up more firmly than most, and perhaps is indeed the key. But it immediately raises two obvious problems. The first is how trees in tropical forests manage to find breeding partners if the nearest individual of the same kind is half a kilometre away. This is indeed a difficulty – which, as discussed in the next chapter, has led to some of the most ingenious and extraordinary evolutionary exercises in all of nature.

But also – if disease causes tropical forests to be so diverse, why doesn't it have the same effect on temperate forest? Temperate trees have plenty of diseases too: Dutch elm disease in elms, chestnut blight in North America, and so on.

The answer is again uncertain (of course), but two grand ideas seem very definitely to be pertinent. One emerged in the 1950s, again from Theodosius Dobzhansky. He pointed out that in the tropics, where life in general is so easy, and so many different creatures can find a niche, the real pressures are biological: in other words, the pressure on any one creature comes from all the other creatures around it. By contrast, in temperate climates, the main problems are posed by the elements – cold winters, whether wet or dry; late frosts; and lack of adequate warmth in summer to stimulate growth. Only the hardy few that can hack such vicissitudes need apply at all. We might further speculate that the cold zaps the parasites too, and so relieves the pressure from them. After all, temperate fruit trees seem to suffer more damage from parasites after a mild winter, when more of the pests survive.

Many a case history offers good support for Dobzhansky's idea that northern creatures must cope first and foremost with the sheer violence of the physical conditions. I will round off this chapter with three examples from North America. But before we look at them we should acknowledge one final joker in the pack. The strongest reason of all why the tropics are so much more varied than the north may be a matter of history.

History

History marches on an infinity of timescales simultaneously. Every living creature or the ancestors that gave rise to it has been influenced by events that happened yesterday, decades ago, thousands of years ago or hundreds of millions of years ago and by the same token, everything that happens in any one moment affects the next second, the next year, and so on into the indefinite future. On the short scale (of years and decades and centuries) all trees everywhere (or their ancestors) have been affected by storms, floods, landslips and fires. On the very grandest timescale, all trees have been affected by the movement of continents, as already outlined. On the intermediate timescale – from centuries to tens of millions of years – the world may experience vast changes of climate.

In particular, the world has grown steadily cooler over the past 40 million years or so, albeit interrupted by a few warm spells, culminating in the ice ages of the past 2 million years. This climatic shift altered the course of all evolution – indeed, it apparently brought our own species into being as the forests of East Africa shrank during a series of cold spells a few million years ago and forced our arboreal ancestors to the ground. More pertinent here, is that the cooling of the earth in general and the ice ages in particular may largely account both for the variety of trees in the tropics, and for the impoverishment of the north.

The reason for the cooling lies ultimately with greenhouse gases in the atmosphere and in particular with carbon dioxide. All kinds of evidence – including cores drawn from extremely ancient ice in Greenland, which still holds bubbles of ancient atmosphere – attest that when carbon dioxide levels are high, then the surface temperature goes up; and this is what seems to be happening now, and is causing global warming. Contrariwise, when carbon dioxide levels are low, the earth cools. Physics theory supports this idea. The basic reason is that greenhouse gases (like carbon dioxide) are relatively impervious to infra-red radiation. The earth is warmed in the day by sunlight, and loses the heat again at night in the form of infrared. But carbon dioxide inhibits the loss of infra-red, and so reduces cooling. This is how the glass in a greenhouse works, which is why carbon dioxide is said to be a greenhouse gas (and so are a number of other gases, such as methane, but carbon dioxide is the one that counts in this context). The earth as a whole has cooled this past 40 million years or so because the concentration of atmospheric carbon dioxide has steadily gone down.

Why should this be so? The explanation was provided most ingeniously by Maureen Raymo, from the Massachusetts Institute of Technology, in the 1980s. It is linked to continental drift. For, as we have seen, about 60 million years ago the tectonic plate that bore the vast island land mass that is now India began to crunch into the south of Asia, and puckered the land in front of it to produce the Tibetan Plateau, with the Kunlun Shan and the Himalayas to the north and south. The plateau and the mountain ranges form, as Raymo puts it, 'a giant boulder'. The wind sweeps over the Pacific to the south and east, picking up water along the way; and as it hits the high land at the top of India it rises, cools, and releases its water, which falls in the rainstorms known as the monsoons, and runs away in eight great rivers that include the Ganges, Brahmaputra, Indus, Yangtze and Mekong. The mass of water feeds both the forests and the farms of Asia (although the land left in the rain shadow includes the Gobi, the Mediterranean and the Sahara).

But the water that falls in the monsoon rains is not pure. Rainwater always contains gases dissolved from the atmosphere, and among the most soluble is carbon dioxide. Thus rain – all rain, everywhere – is a weak solution of carbon dioxide, otherwise known as carbonic acid. As the carbonic acid falls on to the Himalayas and the Tibetan Plateau, it reacts with calcium and magnesium in the rock to form a weak solution of bicarbonates. This is washed away in the rivers and into the ocean, where the bicarbonate salts become incorporated with the ocean bed and eventually (thanks to plate tectonics) are thrust down into the magma beneath. The net result is that carbon dioxide is steadily leached out of the atmosphere – and has been leached, storm by storm, for the past 40 million years. The dwindling carbon dioxide causes what is sometimes known as an 'icebox effect': opposite to a greenhouse effect. This may sound fantastical, but sober calculations based on elementary chemistry and the topography of the mountains suggest that it is all eminently plausible.

There is one more variable. In the early twentieth century the Yugo-slav mathematician Milutin Milankovitch sought to find a relationship between the fluctuations in climate and the changes in the Earth's orbit as it circles the Sun. In general the Earth's orbit is almost circular – but at times it becomes more elliptical. Ellipses are like flattened circles, pointed at both ends; and when the Earth is at the points, it is further from the Sun than it ever is when the orbit is more circular. When the Earth is more distant, said Milankovitch, the amount of sunlight striking it can go down by 30 per cent, and the climate then is obviously cooler. The orbit passes from near circular to more elliptical in cycles, known as Milankovitch cycles, that last around 100,000 years. So, said Milankovitch, we can expect periods of rela-tive cold to alternate with periods of relative warmth every 100,000 years.

When the general level of carbon dioxide in the atmosphere is high, the periodic cooling caused by the Milankovitch effect may not be too disturbing. But when atmospheric carbon dioxide is low, the extra cooling can change the world radically. By about 2 million years ago (thanks to the continued depletion of atmospheric carbon dioxide by the rocks of the Tibetan Plateau), the earth was on the point of freezing over. In practice, over the past 2 million years the earth *has* frozen over, at least partially, every 100,000 years or so. These periodic freezings were the ice ages. In the northern continents, glaciers and mountains of ice in places reached depths of several miles, extending through Europe to southern Britain and France and through North America to what is now New York and beyond. Ice sheets centred on Antarctica encased much of the Southern Ocean, and encroached at least into the south of Australia. During the ice ages, the world was cooler, or course. But it was also much drier, as much of the world's fresh water was locked into ice, and the cold oceans evaporated less freely. The creatures that lived around the poles (particularly in the north) and around the equator were both affected by the combined effects of cooling and drying. But the biodiversity of the north and in the tropics was affected in totally opposite directions.

In the north, the cumulative effect of advancing and retreating ice, in the roughly 100,000 year cycles, was devastating. In North America, Greenland, Iceland and Eurarasia, as the ice encroached from the north it simply obliterated all the creatures in its path that could not move away and caused all the ones that could move to migrate to the south. Reindeer and lynx, archetypal Arctic creatures, came down into southern Europe. Some species of trees were wiped out. Others scattered at least some of their seeds ahead of the ice, and so their descendants lived on, but further and further south. At the end of each ice age, as the glaciers retreated, the remaining creatures could extend north again. The plants generally did so in a fairly orderly sequence. Often the land had been scoured of soil, and the vascular plants had to wait for the hardiest pioneers, like moss, to build it up again. The trees followed in their wake, with the hardiest first: birches, aspens, pines, then oaks, beeches and chestnuts. Records of ancient pollen in the north show this sequence, or 'succession', over and over again. Oaks and ash were typically among the later species to be installed and often became dominant, as in Britain. When any particular group of plants become dominant they and their undergrowth are commonly said to form the 'climax vegetation'. In more recent years, however, the classic concept of 'climax' has been challenged somewhat. Everything is transient, seen on the geological timescale. Oak forest may be ancient, but it is merely the latest instalment in a sequence that's continuously unfolding.

Clearly, the advancing and retreating ice in the north reduced diversity. As the ice came south, it caused extinctions. The trees that followed it north again were a selection of those that had survived the southern putsch: only the tough could undertake the northward journey at all, and the race was to the swift. Whatever diverse creatures occupied the northern lands before each ice age were weeded out. In the past 2 million years the weeding has been repeated at least a dozen times.

In tropical latitudes, by contrast, the cooling was less than devastating. Even in an ice age, the equator was still warm. But tropical creatures were affected by the drying caused by the general cooling and the entrapment of water in polar ice. The vast equatorial forests, which in interglacial times are continuous within each continent, became patchy. Patchiness provides precisely the conditions required to produce more species, as different populations become reproductively isolated one from another, and each evolves along its own lines. Then, when each ice age ended, the patches expanded again, and the newly evolved species from each different place were brought into contact again – to form astonishingly diverse assemblages of the kind we see now in the Amazon. That at least is the theory, and it is highly plausible. We see the same phenomenon among fishes – particularly the cichlid (pronounced sick-lid) fishes of the great African lakes. In glacial times they were reduced to a series of (big) ponds, each developing its own cichlid variants, which then reconverged to form the great inland water masses of today. Various human incursions have caused extinctions, but until recent years Lake Victoria had at least 300 species of cichlid and Lake Malawi had more than 500.

In short, both the northern and the equatorial forests felt the effects of the ice ages. But as the ice came and went, the northern forests lost more and more species, while the tropical forest became more and more diverse.

The ice ages, too, had one more dramatic effect. They caused forests to disappear in some places, but allowed them to flourish in others. Thus if we could look at the world's forests in geological time we would see them racing over the surface of the globe. The rainforest of Queensland seems to have been there for ever. But, like the Great Barrier Reef just offshore to the north-east, it has been there only since the last ice age ended, less than 10,000 years ago. The people of

10,000 years ago were modern, like us. Some people were building cities – Jericho is about that age. Many had long been navigators. Farming was beginning on a settled scale. Doubtless they had priests and paid taxes. Some at least of the stories in the Bible and the memories of the ancient Egyptians and of present-day Australian Aborigines, extend back that far.

It seems to me that all of the ideas outlined above to explain the diversity of tropical forest and the impoverishment of temperate forest could apply at any one time. Each kind of influence could build on the others. Two ideas seem most cogent, however (at least to me). The first is the impoverishment of northern faunas and floras, and the diversification of tropical creatures, by the fluctuations of the ice ages. The second is the grand if simple idea of Dobzhansky: that in the tropics the main pressures are biological, while in temperate and boreal lands the stresses are mainly physical. In the tropics the critical pressure comes from parasites, which make it advantageous for any one species to be rare, and for the individuals to be widely spaced. In the north, the physical pressures are of various kinds, and only the toughest or the most adept survive. The aspens, jack pines and coastal redwoods of North America make the point beautifully.

A TALE OF THREE NORTHERNERS

Canada is the world's second biggest country, at nearly ten million square kilometres, and more than a third of it is boreal forest. Yet this huge northern forest, nearly fifteen times the size of Britain, is dominated by nine species of tree. There are six conifers – jack and lodgepole pines; black and white spruce; balsam fir; and the larch known as tamarack (*Larix laricina*). The three broadleaves are aspen, balsam poplar and paper birch, which sometimes form pure stands and sometimes are mixed in with the conifers. There is more diversity to the south of the boreal forest (including more broadleaves). The beauty is haunting, but life is hard. Only a few species can cope with the northern winters.

Odd though it may seem, a crucial feature of this coldest of all lands is fire. Those species that are especially equipped to cope with it can steal a very large advantage over those that are less adept. One adept is the aspen, sometimes known as the trembling or quaking aspen, *Populus tremuloiodes*.

You would not immediately suspect, if you confronted an aspen in an urban park, that it is among nature's most resourceful trees. It has a languid air, with wanly fluttering leaves on long flat stalks, which in autumn turn a melancholic yellow. Its trunks, at least when young, are ghostly smooth and greenish-white (though deeply grooved near the base when older). Yet for all its bloodless foppishness the quaking aspen has the widest distribution of any North American tree and in large stretches of the far north it is the dominant and at times the only species. How come?

Through a series of neat tricks, is the answer. The aspen's whirling leaves resist the wind by not resisting: they ride the blows, go with the flow. It's wise of the tree to lose them in the winter, when they could do no good. What really counts in the north, though – what ensures that aspens may dominate for hectare after hectare, when trees that are more obviously shaped for the boreal stresses have long since fled – is their ability to bounce back after fire.

For the aspen has long lateral roots which, at intervals, send up suckers that grow into entire new trees. Many other broadleaf trees produce suckers, including the paper birch, another pale and ghostly denizen of the boreal forest. But only the aspen spreads itself so wide: an Alaskan ecologist, Leslie Viereck, has found such suckers more than 80 metres from the parent trunk. Fire inevitably strikes the boreal forest sooner or later and if it occurs in spring or summer it will kill the aspens along with everything else, because it burns the organic matter within the ground, including the aspen's trailing roots. But if fire strikes in winter when the ground is frozen, or in spring when it is still wet, the roots survive. Then the suckers rapidly grow up to form new trees – rapidly because they already have a vast, established root system to draw upon. Thus the mixed forest, probably mostly conifers with a few aspens, may be replaced in months by a grove of aspens.

The overall form of the aspen – lateral roots with suckers – is reminiscent of the calamite trees, primeval relatives of the modern horsetails (*Equisetum*) with their long trailing rhizomes (which are underground roots swollen to form storage organs). All the aspen trees in the grove that springs up around the parent, together with the parent itself if it survives, form a clone: a related group of genetically identical individuals. They all come into flower together, and shed their leaves together. Since they also remain physically joined (by the lateral roots) some biologists have suggested that the entire grove should be regarded as a single organism; and since it may be hundreds of metres across in all directions, such a grove would rank as one of the largest organisms on earth. It's been suggested, too, that some groves date back to the end of the last ice age, since they would have been among the first on the scene after the ice melted. Thus, at around 10,000 years, an aspen clone would also be among the oldest organisms on earth.

This is an intriguing thought, worth musing over, but perhaps not to be taken too seriously. The trees that look like individuals in a grove really are individuals, to all intents and purposes. If some accident or outward-bound gardener were to sever the lateral roots that join the trees to their neighbours, each one would happily live alone – just as young strawberries do when they are cut loose from their runners. What the aspens really illustrate is the power of asexual (or vegetative) reproduction. Most vertebrate animals, including human beings, and many plants, cannot pass their genes on from generation to generation except through the medium of sex. But many organisms can bypass sex, at least some of the time. They simply generate copies of themselves. Human beings can do this only in Greek mythology. But as the aspens demonstrate, asexual cloning can be a useful wheeze.

Two further morals emerge from the tale of the aspen. First, appearance in nature can be highly deceptive. The aspen doesn't look tough, and yet it may flourish where other trees, more obviously stripped for action with needle leaves and pointed crowns that shed the snow, succumb. But then, many of nature's most successful creatures seem to go out of their way to look bizarre and delicate. Sometimes the bizarreness is bravado, intended to attract mates, as with the plumes of the peacock and bird of paradise. Sometimes flamboyance is a warning, as in poisonous lizards and caterpillars. But sometimes, when we look at an animal or a plant, we just get the physics wrong. Moths and butterflies look absurdly frail, fair game for anyone, but they are ubiquitous, and in their season in the Amazon the yellow butterflies come at you in blizzards. Nature moves in mysterious ways, which itself is a moral for all who would presume to take nature in hand.

Secondly, many creatures survive when you might not at first sight expect them to because of just one particular trick. The analogy is with an old and arthritic fencing master who wraps up the young athletic tyro with disdainful ease just because he knows a few wrinkles that the novice does not. The aspen thrives where others fail because of its suckers. It might not seem much, but it works.

But the trick works only if the forest fire spares the lateral roots. If fire breaks out in summer, when the soil is dry and unfrozen, the roots are cooked and the aspen succumbs. Then the jack pine, *Pinus banksiana*, comes to the fore: a quite different tree with a quite different survival strategy.

J. David Henry in *Canada's Boreal Forest* calls the jack pine 'a scrappy tree', which in former times (when the timber of nobler trees like the white pine, *Pinus strobus*, were more available) was often treated as a weed. Now, however, with the alternatives diminished, it is widely used for fence posts, telegraph poles and pit props, and indeed for Christmas trees; and together with the black spruce, says Henry, 'is a mainstay of the pulp and paper industry'.

The jack pine has several qualities that seem to equip it for a life with fire. As in many conifers, the branches nearer the ground tend to die off as the tree grows taller, not least because they find themselves without light. In the jack pine, these dead branches simply fall off: the tree is said to be 'self-pruning'. If the dead branches were allowed to persist, they would provide a 'ladder' for the fire from the ground to the top. The physics of fire is in many ways counter-intuitive. Crucially, a hot fire that burns itself out quickly can be less damaging than one that's somewhat cooler, but lasts longer. Jack pine needles are high in resin and often low in water, especially in the droughts of spring and summer, when fires are likely, and so they burn hot but quick. On much the same principle, the jack pine's bark is flaky. It picks up surface fires but then burns swiftly and does little harm. The stringy bark of eucalyptus in Australia, hanging loosely from the iron-smooth bole beneath, is protective in much the same way. In both cases the discarded bark acts as a decoy, like a hamper thrown from a troika to divert the chasing wolves.

On the other hand, if jack pine bark accumulates on the ground (as it does if there is a long interval between fires) then surface fires – particularly in spring and summer – can be very fierce. Then, most trees of all kinds are killed. But the jack pine is typically the first to spring back. For a very hot fire in the summer burns both the leaf litter on the surface and the organic material in the soil itself, leaving a bare, mineral soil behind. Jack pines germinate well in such soil – and indeed are inhibited by leaf litter: organic matter isn't always everybody's friend. They like bright sunlight, too, and appreciate the open space.

Once germinated, the young trees grow happily in sandy soil that is too dry for other species; and for good measure, the young saplings can tolerate drought of a month or so, and sudden drops of temperature of the kind that for many trees are lethal. They grow swiftly when young – more than 35 centimetres in a year. This is a joke by the standards of tropical trees, some of which reach 20 metres in five years, but good for a land so niggardly in bright sunshine and general warmth. By their fourth or fifth year many of the young jack pines are producing their first cones – which by tree standards is markedly young. The Canadian ecologists Stan Rowe and George Scotter asked why they should be so precocious: why not focus their precious energy on more growth, rather than on reproduction? Forest fires often leave a lot of fuel behind, and sometimes a second fire comes hard on the heels of the first. It seems a good idea to scatter a few seeds before the possible follow-up.

But it's the cones and the seeds of the jack pine that are adapted most impressively and specifically to fire. The cones are hard as iron, their scales tightly bound together with what J. David Henry calls a 'resinous glue'. Many creatures attack cones; but only the American red squirrel will take on the jack pine cone, and even the red squirrel much prefers the easier, fleshier meat of spruce cone. The cones may persist on the trees for many years, and the seeds within them remain viable: in one study more than half the seeds from cones that were more than twenty years old were able to germinate.

The cones do not open *until* there is a fire: it takes heat of 50° C to melt the resin that locks the scales together. Then, they open like flowers. Thus the seeds are not released until fire has cleared the ground of organic matter and of rivals, and created exactly the conditions they need. The output is prodigious. After a fire in the taiga (the northernmost forest that then gives way to tundra), the burnt ground may be scattered with 5 million jack pine trees per hectare.

But although the cone responds to fire, and only to fire, it is remarkably fire resistant. Thus in the early 1960s a biologist called W. R. Beaufant found that the seeds inside would survive for thirty seconds even when the cone was exposed to 900° C – the kind of temperature that potters use for firing. At a mere 700° C, the seeds were perfectly happy for at least three minutes. In short, it takes an awful lot of thermal energy to kill jack pine seeds when they are still in their cones. Trees seem to have evolved cork largely as an adaptation against fire; and jack pine cones contain cork too.

Yet there is more. For as J. David Henry has found, the cone does not respond simply to the presence of fire, like some crude unmonitored mechanical device. As it is heated, it releases resin from within. This oozes to the surface, and 'creates a gentle, lamplike flame around the cone', which lasts for about a minute and half. All in all, says Henry, 'It seemed that, once ignited, the cone was programmed to provide a flame for the right amount of time to open the cone . . . while a forest fire is needed to initiate this process, the cone itself is capable of providing the type and duration of flame it needs to open and disperse its seeds.' He then showed that, once open, the heated cone does not release its seeds until it has cooled down again – which in field conditions may take several days. So the initial opening is controlled; but when the cone is first open, the seeds are held back. They are not sent out like Daniel into the fiery furnace. Henry suggests a mechanism: perhaps the hairs that coat one side of the seed are sticky when hot, and hold the seed in, but lose their stickiness when cooled again. This is speculation, yet to be tested.

In any case, the adaptations are extraordinary. Jack pines belong among a fairly impressive shortlist of trees that not only resist fire, but have become dependent upon it. They cannot reproduce without it. If there is no fire within their lifetime, they die without issue. After a fire, jack pines may flourish and form a monoculture, for hectare after hectare. But without further fire, the jack pine forest fades away.

Yet there is one final twist. In practice not *all* of the jack pine cones need the fierce heat of a fire to open them. In the north, about one in ten open just in the warmth of the sun. In the Great Lake states to the south, where there are far fewer fires, *most* of the jack pine cones are able to open in the sun. Thus jack pines have a 'mixed strategy': the genes that make their cones so special are clearly of two kinds – some that gear the cone to fire, and some that enable it to respond to sunshine. Geneticists call this a 'balanced polymorphism'. Natural selection tips the balance towards the fire-dependants in the north, and to the sunshine-dependants in the south. The jack pine isn't simply the supreme specialist. To some extent at least it is the jack of all trades. Various other pines have comparable fire-resistance and fire-dependence, but none surpasses the scrappy jack pine in its adaptation.

But perhaps the tree that is most thoroughly adapted to the special conditions of the north – not the extreme north, but the central and northern coast of California – is the coastal redwood, *Sequoia sempervirens*.

The coastal redwoods inhabit, or rather they create, temperate rainforest in a discontinuous belt, roughly 15 kilometres wide, from Big Sur south of San Francisco northwards to the Oregon border. They are of course magnificent: the height of a cathedral spire, 60 to 70 metres. The tallest, known prosaically as 'The Tallest Tree', is in Redwood National Park and is 111 metres high. Its trunk is 3 metres in diameter – although this is quite slim by redwood standards. They often go to up 5 metres. Many live to 1,000 years and some reach more than 2,000.

All forests can be peaceful (you can sleep the sleep of the just in Amazonia without being carted off in pieces by voracious ants), but nothing compares with the tranquillity of redwood forest. The columnar trucks reach far higher than in a tropical forest, where 30 metres is more standard. Tropical forest is a mass of small trees, battalions of poles, with just a few trunks of respectable garden size and only the occasional giant glimpsed through the gloom; all festooned with climbers and epiphytes. But the coastal redwoods for the most part are decorously spaced – except where they form little circles, so-called 'fairy rings'. They are an army of giants; the biggest on earth. The ground beneath them is littered with the delicate, yew-like branchlets that the trees shed every few years, chestnut brown after death. The mosses and ferns, herbs and shrubs, dotted here and there, stay at the feet of the great trees. There is no importunate clambering. There are few birds. You rarely hear such silence as in a coastal redwood forest. The light is green. The sun shines through in sharp bright shafts. On a warm, late afternoon the Pacific rainforest of coastal redwoods is perhaps the most serene of all earthly environments.

But it's not always like that.

The first problem is flood. Redwoods like moisture; a 'mild maritime climate' as it is sometimes described (though they don't tend to like salt spray). Indeed they go to great lengths to capture and condense the thick fogs of the cool Californian night and morning in their leaves. It falls as 'fog drip', and in the rainless summers, may add 30 centimetres of extra water. So they make their own climate, humid and shady.

But you can overdo the water. In winter, there may be 250 centimetres of rain. Storms are frequent; and with storm comes flooding. In Humboldt County, redwood country, there were severe floods in 1955, 1964, 1974 and 1986. The floods of 1955 swept away sawmills, farms and whole communities along the Eel, Klamath and Van Duzen rivers. Buildings were buried deep under mud. More than 500 redwoods were swept away along Bull Creek, a tributary of the Eel. Elsewhere the forest floor was buried under 1.3 metres of silt.

Then, with the aid of radiocarbon dating, Paul Zinke showed that Bull Creek had often suffered such insults in the past. In fact, a study in 1968 cited by Verna R. Johnston (see Notes and Further Reading) showed that there have been fifteen major floods in the past 1,000 years, and between them they have raised the level of the whole surrounding area by more than 9 metres – the height of a three-storey house.

In short, over time along these north Californian coastal rivers, the banks and surrounding areas are eroded in some places and built up in others: the kind of pattern that is seen for example around the coast of Europe, as the North Sea picks up entire beaches from some places and dumps them somewhere else. The cartographers of eastern England have been particularly busy this past few centuries, and surely will be even busier as global warming strikes.

This is where the redwoods reveal their own set of tricks. Of course, if the ground is removed from around them all together, then they are swept away. But if they are merely buried, to a depth of a metre or so, they are untroubled. Most trees *do* succumb to such treatment. They are suffocated. But redwoods send up roots, vertically, from

Coastal redwoods re-root themselves as the silt piles up around them

their buried lateral roots, into the silt above; and these verticals grow so quickly they sometimes come bursting through the surface.

These rapid-growing verticals, however, are merely the emergency procedure, the front-line troops. Before long, new lateral roots grow from the buried trunk, just below the surface of the newly deposited silt; generally speaking bigger and broader than the previous roots at the lower level. Thus an old redwood, that's survived many floods – and those a thousand years old or more must often have survived more than a dozen – finish up with a multi-layered root system, like an inverted pagoda; a fairly small set of roots deep down, then a bigger set higher up, and so on, each set corresponding to some earlier flood. The net result is a truly remarkable anchorage. Thus they hold their otherwise precarious trunks steady for a thousand years. Incursions of silt kill most trees, but redwoods have made a virtue of it. Here again is nature's opportunism.

Human beings are managing to queer the pitch, however. Forest clearance exposes the survivors to stronger winds, and may blow them over despite their anchorage. Roads alter the natural drainage from the hills, and so alter the pattern of flooding, leading to landslides, which nothing can resist. Added silt builds up in the middle of rivers, forcing the currents to the sides and undercutting the trees along the banks. More than 100 ancient trees in California's Avenue of Giants were killed this way in 1986, when 58 centimetres of rain fell in nine days. And although the redwoods have turned partial burial to their advantage, their roots are damaged if the soil above them is compacted – which it can be by the dedicated tractors (skidders) that are used to extract logs, and by other traffic. In great gardens, as at Kew, soil that has been compacted by the feet of thousands of visitors coming to admire the trees is loosened by pumping in nitrogen gas under pressure. In some of New Zealand's forests, visitors walk along catwalks a foot above the ground, with little bridges over exposed roots. Humanity needs wild trees. But sometimes we need to tend the wild as carefully as any garden. 'Managed wilderness' may seem a paradox, an oxymoron; but it is the reality we have to come to terms with.

Redwoods are wonderfully fire-resistant, too – and show yet more survival strategies. Their bark is a fireman's dream: tough, fibrous, and up to a foot thick. Their timber retains a lot of water. They produce a pitch that is almost non-flammable. A severe fire will break through the bark, and thus allow access for subsequent fires. But the charred wounds on the boles of many old redwoods merely attest that they have endured.

Like aspens and jack pines, too, coastal redwoods have a wonderful ability to recover after fire. Almost uniquely among conifers, redwoods can sprout new stems and roots from trunks that have been burnt, cut down or blown over. From the earliest age – indeed from the seedling stage – redwoods carry latent buds around the base and along the trunk. If they are damaged, or irritated, these dormant buds develop into shoots or roots and then into whole new trees. When the parent tree is damaged the buds around the base sprout to form a 'fairy ring' around it, of up to a dozen scions; and when the main tree dies they persist, in conspiratorial circles, that indeed seem enchanted. Sometimes the process is repeated: the trees of the inner circle are damaged and more grow up around them. By such means redwoods may return even to areas that have been heavily logged.

The many thousands of cones on each redwood produce prodigious harvests of seeds. They do not germinate well, and redwood groves generally contain few seedlings. But, as with jack pine, the seeds fare best on bare, mineralized soil that has been stripped by fire or newly deposited as silt. Again, then, the redwood is equipped to return after fire – and for sexual reproduction at least may be said to be dependent on it. Again, the principle is common throughout nature: that organisms often delay reproduction until conditions turn nasty, and a new generation is needed to carry the day. Many organisms reproduce only *in extremis*.

Despite their solitary splendour in the western forests, coastal redwoods are not innately standoffish. They do not set out, for example, to poison their putative competitors, as some plants do. On the slopes, away from floods, they may mix well enough with Douglas fir, tanoak, grand fir or California bay. Tanoaks too sprout vigorously after fires, and tolerate shade, and are in other ways resilient, with a bark rich in tannin that protects them against insects and fungi. But (like Douglas fir, grand fir and California bay) the tanoak succumbs to flooding and silting; so on the floodplains, the redwood reigns supreme.

Overall, I suggest that the tropics show what nature can really do: how various it becomes when conditions are easy, or when the pot is stirred now and again, as by the drying of the ice ages, in a creative way. By contrast, in the high latitudes, which mainly means in the north, nature was repeatedly filleted by the ice ages, and what we see now are the few brave and specialist survivors.

Yet, so far, this book has dwelt mainly on the basic cast list: who's who, where they all are, and why. Ecology begins when we ask how all the different players interact. That is the subject of the next chapter.

lives without offspring. But if any lineage of creatures abandons reproduction all together, it dies out. Yet reproduction is costly. It takes energy to produce gametes. It takes even more to produce viable, fertilized eggs, or live young. Many creatures die after a single bout of reproduction, including most annual plants and even some mighty trees like the big bamboos and some of the greatest palms. Even a few eudicot trees follow such a strategy. Over eighty years or so in the forests of Panama, *Tachigalia versicolor* of the Fabaceae family grows into a tree imposing enough for any royal park – then blossoms, sets seed, and dies. I have stood at the feet of one already dead. If parent organisms do not die (as most do not), then their offspring may become their assistants, as in many bird families, and in our own societies when they are working well. But children also become rivals. Contrariwise, the greatest threat to the offspring often comes from its parents – and so as we have seen, tropical trees that aspire to grow too close to the mother tree may perish from its diseases. The offspring of *Tachigalia*, presumably, are better off as orphans. The theme of parent–offspring rivalry runs through all of literature, beginning with the Bible and the Greeks (and doubtless through pre-literate cultures too) – and the greatest themes of literature are those of nature. Reproduction is necessary, to be sure. But it takes a heavy toll.

Sex adds another layer of complication. Sexual reproduction is *less* efficient than the many asexual strategies seen in nature. It takes two parents to produce even a single offspring by sexual means, whereas cloning requires only one. Yet most creatures practise sex – including all the trees that I know about, even though many of them also reproduce asexually. In truth, sex has nothing directly to do with reproduction. Natural selection has favoured sex for another reason – because it mixes genes from different organisms, and so produces variation: offspring produced sexually are all genetically unique, different from either of their parents, and from their siblings. This offers long-term advantage, since variation is a key ingredient of evolutionary change. But natural selection does not look to the future. Sex must bring short-term advantage – for if it did not, then sexual beings would lose out to asexual ones, who in principle can reproduce twice as quickly. What is the short-term advantage? There are two main hypotheses: one (from the American biologist George Williams) says that if all the offspring are different, then this increases the chances that one, at least, will find itself in favourable conditions, and live to pass on its parents' genes. The other (from England's Bill Hamilton) says that short-term variation keeps parasites on the back foot. If all members of a population are the same, then any parasite that can attack one of them will be able to defeat all of them. If they are all different, then each individual poses new problems. There is direct evidence to support this idea. Certainly, modern crops grown as genetically identical monocultures are especially vulnerable to parasites.

For sex to work, each creature must find a mate. For animals this can be dangerous – many a lion and stag has died from battles with rivals; many a male spider has died in his chosen one's jaws. Trees, rooted to the spot, must find a way to transport pollen from male flower to female. The female flowers (or the female parts of hermaphrodite flowers) must in turn be able to capture pollen. Yet they must not be too promiscuous, and allow themselves to be entered by pollen of the wrong type. In general they reject pollen from other species. Many trees, including domestic plums and apples, also reject pollen from other trees that are too genetically similar to themselves: by such self-incompatibility they avoid the genetic perils of incest. Growers of plums and apples must grow varieties side by side that complement each other. Again, the tensions between males and females are a principal theme of literature, and always have been, running as strongly through the Bible and the Koran as anywhere else.

Yet the relationships between creatures of the same species – parents and offspring, friends and rivals, males and females – are only a part of life's complexities. Each creature must perforce interact with all the other species that share its environment, and especially for those that live in forests, the catalogue of cohabiting species is very long indeed. Sociology merges with ecology. It is not true (as Tennyson's line implies) that each individual or each species is inevitably in conflict with all the others. No species, to extend Donne's metaphor, can ever be an island. Cooperation is often the best survival tactic, as Darwin himself emphasized – and so it is that many pairs and groups of different species are locked in mutualistic relationships that are vital to all participants. Yet even here there is tension. Figs need fig-wasps, and fig-wasps need figs. Their interdependence is absolute. But as we will shortly see, no once-for-all peace treaty has been signed, or ever can be. The relationship is always liable to break down as each partner begins to take advantage of the other – and 'freeloaders' (a technical biological term) cash in on both. Machiavelli spelled out the intricacies of such relationships in *The Prince*. The themes of literature are indeed the themes of nature. Still, though, it is not true as has often been argued of late, that human beings need to reject their own biology in order to behave unselfishly, as moral beings. Cooperativeness and amity are at least as much a part of us as viciousness. The point is not to override our own nature, but simply to give the positives a chance.

So what does all this mean for trees?

To begin with, for a tree to reproduce sexually, it must transfer pollen from anthers to stigmas. In theory, a hermaphrodite flower might achieve this easily enough by self-fertilization, but in reality this is rare. Most wild trees are 'out-bred'. Pollen travels from the male flowers (or male parts of hermaphrodite flowers) of one tree, to the female flowers (or female flower parts) of another tree. Since trees do not move, they must employ couriers. In mangroves, water sometimes serves as the vehicle. In temperate forests, where any one tree is liable to be surrounded by others of its own species, and the weather in general tends to be breezy, the wind does the job. This is hit-and-miss, of course, and trees that do reproduce by wind tend to produce prodigious quantities of pollen. Flick a young male pine cone and the pollen swirls out like orange smoke. One early summer in Oxfordshire I watched two wood pigeons jostle for position in a birch tree. From a field away I could see the thick yellow puffs of pollen that they dislodged – illustrating, incidentally, that wind-pollination may be animal assisted.

In tropical forest, however, where any two trees of the same species may be half a kilometre apart, with thousands of trees of other species in between, it just will not do to scatter pollen literally to the four winds and hope for the best. In the tropics, only some trees of the open savannah or the Cerrado practise wind pollination – apart from a few like *Cecropia*, which grow only in forest clearings. Most tropical trees rely on animals to carry their pollen. This has led to some of the most spectacular examples of mutualism in all of nature.

ANIMALS AS GO-BETWEENS

Transmission of pollen by animals requires many layers of co-evolution. The flowers, both male and female, must have the shape and colour that the chosen pollinator will respond to, and they must be displayed appropriately. The pollinator in turn must be geared to the plant's signals. When bees are at work in a rose garden it all looks simple enough, but in a tropical forest the bee or wasp or fly or bird or bat must seek a particular glint of colour, shape or whiff of scent among a cacophony of colours, shapes and scents, as a million different organisms send a million different signals to their potential mates, allies, predators or prey; and there are many other smells besides to sow confusion, including those of general decay.

How is this possible? It's as if any of us could pick out the reedy squeak of the oboe when the Berlin Philharmonic was in full spate – but then of course we can; or at least, the conductor does. Some biologists have suggested that the ability of a wasp to detect its particular fig, or of a night-time moth to pick up the ultrasound pulse of a bat, implies an advanced ability to filter out all extraneous scents or sounds, as any of us can do (up to a point) when chatting at a cocktail party. But perhaps the truth is the other way around. Perhaps particular animals are geared *exclusively* to the particular sights, smells or sounds of the flowers they feed from or the predators they seek to avoid, and register nothing else. In the same kind of way we see light, and yet are not at all fazed by the radio waves and ultra-violet and cosmic rays, not to say the swarms of neutrinos, that assail us all the time. Our senses are simply not aware of them. No filtering is required.

We see, too, that co-evolution is an exercise in give and take. The tree must buy the animal's help. Many primitive plants, such as waterlilies and the trees of the custard-apple family, the Annonaceae, allow or encourage the pollinating insect to eat great chunks of the flower itself. Others offer custom-made especially attractive food, which commonly but not always takes the form of nectar, both sweet and nourishing. Sometimes they lure the pollinators with aromatic oils. Often the pollinating animals eat a lot of the pollen itself. In short, trees that seek insect help must pay a double price. First they must produce the pollen and ovules, and all the supporting apparatus of petals and sepals – but then they must make a surplus, to bribe the pollinators.

Some of the relationships between trees and their pollinators are somewhat loose: many trees solicit the help of several or many animal helpers, and many animals seem happy to pollinate many different trees – domestic honey bees, for example, are generalist pollinators. But often the relationship is specific. Often a particular plant is completely committed to one particular pollinator (a bee, a moth, a wasp, a hummingbird), while each pollinator depends absolutely upon the particular tree. Generalism spreads the options, but reduces precision. Specialization improves the accuracy, but also means that the fate of any one creature is linked absolutely to the fate of another. Lose the pollinator (for example through some over-enthusiastic attack with insecticide) and you lose the plants that depend on it.

Insects and birds are the chief animal pollinators and for them (unlike mammals, which tend to be colour-blind) colour is critical. They each have their preferences. Beetles were probably the world's first animal pollinators (they pollinated cycads long before flowering plants came on the scene), and beetles prefer white flowers. They ignore red. Bees too prefer white – although bees are perhaps most alert to ultra-violet, which we don't see at all: often the flowers that seem to us to be plain white turn out to be ultra-violet coloured (and intricately patterned – sometimes with a road map to the nectaries).

Red or purple flowers attract butterflies – and may repel all the insects that prefer white, including the potential freeloaders. Moths are closely related to butterflies and yet, like beetles and bees, they prefer white. The colour preferences of butterflies and moths are reflected in the related Amazonian trees *Hirtella* and *Coupeia*. (They are both in the family Chrysobalanaceae, in the Malpighiales.)

The many species of *Hirtella* are geared up perfectly to butterflies. They open by day; they are pink or purple (rarely white); they have hoods, which provide the butterflies with a place to land; they reward their pollinators with copious nectar; and they have only a few stamens (the organs that bear the pollen), which are neatly and widely spaced. Butterflies of many species, guided by sight as well as scent, land at leisure on the flowers of *Hirtella* and feed decorously, coating themselves in pollen as they do so. *Coupeia* puts its trust in moths – particularly big hawkmoths, which fly by night and hover like hummingbirds to feed. *Coupeia* flowers open at night, just a few at a time, and are always white. They have a great many stamens – from 10 to 300. They coat the hovering hawkmoth (and occasional hummingbird) with liberal quantities of pollen as it probes among the tangled stamens for the nectar – which *Coupeia* provides even more generously than *Hirtella*.

As a relative of the custard apple, the Amazonian tree *Annona sericea* is pretty primitive; and it is pollinated mainly by beetles, which as insects go are primitive too. But 'primitive' does not mean 'simple', or merely 'prototype'; and the degree of co-adaptation between *A. sericea* and its pollinators is extraordinary. To be sure, the flowers of *A. sericea* are simple: three fleshy petals that never fully open, grouped around a central conical knob which bears both the stigma (female) and the stamens (male).

At about seven o'clock in the evening – soon after dark in the tropics – the flowers begin to warm up, to about 6°C above ambient. You may well find this surprising: after all, we all learn at school that only mammals and birds are 'warm-blooded', able to raise their body temperatures just for the sake of it. In truth, though, many creatures can do this – probably including some dinosaurs, and certainly some modern insects and some sharks. Some flowers can do it too. The rise in temperature helps to intensify their odour. The flowers of *A. sericea* do not breathe out the sweet smell of violets and honeysuckle that entranced lovers in Shakespeare's comedies but (says Ghillean Prance) a perfume more 'like chloroform and ether'. But it serves its purpose. Beetles (of the particular kind known as chrysomelids), and also some flies, come flocking in.

The beetles squeeze their way past the fleshy petals to the cone of male and female organs within. This is a common device among insect-pollinated flowers: provide an obstacle, which only the desirable insects can overcome. The stigmas at this time are ready to receive pollen, but the anthers, which provide pollen, are still closed. Any pollen that the beetle has about its person may thus be transferred to the stigmas. But the beetle, at this stage, cannot obtain pollen from the same flower that it is pollinating; so there can be no self-fertilization. The beetles often stay to copulate within the flowers and as they mill about they transfer even more pollen.

When the beetle has transferred its pollen, the stigmas at the top of the central cone fall off. Then the anthers become erect and release their pollen, and so the beetles become coated in it. Then the stamens drop off. The beetles eat the bases of the petals and then the petals fall off. Then the beetles can escape (they could escape before the petals fall off but they generally do not) and fly to another flower – now carrying pollen from the flower they have just pollinated. The flies that may visit also serve as pollinators in passing, and may lay their eggs on the sepals, but the beetles are the main players. Note, in this account, that the flower is seriously damaged, not to say destroyed by the beetle: the flower sacrifices bits of itself to bribe the beetle. But so what? The flower is only a lure. Once pollination is effected, its job is done.

Flies are only bit players in the life of *Annona sericea*, but flies including midges are the prime pollinators of many a plant, including many a tree – and including the cocoa tree, *Theobroma cacao*. Again, the cocoa flower sacrifices part of itself – sterile parts of the flower – to encourage the midges: and again, the flower is organized in such a way that as the midge feeds it is brushed with pollen. The flowers are produced on the trunks and branches (this is 'cauliflory', so typical of tropical forest trees), and the midges breed mainly in the fruit pods which fall to the ground and decay. If the cocoa grower is too tidy, and clears away the pods, the cocoa loses its pollinators. Here, as in all of life, too much hygiene doesn't pay. Many flowers that are pollinated by flies smell horribly, incidentally, imitating the rotting flesh that flies prefer; and some of them heat up to make it worse. (Most famous is the world's biggest flower, *Rafflesia*, from Indonesia.)

But the best-known insect pollinators and probably the most important, are the bees. Some bees are very small. Some are extremely large, like the carpenter bees and bumble bees. Others are in between, like honey bees and the long-tongued ('euglossine') bees. Many are solitary, some live in small colonies like the bumble bee, and some in very large colonies like the honey bee. Many are generalist pollinators but some – especially among solitary species – are adapted to pollinate particular flowers, which in turn are highly adapted to them. Bees are strong fliers: studies in the Amazon in the early 1970s showed bees of the genus *Euplasia* returning to their nests when released from a distance of 23 kilometres. In the normal course of foraging they commonly fly many kilometres in a day, following a regular route from flowering tree to flowering tree according to the strategy known as 'traplining'.

Some trees attract many species of bee: one study in Costa Rica in the 1970s showed that one leguminous tree (*Andira inermis*) attracted seventy different kinds, from middle-sized long-tongued bees to big carpenter bees. By the same token, many species of bee visit many different species of plant. But a few – particularly solitary bees – do have very close, specific relationships with particular trees, requiring a great deal of co-evolution between the two.

If a bee, on any one foraging trip, visited many different kinds of plant indiscriminately, then it would not be much good as a pollinator. It wouldn't help England's wayside roses, for instance, if visiting bees flew off and distributed their pollen among the local clover. But it turns out that on any one trip, most bees (more than 60 per cent in one study in the Amazon) focus on only one plant species and very few (only 15 per cent) visit three or more species per trip – and this of course makes them far more efficient as pollinators. Perhaps this focus reflects an 'optimum foraging strategy': a method by which feeding efficiency is maximized. Optimum foraging strategy has been studied most closely in birds. For example, if a pigeon is given a lot of barley with a few peas, it ignores the peas altogether. Confronted with a lot a peas and a few grains of barley, it ignores the barley. It pays a forager to get its eye in. By focusing on whatever food source is commonest, and is known to be reliable, the pigeon does not have to waste time in wondering whether any one item is a pea, or a barley grain, or a pebble. By the same token, if a bee once establishes that roses are in bloom, it tends to stick to roses. Let others focus on clover.

Furthermore, other studies have shown that once back in the hive, colonial bees exchange pollen with each other – not deliberately, but just as they mill about. So a bee that picks up pollen 5 kilometres to the east of the hive, may pass some pollen to another that is foraging up to 5 kilometres to the west – and trees that are 10 kilometres apart may thereby find themselves exchanging pollen. Of course, at different times of year the generalist bees switch from one species to another, as each comes into bloom, and thus ensure a year-round supply (or season-round, in temperate latitudes). Then again, although the bees may be generalists, individual species of tree seem to adjust their flowering strategy to the needs of particular types. Thus, in the Amazon, some trees flower in a 'big bang' fashion – a brilliant show of flowers in a short season; which ensures that bees in general will notice them. Others, however, favour the 'steady state' approach. They produce only one or a few inflorescences per day over a long period – and this tends to attract the kinds of bees, like the carpenter bees, that habitually fly long distances, and follow the same kinds of routes every day. This, then, seems to be a particularly good strategy for trees that are very widely spaced; but of course it relies on the regular habits and industry of a few species of bee.

In the forests of Amazonia the pristine ecology has been much interrupted and to a large extent pre-empted this past few decades by bees imported from Africa: the so-called 'killer bees'. These are simply an African race of the familiar honey bee; and very good honey-makers they are too, favoured by many bee-keepers. They got into South America in the 1970s from a research laboratory in São Paulo in the south of Brazil, and spread at more than 200 kilometres per year. By 1982 they were already crossing Colombia, thousands of kilometres to the north of São Paulo. 'Killer' is well over the top, but they are certainly aggressive both to people and to other insects. Thus in 1973 Ghillean Prance, near Manaus, observed the insects that came to pollinate *Couroupita subsessilis*, a relative of the Brazil nut. The visitors included wasps and bees. The only visitor the following year was the honey bee; not necessarily, but very probably, the African interloper. Presumably what Professor Prance saw in Manaus is common all over South America. Presumably, too, the African bees will sometimes do a good job; but in general, it seems unlikely that any one generalist, however aggressive, can pollinate the trees of the neotropics as efficiently as the droves of insects that have evolved specifically to the task.

Yet the Brazil nut itself might well be a beneficiary. Brazil nut trees are wonderful. They are emergent species, half as tall again as most canopy trees. Of course, too, their nuts are extremely valuable and so the Brazil nut is among a shortlist of Amazonian trees that it is forbidden to fell. So it is that when forest is cleared, the pastureland that is sown or grows up in its wake is punctuated by isolated Brazil nut trees, rather like the big solitary oaks in England's stately parks, although the parkland oak trees are to the manner born and spread themselves most opulently while the Brazil nut trees, deprived in middle age of their companions, seem forlorn, magnificent but somewhat haggard. Furthermore, although the Brazil nut trees have been conserved mainly because of their nuts, when they are in the middle of nowhere they are liable to remain unfertilized. Most of their pollinators won't fly over big open spaces. But isolated oaks in English parks, served by wind that's laden with pollen from surrounding woods, have no comparable problem.

Enter, though, the African bee. Its aggression is matched by its energy, and it does apparently make the long trek out to the isolated Brazil nut trees. Introduced species on the whole are a bad thing – in fact, 'exotics' seem to be the chief cause of extinction of native species, apart of course from gross loss of habitat. But here at least, just for once, there is some compensation. 'It's an ill wind', as the adage has it.

Birds are great pollinators too – notably, but by no means exclusively, hummingbirds. Like butterflies, they typically prefer red, which they see best. Mammals also. Some, like the honey possums in Australia or the desert rats in South Africa, are highly specific. Others, like the capuchin monkeys of Amazonia or giraffes in Africa, may pollinate their prey trees inadvertently (but nonetheless usefully) as they forage. But among the mammals, as with all creatures, the most efficient pollinators are the fliers: and with mammals that means bats.

There are well over 800 species of bats – among mammals, only rodents have more species. They are of two main kinds. The microbats live all over the world. Many – like the familiar bats of temperate lands – eat insects; but others live on mammalian blood (the vampires) or on frogs, or nectar and fruit. The mega-bats, which include the flying foxes, live exclusively in the Old World. For the most part they are big and live largely or exclusively on nectar and/or fruit. They rely heavily on scent to find their food, as most mammals do. By contrast,

Trees (including many cacti) hold their flowers high for bat pollination

the micro-bats locate their prey, whatever it may be, by echo-location: sending out a high-pitched squeak and analysing the echo. Both groups of bats are nocturnal. If they fly by day – as they sometimes are obliged to do, particularly in cold weather when there are too few insects flying at night – they are quickly picked off by hawks. Some studies have shown that day-flying bats rarely survive for more than a few hours. Mammals and birds both flourished in the wake of the dinosaurs and flying reptiles. But while the mammals came to dominate the ground, birds very definitely dominate the air. Bats are very successful – ubiquitous and various – but they are advised to stick to the hours of darkness. Even then, they must avoid the owls.

Trees that would be pollinated by bats must adapt to them – just as they must to specialist bees, butterflies or moths. None has adapted to them more spectacularly than a relative of the mimosa:

the Amazonian tree *Parkia*. *Parkia* has been studied in particular by Dr Helen Hopkins but it is everybody's favourite, like a favoured sister, endlessly endowed. It is beautifully shaped, like a great flat-topped umbrella. Its leaves are doubly compound, like feathers. Its bark is smooth but not too smooth, in some species dotted with red. But most glorious of all are the flowers, pompoms of stamens and styles, bright red or pale yellow or bronze depending on species, that hang from on high on long thin threads like Christmas baubles.

The topmost flowers in each inflorescence are sterile, with copious nectar, which begins to flow early in the morning. By mid morning the first flowers begin to open, and all of them are open by mid afternoon. At dusk, the flowers begin to release their pollen. Then the bats come. By morning all the nectar has gone, the filaments that bear the anthers have wilted, and the flowers have faded. One of nature's most glorious shows lasts only one night. Showbiz is not the point. Replication and multiplication is the point. One night is enough: as we have seen already, animal pollination can be remarkably efficient. The would-be pollinators are constantly alert.

So there are many specific, mutualistic relationships between many different kinds of tree and many different kinds of animal pollinators. But the most stunning relationships of all are between the world's figs and their pollinating fig-wasps. Here we see, writ large, all that is involved: the exquisite precision of co-evolution; the sacrifice that the figs must make to entice the wasps; the constant temptation on both sides to renege; and the inevitable imposition of freeloaders (including parasites).

OF FIGS AND WASPS

All organisms are different but some, as George Orwell might have said, are more different than others. There is nothing quite like an octopus. There is nothing quite like a human being. And there is nothing, absolutely nothing, like a fig. Figs, as described earlier, are of two main kinds: the kind that grow as other trees do, when planted in the ground; and the kind that begin as epiphytes. The epiphytes in turn may be stranglers, encircling and throttling their hosts, or non-stranglers which, like banyans, begin aloft but contrive to make contact with the ground without crushing their hosts. (In the Solomon Islands, too, a few kinds of fig grow as climbers.) In all kinds, the young fruits take the form of 'syconia' (not 'synconia' as often printed). They are like fleshy cups, though almost closed at the top. The flowers, often hundreds of them, are borne inside the cups, facing inwards. The effect is womb-like, as D. H. Lawrence was not slow to observe. Each female flower contains only one ovule which, when the flower is pollinated, develops into a seed. If the flowers are not pollinated, then the entire fruit aborts – or that at least is what usually happens; for as we will see, there are twists. (Strictly speaking, a syconium does not become a fruit until the seeds are fertilized. But it is convenient to use the term 'fruit' to mean syconium.)

The flowers inside their syconia are pollinated by small black fig-wasps. A female fig-wasp enters the syconium through the hole at the top, and lays her eggs in some (up to half) of the flowers. A wasp that colonizes a fruit establishes a new generation of wasps; and so she is called a 'foundress'. After she has laid her eggs, she dies. The fruit becomes her sepulchre. Each egg hatches into a larva, which feeds upon the developing seed within its allotted flower (which of course kills the seed). Then the larvae pupate, each one still in its flower, until they emerge as adults. At least half, and often more than 95 per cent, of those new young wasps will be females; less than half, and often fewer than 5 per cent, will be male. The males are wingless. They emerge before the females, chewing their way out of their respective seeds. Then they chew their way into seeds containing females, and mate with them while the females are still inside. Then they too die. The newly-emerged females, already with sperm on board, now fossick around inside the syconium, picking up pollen from the male flowers, although they do not apparently feed on it. Then they fly off to found a new generation in a new syconium. These young females are laden with pollen and so they fertilize the flowers of the next syconium.

Some species of fig-wasp are 'active' pollinators, and some merely 'passive'. The active pollinators have specific adaptations that include special sacs for carrying pollen, which they conscientiously fill; and when they reach the new fig tree, they equally conscientiously place the pollen on the stigmas of the female flowers within the syconium. The passive pollinators merely mill about before they leave the syconium of their birth and so become covered in pollen; and scatter it equally haphazardly in the next fruit they visit in the next fig tree, in which they lay their own eggs. However, some apparently passive pollinators also have specific pollen sacs – suggesting that they are descended from ancestors that were more meticulous.

The syconium (fruit) of a fig is both womb and sepulchre

The net result is a wonderful mutualism. The figs benefit, because they get their flowers pollinated, and so are able to set seed. The wasps benefit, because they are supplied with a good, safe place to lay their eggs, and a food supply for their larvae. Without the wasps, the figs could not reproduce, and so would die out; and without the figs, the wasps could neither reproduce nor feed. Although the wasps are small they fly prodigious distances, so any one fig tree may spread its pollen over 100 square kilometres. Indeed, in the gallery forests of Africa it seems that individual wasps may carry the pollen from some trees hundreds of kilometres: presumably the wasp rises into the upper air currents, and is whisked along (though how it knows when to descend again, and how it does so, is not at all obvious). This has been discovered not by studying wasps directly, but by DNA studies of the figs, which reveal that any one fig may have daughters living a very long way away.

Figs are immensely successful. More than 750 species are known –
more even than oaks or eucalyptus – and they thrive throughout the
tropics, both Old World and New, and into the subtropics. Extraordinarily – indeed it all starts to seem miraculous – *each* kind of fig has
its own species of symbiotic wasp. Each wasp co-evolved with the fig
it pollinates.

This, at least, is the basic story. It was beginning to unfold even by
the 1940s, and has been steadily expanded since by generations of
biologists. The research has been outstanding. It is difficult to do, but
figs are of tremendous ecological importance – the sole or the stand-by
food supply for hosts of forest creatures; and the study of mutualisms,
as Darwin himself first noted, is a rich source of evolutionary hypotheses which also offers the means to test them. Modern techniques,
including molecular methods that enable researchers to explore the
evolutionary and genetic relationships between different figs and different wasps, are enriching the insights month by month. The following account is based mainly on the extraordinary studies over the past
two decades by Dr Edward Allen Herre and his colleagues at the
Smithsonian Tropical Research Institute (STRI) in Panama (which I
was privileged to visit in 2003).

To begin at the chronological beginning: How did the precise
relationship between fig-wasps and figs – basically one fig per wasp,
with hundreds of species of each – first evolve? It is impossible to
reconstruct history with certainty, but modern investigations are
providing many intriguing insights.

First, genetic and anatomical studies show that all the wasps that
pollinate figs are descended from a common ancestor. Comparable
studies suggest that all modern figs are also descended from a common
ancester. The two facts together imply that the present-day mutualism
of figs and their pollinating wasps *evolved only once*. All the 750
species of fig and their corresponding wasps are descended from one
kind of fig and one kind of wasp that first appeared a very long time
ago. Furthermore, as each lineage of figs divided to form two daughter
species, so the wasps that pollinated them also divided into two new
species: a new kind of wasp for every new kind of fig.

What was the first of these wasps like? Genetic studies suggest that
of all the several genera of present-day fig-wasps, the most ancient is

Tetrapus. This is a passive pollinator, and its various species pollinate a group of closely-related figs that live in South America. Perhaps, then, the first ever fig–wasp mutualism was established between the ancestor of those South American figs and a *Tetrapus*-like wasp; and presumably it was first established in South America, where the most ancient kinds of fig-wasps, and their corresponding figs, still reside.

When did this fateful liaison take place? A few years ago such a question would have led only to the waving of arms, but modern studies of DNA provide at least the outline of an answer. For DNA changes little by little over time – not simply because of pressures from natural selection, which cause it to evolve in quite new ways; but spontaneously, simply because, as it is copied and re-copied, generation by generation, mistakes creep in. The point of DNA is to provide the code on which proteins are based. But in plants and animals (and fungi and protozoans and seaweeds) *most* of the DNA does *not* seem to code for protein. Indeed it seems to be more or less functionless, or at least its function is unknown. Small mistakes in the 'non-coding' sections of DNA seem to make little or no difference to the life of the creature, and natural selection fails to weed them out. Thus over time such mistakes accumulate. They do so, furthermore, at a fairly regular rate – so that by measuring the difference between the non-coding bits of DNA of two different creatures, it is possible to see when they last shared a common ancestor. The (fairly) steady change of DNA over time is sometimes said to provide a 'molecular clock'.

The differences observed in the DNA of existing fig-wasps suggest that they last shared a common ancestor around 90 million years ago. The same is true for existing figs. So that, presumably, is when the ancestor of today's fig-wasps (who presumably up to then had been making a living in some other way) began to lay its eggs in the developing seeds of the figs' first ancestor. Ninety million years ago was when the last wave of dinosaurs and the great marine and flying reptiles were still in their prime.

So figs and wasps have depended on each other, and have co-evolved in step – each new kind of fig accompanied by a corresponding, new kind of wasp – for tens of millions of years. Clearly the partnership works very well: both figs and wasps benefit. Clearly (we might suppose) it is in the interests both of fig and wasp to keep the alliance intact. Thus biologists traditionally argued that in such relationships as this, natural selection must favour true and stable mutualism. Peaceable, not to say amiable mutualism has often been portrayed as the natural end-state that would inevitably come about sooner or later. But modern theory suggests that it could be in the short-term interests of either party to cheat; and since natural selection does not look ahead, we might expect that short-term betrayal would indeed take place. Modern studies by Dr Herre's group and others elsewhere now show that in reality the relationship between figs and wasps has often been reneged upon and otherwise flouted, in various ways and to varying degrees. So biologists must now ask how it is that wasp and fig have served each other so well for so long *even though* it would pay the wasps to cheat (and the figs too, although probably to a lesser extent).

The relationship between the two can be analysed in the vocabulary of game theory, or indeed of cost–benefit analysis. The bottom line is that the figs and the wasps must each pay a price for the services of the other, but neither can afford to pay too much. The figs sacrifice a lot of their would-be seeds – up to half of the possible seeds in each syconium feed the young wasp larvae. The wasps, on the other hand, seem to exercise restraint – for if they lay their eggs in half the seeds, then why not in all of them? To be sure, this would kill off the figs, the geese that lay the golden eggs. But such things happen in nature – precisely because natural selection does not look ahead, and the long term is sacrificed to the here and now. Alliances of wild creatures are known to break down sometimes, and have led to mutual extinction. In the same way, human societies have often broken down as agreements are reneged upon, to the short-term advantage of some but sometimes to the total destruction of all. Why, then – since it would apparently pay them to cheat – do the wasps exercise restraint?

Just to stir the pot a little more: wasps benefit figs by dispersing their pollen – and as we have seen, the active pollinators are especially adapted to do so. But why? What's in it for the wasp? In the long term they benefit of course – fertilizing seed to provide a future genera- tion of figs. But since natural selection cannot consider the long term we always have to ask, what is/was the short-term advantage in any particular mode of behaviour? In this case, why does it benefit a pollinating wasp to play an honest game?

In truth, the fig could well have evolved mechanisms to prevent cheating. Perhaps the seeds in any one syconium are not all edible. Plants are remarkable chemists; and virtually all organisms are capable of some degree of polymorphism – producing some offspring of one kind, and some of another. Figs could well produce some would-be seeds in a tasty form, which serve as sacrificial offerings to the essential wasps, and others that the wasps find foul, and leave alone. There is some preliminary evidence to suggest that something like this is happening, but the picture is not yet clear.

Alternatively, the female wasp that enters a new syconium may simply be unable to lay enough eggs to take over all of the flowers. This is eminently possible. As we will see, however, whereas some fruits in some fig species are generally colonized by only a single female wasp, in others – generally the larger fruits – several different wasps invade. Between them, we might suppose, they could colonize all the flowers. So some form of positive repellent does seem likely, to stop them doing so.

But why (in the short term) should the wasps – the active pollinators in particular – go to such lengths to pollinate the fruits they invade? Why not simply pinch the ovules in the syconia they colonize without bothering to fertilize the ones they leave alone? The answer seems to be that any syconia whose ovules remain unfertilized are aborted. No pollination, no nourishment. There is, as the adage has it, no such thing as a free lunch. In short, the fig seems to have rigged the game so that the wasp *does* gain short-term benefit from playing honestly, and would be punished even in the short term if it cheated.

Such mechanisms seem to have kept the whole system on course. Yet (so the Smithsonian studies have shown) it clearly is not quite so stable as has been supposed. Thus it looked for a long time as if the simplest rule applies: each species of fig has its own particular species of wasp; each wasp is adapted to only one kind of fig. But modern techniques enable biologists to explore the relationships between different creatures by examining their DNA. DNA really is a most obliging molecule. Some bits of it – notably some apparently non-functional bits known as 'micro-satellites' – change so rapidly that differences between them reveal relationships even within families: who is whose sibling, or parent, or offspring. (These are the kinds of studies used in legal paternity cases.) Other bits of DNA change more slowly, and significant differences between these bits indicate that different individuals belong to different species. This of course is especially useful when two or more different species look very similar. Thus DNA studies in recent years have revealed that various populations of owls, mice, monkeys and bats that were each thought to represent just one species, sometimes should be divided into two or more. When species can't be told apart except by their DNA they are called 'cryptic' species.

By DNA studies, the Smithsonian biologists have shown that fig-wasps include many cryptic species. They even found cryptic species within the ancient wasp genus *Tetrapus* – suggesting that rule which says 'one fig: one wasp' has long been 'routinely violated', as Allen Herre puts the matter. We might reasonably guess that when two species of wasp pollinate the same kind of fig, this is simply because some ancestral species of wasp, which served that particular fig, had split to form two species. But DNA studies show that sometimes two different kinds of wasp sharing one fig are *not* close relatives. This means that one of the two must have come from some other kind of fig. As seems to follow, it also turns out that some kinds of wasp colonize more than one kind of fig. So fig-wasps can be more like honey-bees than had been supposed. When two kinds of wasp meet in one kind of fig, they might in theory hybridize – and DNA studies show that this sometimes happens; although the hybrids do not themselves seem to spread. When one kind of wasp pollinates more than one kind of fig, the figs could be hybridized – and as we have already seen in willows, hawthorns, poplars and many others, including the London plane, hybrid plants of all kinds can and do evolve into new species (and perhaps this is one reason why there are now so many different species of fig).

In the Smithsonian's Panama studies, whenever more than one kind of wasp colonized the same syconia, both wasps may serve as perfectly good pollinators. But this is not always so. A study in Africa has shown that at least in one case, one of the cryptic wasps that colonizes one particular species of fig tree behaves simply as a parasite. It lays its eggs in the flowers and so feeds its young, but it does no pollinating. It is a cheat: an archetypal freeloader. Game theory predicts that freeloader fig-wasps might exist – it's a possible niche – and so they do. Again we see that the simple one-to-one relationship between figs and wasps, evolved over millions of years into perfect mutualism, is not quite so cosy as it has seemed. The relationship, like all mutualisms, is dynamic. It is always prone to decay.

Often, too, any given fruit may be colonized by more than one foundress from the same species. This raises another set of complications – complications which again have been predicted by modern evolutionary theory, and which again (very satisfyingly) have now turned out to be what actually happens.

Let me refer you back (as Perry Mason would say) to an earlier comment: that the proportion of young male wasps born within a given syconium varies from around 5 per cent (one in 20) to around 50 per cent. The preliminary point is that many creatures can adjust the sex ratio of their offspring. Humans cannot do this, but we do not shine at everything. Such adjustment is especially easy for wasps (and bees and ants) because in these insects, the females all develop from fertilized eggs, while the males develop parthenogenetically, from unfertilized eggs. The mother wasp (or bee or ant) keeps the sperm separate from her eggs until the time comes for laying, so she can decide in the light of circumstance whether or not to fertilize them before laying. Again the mechanism seems so subtle that it beggars belief, and yet it is the case.

Theory predicts that when only one foundress colonizes a particular fruit, the ratio of males should be low: one in twenty rather than one in two. But the more foundresses there are, the more we would expect the ratio of males to increase. The point is that each female strives to pass on as high a proportion of her own genes as possible. If all the larvae within any particular fruit are her own offspring, then all her daughters are bound to be mated by her own sons (if they are mated at all). The shortcomings of incest and inbreeding are apparently outweighed by the advantage that the mother thereby passes on her genes both via her sons and via her daughters. So she needs only enough sons to ensure that her daughters are all fertilized – and one son per twenty daughters seems enough. It is good, of course, to focus on daughters because they are the ones that lay the eggs that supply the grandchildren generation.

But if there is more than one foundress per fruit, then the young males find themselves with rivals who are not simply their brothers (who would be genetically very similar) but come from a different lineage (albeit of the same species). In such circumstances, we might suppose that it could pay a female to produce sons exclusively – provided those sons are big and tough enough to mate all the daughters of all the other foundresses. The theory shows, however, that it never pays to produce more sons than daughters. A 50:50 ratio of sons and daughters is the maximum. Again, natural history supports the theory. Dr Herre and his colleagues have shown that as the number of foundresses per fruit rises to about six, so the proportion of males rises to around 50 per cent.

But here comes another twist. The males have one function only: to mate the females. Apart from that, they are a dead loss – both from the wasp's point of view, and from the fig's. After all, the fig has to sacrifice a seed for every young wasp that is born. The youngsters that matter to the fig are the females, which fly off to pollinate other figs. As far as the fig is concerned, the fewer males, the better. This, in turn, implies that figs should encourage wasps to enter their syconia one at a time: that they should evolve some limitation on access (and there are many comparable examples in nature). As things are, small syconia are much less likely to attract multiple foundresses than are large syconia – so we would expect natural selection to favour small syconia. In reality, while some fig species do have small syconia, others have larger ones. So why does natural selection ever favour big syconia? This question will be raised twice more as this narrative unfolds, in two quite different contexts. Whichever way you look at it, big syconia seem like bad news. Yet there is an answer, which will be provided later in this chapter. Patience, gentle reader.

As if the game between figs and wasps were not convoluted enough, there enters now a third set of players: parasitic nematode worms.

ENTER THE NEMATODES

It has been suggested that every species of creature on earth above a certain size has its own specialized nematode parasite; and if this were so, it would mean that the total number of species on earth is equal to the number of non-nematodes, times two. Whether this is so or not, it does seem that every species of pollinator wasp *does* have its very own species of nematode parasite. All the nematode parasites of Panamanian fig-wasps belong to the same genus: *Parasitodiplogaster*. Since STRI where Dr Herre works is based in Panama, this is the genus they have studied most.

The life-cycle of *Parasitodiplogaster* nematodes is superimposed on that of the wasps they attack. Not all figs are infested with nematodes, but in those that are, the nematodes will have reached the immature, dispersal stage by the time the young female wasps are emerging from their flowers. The worms then enter the wasp's body cavity, and begin to consume it from within. Their efforts are not immediately fatal, however, and so their host carries them on to another syconium. When the infested wasp finally dies, generally after laying her eggs in the next syconium, up to twenty or even more adult nematodes crawl from her body, then mate, then lay their eggs. The young nematodes hatch before the young wasps emerge – and so they are ready to invade the young wasps and begin the cycle afresh.

In general, the relationship between parasite and host is as delicate as that between partners in a mutualistic relationship. The aim of the parasite is to grow and reproduce, and for this it must feed upon its host. If it feeds too vigorously, it is liable to kill the host. If it is too decorous in its approach, it loses out to rival parasites who are more vigorous and so breed more quickly. In general, then, it pays parasites to be as vigorous – 'virulent' – as possible, but without overdoing it.

Now a further twist. Theory predicts that if nematodes infest wasps that occupy fruits on their own – one wasp per syconium – then they should be less virulent. After all, if they are too virulent, and kill their host wasps, they have no chance at all of being transmitted to a new fruit to lay eggs of their own. But if the nematodes attack wasps that invade fruits more than one at a time, they can afford to be more virulent. It doesn't matter too much if some of the young host wasps are killed off, since there are liable to be others which are not killed, and will carry the nematodes to pastures new. The Smithsonian scientists found that this prediction stands up. Wasps that invade fruits singly generally manage to fly off to new fruits even when they are infested with nematodes. But in fruits that entertained more than one foundress, a proportion of infested foundresses perish before they leave the syconium of their birth.

Nematodes are clearly bad news for the wasps; and particularly virulent nematodes are bad for the figs, too. After all, the fig has to sacrifice one of its would-be seeds for every wasp that is produced, and the sacrifice is wasted if the wasp then dies from nematode attack. Again, then, it seems that figs would be better off producing syconia that attract only one foundress. Again, small-sized syconia seem advantageous – because, in general, the bigger the syconium, the more foundresses it is liable to attract. So the question is prompted again: why do some figs continue to produce large syconia?

Finally, neither the fig-wasps nor the nematodes have the field to themselves. A whole number of other creatures – including other kinds of wasp – also feed on the syconia, though purely as parasites; offering nothing in return by way of pollination. Particularly intriguing is a group of wasps in the New World that thrust their ovipositors into the syconium from the outside, and lay their eggs in the outer flesh or in the flowers within. The syconium responds by producing a gall – a mass of tissue that may be seen as a defensive response but also provides a very convenient home for the developing wasps. Galls are common in plants, and particularly in trees which live a long time and are attacked many times. A syconium attacked by a gall wasp fails. Its flowers do not get pollinated.

Yet as we have seen, figs normally abort syconia that fail to become fertilized, so the syconia with galls should simply be shed. But they are not. Evidently the gall wasp stops this happening. The fig mobilizes hormones that cause changes in the stalks of failed syconia, which cause them to fall off – but the wasp apparently produces chemical agents of its own to subvert or otherwise block the fig's hormonal signal. The wasp, like a hacker, has cracked the fig's protective code. Here is a prime example of the kind of 'arms race' (another technical term in biology) that we commonly see in nature as predators and prey, hosts and parasites, evolve in each other's presence. But – as with the figs and the pollinator wasps – we can be sure that the race is not over yet, and indeed will never be over. We should come back in a million years or so and see if the figs have thought of a new way to overcome the gall wasps' knavishness. Probably not, for various reasons. But it would not be surprising if, at some time in the future, gall wasps had to up the ante once more. Incidentally, the gall wasps are in turn parasitized by smaller wasps. Jonathan Swift's observation that fleas 'have smaller fleas to bite 'em,/And so proceed *ad infinitum*' seems to apply just as cogently to wasps.

There is more.

COOL FIGS AND HOT FIGS

Although figs surely have no love for gall wasps, they do go to great lengths to protect the vital pollinator wasps. In particular – as again revealed by the Smithsonian studies – they maintain a temperature within the syconia that allows the young wasps within to develop.

As a preliminary observation, the scientists showed that when the temperature is only 5° to 10° C higher than the ambient temperature at midday, the pollinator wasps of Panama (or at least two species of them) are incapacitated or die. But, say Dr Herre and his colleagues in a paper published in 1994, 'Such lethal temperatures would be expected in objects exposed to full sunlight.'[1] And 'such objects' include the syconia of figs, hanging on their trees. So they measured the temperature inside syconia – and found that they stayed more or less as ambient: still comfortable for the young wasps developing within them. Even on the fiercest days, the temperature within small syconia never rose above 32°C, which wasps find perfectly acceptable.

Yet there was a greater oddity. For although the physical theory is complicated, it suggests that small fruits should find it easier to stay cool than large fruits do. But in fact, the larger fruits were often even cooler than the small ones. So how do figs in general stay cool? And how is it that the large ones – apparently in defiance of physics – tend to be the coolest of all?

Perhaps, the scientists surmised, the large fruits cooled themselves by evaporation, as leaves do, or as mammals do when they sweat. Evaporation would be effected, as in leaves, via holes (stomata) in the syconium surface. To test this idea, the scientists simply covered the figs in grease, to block the stomata. Sure enough, the temperature inside the big syconia then rose by about 8°C. When the outside temperature was at 29°C (which is common enough), the temperature inside the big greasy fruits rose to around 37°C – hot enough to kill the wasps inside within about two hours. Small fruits do not need such refinements. They have no stomata, or very few.

So the big fruits can keep themselves cool – but only at a cost. They have to waste a considerable amount of water to do so. We have already seen two reasons why small syconia seem preferable to large ones. There are more male wasps in the big syconia (because there are more foundresses), which is wasteful. The nematodes are more virulent in the big fruits (because there are more foundresses), which is wasteful again. Now, to cap it all, the big fruits have to waste water, a precious commodity, just to keep themselves cool. So why do any figs have big fruits? How could natural selection have favoured such an apparent absurdity?

I must delay the answer still further. It lies under the heading of seed dispersal, the generalities of which we should look at first.

SCATTERING OF SEED

Many plants, temperate and tropical, rely on animals to disperse their seeds. As with the pollinators, the relationship is mutualistic, with give and take on both sides. The tree gets its seeds dispersed, to be sure. But animals cannot afford to run charities, and they must have their quid pro quo. Sometimes they expect to eat a proportion of the seeds, and so squirrels typically eat at least as many acorns as they scatter. When trees produce fleshy fruits, animals may simply consume the pulp and then either spit out the seeds (as monkeys may often be seen to do with machine-gun efficiency) or else allow them to pass through their guts (whereupon they are deposited with their own consignment of fertilizer).

Always, though, and inevitably, there is tension. If a particular tree evolves to become dependent on a particular disperser, and the disperser disappears, then the tree might disappear with it. Thus many a seed seems simply to languish in tropical forests – though perhaps in the past dispersed by long-gone dinosaurs or some extinct giant mammal. On the other hand, if the dispersers become too common then they may eat too many of the seeds, and then the tree is also liable to die out. Balance is all. Many thousands of examples could be cited, but a couple must suffice.

The first is the almendro tree, *Dipteryx panamensis*, from the Fabaceae family, which grows on Barro Colorado Island in the heart of Panama, which for many years has been studied by scientists of the Smithsonian Tropical Research Institute. Egbert Leigh, who has worked on the island for the past thirty years, introduced me to the almendro one very rainy morning. It is indeed lovely, with bark the colour of pale pink salmon and a trunk that forks and forks again to produce, says Dr Leigh, 'a graceful, somewhat hemispherical crown of compound leaves spiralled around its twigs'.

There is only about one almendro per hectare on Barro Colorado: but that is a fairly typical number for a tropical forest tree. Come June and July, it produces fine bunches of pink flowers at the ends of its twigs. These are apparently triggered by the onset of the rains, in late April and early May. Certainly if the start of the rainy season is not clearly marked – if, for example, the previous dry season is not as dry as it should be – then the almendro produces far fewer flowers, and so far fewer fruits. This is bad news for Barro Colorado's animals, for the almendro is a serious food tree. Thus small quirks of weather can have far-reaching effects. As global warming continues to bite, we can expect the weather to become quirkier and quirkier.

The fruits, as befits a legume, are produced in pods: a hard wooden pod covered in a thin layer of sweet green pulp, with a single big seed inside; twenty or more fruits per square metre of crown in a good year. This is a prodigious crop and, says Dr Leigh, 'swarms of animals flock to the feast'. Some take the fruit directly from the trees. These include some carnivores like the kinkajou and coati (many carnivores are omnivorous – notably bears), and also monkeys, bats and squirrels. Some take the fruits from the ground, including agoutis and pacas, which are big relatives of the guinea pig (and resemble small antelope or deer), spiny rats (also related to guinea pigs, rather than to rats), peccaries (New World pigs), and the occasional tapir. Many of these feasters simply eat the sweet pulp around the wooden pods. But some – notably peccaries, squirrels, spiny rats and agoutis – gnaw through the hard casing as well, to the bean inside.

Most of the feasters are bad news for the almendro: they eat, but they do not disperse. Squirrels eat but are poor dispersers. Monkeys can be useful: sometimes they eat the fruit where they find it, but sometimes they carry it away from the tree. A young almendro, as is commonly the way with tropical forest trees, will not grow close to its parent. Wide dispersal is necessary. For this, the most important disperser by far is Barro Colorado's largest fruit-eating bat (although it still weighs only 70 grams). Fruit bats do not hang around on fruiting trees. If they did, they would be picked off by the meat-eating predators that also lurk in trees (waiting for the fruit predators), or by owls. Instead bats carry the fruit some distance away to a quiet roost where, says Dr Leigh, 'they can chew off the pulp in peace'.

But mere dispersal is not enough. The seeds of the almendro also have to be planted. Bats do no planting. But when they drop the pods (they are interested only in the pulp around the outside), these are found by agoutis, which eat some of them with the seeds inside, but also – like temperate squirrels with acorns – bury some against leaner times. In some years almendros bear fruit while other trees bear very little, and then the agoutis eat all the almendro fruit. If other fruit are available, then some almendro fruit pull through.

Clearly, this process of seed dispersal is extremely chancy. The fruits and seeds of the almendro must first run the gauntlet of a whole range of animals, most of which simply gobble them up. Eventual success depends on the good offices of two very different kinds of animal: the fruit bat in the air, and the agouti on the ground. The bat does not eat the seeds themselves, and so is a reasonably safe ally. But the agouti does eat the seeds and is useful to the tree only because it sometimes fails to eat all of them. Partly this may be because it is simply forgetful (although it is always likely to find the young germinating almendros). Agoutis may fail to recover all their buried booty too because, between the burial and exhumation, they are themselves eaten, notably by ocelots, the mid-sized spotted cats of South America. But the almendro also contrives to satiate the agoutis – to produce more seed in a given year than the agouti ever gets around to eating. Big crops matter. This is why the almendro and other such trees *need* to produce good crops. A poor crop (caused by quirky weather) means total wipe-out for the particular year.

But, says Dr Leigh, the almendro has not been replacing itself on Barro Colorado. There are very few young trees. Perhaps, he says, this is because the island has too few ocelots, and so has too many agoutis: too much of a good thing. So perhaps we should say that the safe dispersal of almendro seeds requires three kinds of animal – fruit bats, agoutis, and cats to keep the agoutis in check. It seems a very precarious existence. But up to now it has clearly worked, or there would be no almendros – and this year (2004) there are some saplings since such ocelots as there are have apparently reduced the agoutis.

Clearly, though, overall diversity is necessary for the survival of any one species. The elusive concept of natural balance matters. Trees are not adapted simply to the presence of particular animals. They are adapted to their whole environment – climate, flora and fauna. But among the whole, they are particularly reliant on particular allies.

My second tale is an anecdote from a different continent – but it again shows how the fate of trees depends so much on the caprices of environment, and (increasingly) on the whims of human beings.

At the magnificent Forestry Research Institute in Dehra Dun, near the foothills of the Himalayas, Dr Sas Biswas likes to show his students an impressive row of *Chukrasia velutina* that form one side of an avenue along one of the main streets across the institute's huge campus. *Chukrasia* is a relative of the mahogany, and these trees grow tall and straight – and, along this roadside, they are perfectly evenly spaced. The question he puts to the students – and to me – is, 'Who do you think planted them?' All who are asked pluck various plausible bigwigs out of the air while Dr Biswas looks on with mounting glee. Pandit Nehru? Gandhi himself? Some passing British royal? Finally when the students run out of steam he reveals the answer: 'Ants!'

How can it be? It is easy to see how ants might help to plant a tree. They could carry the seeds to their nests if the seeds are small enough. Those of *Chukrasia* are only a couple of millimetres long and ants are prodigiously strong. But how could they space the seeds so neatly? Colonies of ants are often compared to armies, yet they receive no military training. They do not naturally distribute themselves so evenly, or in lines as straight as the cavalry's tents at Balaclava.

On the sites where the trees now stand, there once were small beds of the white-flowered *Tabarnaemontana coesnana*, a relative of the oleander, planted for decoration in brick containers and regularly spaced. Sas Biswas remembers them from the 1970s as he rode past them every morning on his bicycle (people tend to stay a long time at the FRI). This plant repels most insects. Only the ants have learned to live with it. So they did carry seeds to the beds of flowers – from an old, big, mother tree that's still growing on the other side of the road – and the seeds that the ants didn't eat themselves escaped the attentions of other insects too. So the surviving seeds germinated – one or two within each small bed. And so, within the working life of Dr Biswas, they have sprung up – 'Before my eyes!' he says, with a huge smile.

But for one of the most intricate stories of seed dispersal we must return to the figs.

WHY SOME FIGS HAVE BIG FRUITS DESPITE EVERYTHING: THE MYSTERY SOLVED

The syconia of figs qualify as bona fide fruits when the seeds within are mature and ready for scattering. Many animals come to prey upon them. Many, like monkeys, may do some dispersing at the same time, but on the whole are destructive. The figs' main allies at this stage of its life, dispersing its seeds without taking more than their fair share, are various birds, and fruit-eating bats.

Each species of fig produces either red fruit, to attract birds, or green fruit, to attract bats. Fruit-eating birds hunt by day and rely on vision, and bright red does the trick. Among the birds, manakins are the main specialist fruit-eaters, while tanagers, tyrant flycatchers and woodpeckers are opportunist feeders, taking what's on offer but happy to eat other things as well. Bats hunt by night, and for them plain green will do. In fact, fruit-eating bats are less active on nights of the full moon when there is more light, than on dark nights. On moonlit nights they are picked off by owls: a gothic encounter indeed. The birds that dispense figs are all of roughly the same size and so, accordingly, are the red fruits that have evolved to attract them. The fruit-eating bats are of various sizes. So, correspondingly, are the fruits they eat. Bats, unlike birds, carry fruit away from the tree to eat it at leisure elsewhere, in some more private roost. Perhaps this too is a defence against predators. Otherwise owls (or civets, or leopards) might simply hang around the fruit trees in wait.

So here at last, or so it seems, comes the answer to the puzzle – why figs bother to produce big fruits, which seem to cause them so much trouble. Big bats fly further than small bats. So on average, big fruits are dispersed more widely than small fruits. Dispersal, in tropical forest, needs to be as wide as possible. Big fruits seem to lose out at every turn but in the end, they are worth it.

This, at least in outline, is the fig story so far. Several morals are attached to it. The point that Allen Herre emphasizes himself, is how hard it is, in biology, to relate theory to what happens in the field. The basic interplay of figs and wasps is complicated enough; but then the complications multiply and multiply again as you stir in the cryptic wasps that look like pollinators but in fact are parasites, and the nematodes, whose virulence depends on the number of foundresses, and so on. However much we know about nature, we can never know enough. Science is wonderful – the studies that have led to the present understanding of fig-wasps are breathtaking: a brilliant amalgam of natural history, persistence, imagination and intellect – and yet the more that's known, the more it seems there is to know.

Already it is clear, though, that if we do anything to interrupt the lives of the wasps – are too free with insecticide, for instance – then we will kill off the figs, or at least ensure that the present generation is the last. The fruits of figs are essential provender not only for bats and birds, but for a host of other creatures too. In Panama, figs of various kinds are in fruit all through the year while most other fruits are far more seasonal. There are times when figs are all there is. Take away the figs, and half the fauna could be in serious trouble. The whole ecosystem balances on a pin point – and we could tip it into oblivion without thinking; or indeed, we could let it slip through our fingers even if we were trying very hard to save it. On the other hand, precarious though it seems, figs and wasps have maintained their relationship in one form or another without interruption for more than 40 million years. There is robustness in the system. If only we can work with it, it might pull through yet.

Finally, given that so many trees rely so heavily on animal pollinators and dispersers, we might ask what happens to them if their allies disappear. One highly intriguing answer comes from the island of Mauritius, in the Indian Ocean.

THE DODO AND THE TAMBALACOQUE: A SAD TALE WITH A FAIRLY HAPPY ENDING

On the island of Mauritius lives *Calvaria major*, known as the tambalacoque; one of the vast tropical family Sapotaceae. But in 1977 Dr Stanley Temple of the University of Wisconsin reported in the journal *Science*[2] that the only tambalacoque trees left on Mauritius were all over 300 years old. There were no young ones. Since the tambalacoque lives exclusively on Mauritius, these ancients were the only ones left in the world. Yet the tambalacoque used to be common – common enough to be used for lumber. The remaining trees were fertile and were clearly pollinated, for each year they produced plenty of seed. So what was going wrong?

Perhaps, Dr Temple suggested, the tambalacoque had relied for dispersal upon the dodo, which was extinct by 1681, barely 200 years after European sailors first landed on Mauritius. For tambalacoques produce very big seeds – about 5 centimetres across – surrounded by an enormously hard and woody husk, up to 1.5 mm thick: too thick, apparently, to allow the young seedling to emerge unless the walls are first weakened. The tree, Dr Temple said, evolved such a stony seed to cope with the dodo.

The dodos ate the fruit of the tambalacoque. They digested the pulpy exterior, and the big wooden pip passed to the gizzard – the extension of the gut which birds pack with stones and use to crush seeds. Tambalacoques evolved seed-coats that were thicker and thicker, in response to the dodo gizzard's enormous crushing strength. Eventually they became so thick that the seeds could not germinate at all *unless* they had first been eaten by a dodo. Of course, the seeds would fail if they were crushed in its gizzard. But if the pips were merely abraded, or 'scarified', they would germinate much better than if not: indeed, they needed the scarification. Here, then, was another example of co-evolution. Dr Temple tested his theory by feeding tambalacoque seeds to turkeys, which are not related to dodos but are ecologically equivalent. Wild turkeys eat hickory nuts. The turkeys crushed some of the tambalacoque seeds – seven out of seventeen – but after six days or so they gave up on the other ten, and either coughed them up or passed them through their guts. Those ten seeds, duly abraded, *did* germinate.

As we have already seen, many other seeds benefit from some pre-treatment from animals in one way or another. In India, teak seeds are sometimes prepared for sowing by laying them out on the forest floor where the termites can get to them. They germinate better after the insects have nibbled some of the seed-walls away (though it's important not to leave the seeds out too long). All in all, the tale of the dodo and the tambalacoque has all the elements of a classic.

Indeed the story has only one shortcoming. It does not seem to be true. Areas of forest in Mauritius have now been fenced, and cleared of the pigs, deer and monkeys that meddlesome Europeans have introduced to the island over the past three centuries – and lo, in the cleared areas, young tambalacoques have been springing forth. Evidently, it wasn't the presence of dodos they required, but an absence of imported herbivores. This is excellent news for the tambalacoque. But it is pity indeed to kill off such an excellent story.

In general, then, trees like all living creatures have a mixed relationship with their own kind, and with all other creatures: part war, part peace, and part uneasy truce. This is true even of the creatures that eat them and cause diseases.

LIFE'S TORMENTS – AND AUTUMN COLOURS

All trees, like all plants, are beleaguered from the time they are seeds to the time they return to the earth by predators and parasites. Predators in this context means big herbivorous animals, from cattle and squirrels to leaf monkeys; and parasites are loosely defined here to include the viruses, bacteria and fungi that are commonly known to cause disease, and all the animals such as worms, insects and mites that burrow into them, and indeed all the insects and other creatures commonly classed as pests. Old-fashioned accounts of ecology tended to pass over the parasites as if they are mere accidents. Yet they are major drivers in all of nature, that may determine the shape and direction of an entire ecosystem. We have seen the role of nematodes in the fig–wasp relationship. More profoundly, the need to avoid parasites may largely explain the huge variety of trees in tropical forest: no tree can afford to be too close to another of the same kind, for fear of infection. More cogently still, it may be that if there were no parasites, there would be no sex, and the transformation of all life would then be absolute. It isn't simply that creatures would live their lives very differently. Without sex to mix the genes, creatures like us (and oak trees and mushrooms) would not have evolved at all. It seems indeed that we are as we are, and trees are as they are, *because* our respective ancestors had to cope with disease.

In truth, parasites and other pests do a deal of damage, and trees seem particularly vulnerable because they must stay in the same place for so long – not like annual plants, which metaphorically speaking are here today and gone tomorrow. Most tree diseases pass most of us by most of the time, but everyone in Britain became aware of Dutch elm disease, caused by fungi of the genus *Ophiostoma* and carried by various bark beetles. Elms had been one of Britain's most characteristics trees: the ones most likely to persist in hedgerows, where traditional farmers were happy to retain them for shade and as a future source of timber – a casual exercise in agroforestry. Elms feature strongly in the landscape paintings of Constable, from Suffolk, and they also grew so rampantly in the west country that they were known as the 'Wiltshire weed'. But within about a decade, between the 1970s and 80s, English elms above the size of a small shrub were all but eliminated – one of the most dramatic extinctions in historical times.

Of course all trees suffer from pests and diseases to some extent. The average oak in Britain loses roughly half of its leaves each year to insects. Caterpillars sometimes take virtually all of the first crop of young leaves in spring, whereupon the oak may respond with a second flush in May and June, known as 'Lammas growth'. (Although Lammas, meaning 'loaf mass', is a Christianized pagan festival that falls on 2 August. Hmm.) Periodically we read of threats of various kinds to oaks or chestnuts in Europe and the United States from fungi or viruses or whatever, until it seems we will soon be lucky to have any traditional species at all.

The world's two most valued tropical hardwoods, teak and mahogany, both have their dedicated pests that beleaguer them in the wild and hugely affect their economy in plantations. Teak suffers primarily from the defoliator moth, *Hyblaea puera*, whose caterpillars may strip the leaves completely almost every year, soon after they emerge. This leaves the trees gaunt and skeletal – teak trees are often a sad sight – and also means they take much longer to reach harvestable size. Thus traditional plantations in India typically raised teak on an eighty-year cycle. Modern selection and cultivation has brought this down to thirty years. But in Brazil, where the defoliator moth mercifully remains absent (the trees left it behind in their native Asia) the cycle of harvest is down to eighteen years (or so Brazilian foresters are hoping). New research in India on biological control promises to deal with the moth at last, but we have yet to see whether it works. Mahogany is plagued in particular by caterpillars of shoot-borer moths, which burrow into the growing tip and destroy it. The tree does not die, but instead of growing straight and true as a prestige timber tree should, it sends out a mass of branches below the ravaged tip, like a bush. The reasons are as described in Chapter 12: the growing bud normally sends out a hormone (auxin) to suppress such unruly behaviour. With the source of the hormone gone, the lesser buds beneath are given free rein. Many other valuable trees worldwide (the cinnamon plantations come to mind, on Madagascar) have their own particular murrains that are of huge economic importance.

Trees, like all living creatures, contrive in various ways to make life difficult for their parasites. Commonly, tree leaves are low in nutrients: the parasite has to work prodigiously hard simply to get enough to eat. All trees present physical barriers to would-be predators and parasites, including thick waxy cuticles on their leaves that inhibit the entry of fungi or bacteria, while deciduous trees plug the scars left by their falling leaves with cork, like Elastoplast. Finally, trees are fabulous chemists. In addition to the proteins, fats, carbohydrates and other materials they need to synthesize for the everyday tasks of staying alive, they also turn out a huge range of recondite molecules known as 'secondary metabolites'. Clearly these are not essential for day-to-day living. Some trees produce some kinds of secondary metabolites, and some produce other kinds, and some seem to produce very little at all. In times gone past botanists wrote them off as waste or by-products: things the tree produced apparently through carelessness. That is how plants might have produced them first of all, in the deep evolutionary past. Now it is clear that secondary metabolites play many vital roles in the life of the plant – and paramount among them is the repulsion or destruction of would-be predators and pests.

But although pests and predators clearly do cause huge problems, the relationship between trees and their tormentors is not a simple battle. The subtleties are far from understood – the research is difficult, and most studies so far have focused on the pests of herbaceous crop plants, which are easier to work with than trees, and offer quicker financial returns. But already we can see that between trees and their parasites there is the same counterpoise of antagonism and collaboration – war, peace, and uneasy truce – that we find in all ecology. Over time we can discern co-evolution, as each player in each relationship adapts more and more minutely to the other. When the relationship is antagonistic, this co-evolution becomes an arms race, with predators or parasites and prey each upping the ante as the centuries pass. When it is cooperative, the relationship tends to become more intricate with time, until the various players become totally interdependent. The little that is so far known about trees and their parasites already reveals relationships of endless subtlety.

Upping the ante is the first sign of an arms race. So it is that many trees have spines and prickles. But spines and prickles (like cuticles and corky plugs for leaf scars, and all the secondary metabolites) require a lot of energy to produce. So we find that in various ways, trees contrive to be minimalist. Thus as we noted earlier, many species of palm that live in continental forests where predators abound are spiked as fiercely as a medieval prison, while related types, on islands free from abuse, are spikeless. So we find too that the leaves of holly are spiny on the lower branches, where they might be browsed by deer and cattle, but tend to be spineless higher up. In general, a plant that can do without spikes and such adornments has energy to spare for other things, like rapid growth – and in a competitive world, other things being equal, it doesn't pay to waste energy on things that are not necessary.

In their secondary metabolites, too, we see on the one hand a continuous upping of the ante – the trees becoming more toxic, the predators and parasites evolving new ways to cope – but also the constant need, on both sides, to economize.

Among the commonest of the secondary metabolites – very evident in oaks, for example – are the tannins. Tannins bind with the proteins of animals and in various ways disrupt their feeding – and are used for tanning leather, making it tougher and more waterproof, which is where they got their name. Heartwood rich in tannins is evidently less prone to rot than wood without tannins – although, of course, old oaks tend eventually be hollow; and in truth (for nothing is simple) trees that are only partly hollowed (but not so much that they fall apart) may be *stronger* than those that are still solid, just as an iron pipe may be stronger than a solid rod. Cattle, deer and apes are among the creatures known to be put off their feed by too many tannins but rodents and rabbits have joined the arms race and have adapted to them. They produce an amino acid (proline) in their saliva which binds with tannins and blocks their activity. Other mammals are attracted to the astringency of tannins – and so it is that human beings like tea, and tannin-rich red wines. But then, for mammals at least, tannins are not all bad. Evidently they block the chemical signals that cause blood vessels to contract. Red wine is known to protect against heart disease – and this may be in part because the tannins help to dilate the coronary blood vessels that feed the wall of the heart. Tea, a cardiologist now assures me, has the same effect; pleasing news indeed.

Insects in general are put off by tannins – but, as part of the arms race, some at least have evolved ways of coping. Leaves tend to focus first on growth, and only then have energy to spare to create physical defences and secondary metabolites; so pests such as the moths whose caterpillars feed on oaks commonly focus on the youngest leaves. Deciduous trees in turn seek to outwit the moths by producing their springtime leaves with tremendous speed. Thus the buds of oaks seem to unfold before your eyes. Still, though, the moths are liable to win because they have already laid their eggs on the oak's buds. The caterpillars emerge just before the leaves, and so are lying in wait. How the eggs *know* when exactly to hatch, is unknown. Do they simply respond to the same climatic signals as the oak buds do? Or do they pick up some chemical signal from the oak itself?

Many trees and other plants produce secondary metabolites known as 'terpenes' that are specifically insecticidal. Among the best known terpenes are the pyrethroids, which human beings have extracted in particular from African daisies of the genus *Chrysanthemum* (which is not the same as the 'chrysanthemums' of the florist), and adapted as commercial insecticides. Pines, firs, and many other conifers harbour similar agents in their resin ducts. Of course, it is expensive for the tree to produce such chemical agents. But the conifers economize by not producing more than they really need: at least, when they are attacked by bark beetles they produce more terpenes in response.

The terpenes also include the 'limonoids' found in citrus fruits: the skins of oranges and lemons also repel insect predators. The most powerful insect repellent known is also a limoinoid: this is 'azadirach-tin' – produced by the all-purpose medicinal neem tree, *Azadirachta indica*. Azidarachtin will repel insects in astonishingly low concentrations (fifty parts per billion) and also has other toxic effects – yet it has no toxicity to speak of in mammals. So, like the pyrethroids, it is favoured as a commercial insecticide (and modern politics being what it is, the neem tree is now the cause of international disputes).

All the chemical agents cited so far tend to stay within the plant that produces it. But many others are 'volatile', meaning that they rapidly evaporate and float off in the wind. Some of these volatiles, particularly those known as 'essential fatty acids', are highly scented.

Hence the fragrance of sage, mint, basil, and other such relatives of teak. Hence, too, the powerful medicinal niff of the eucalyptus. Chemical repellents that are not volatile have no effect until the predator has taken its first bite – but the volatile ones warn insects and other creatures to stay away *before* they attack: a more sophisticated measure altogether. But, as with tannin, these essential oils often prove agreeable in small quantities, and human beings, ever opportunist, extract the oils of eucalypts (and many other plants) for perfumes and medicines – simply by boiling them up and then distilling the oil.

Then again, a few specialist animals, equally opportunist, have developed ways of coping with such repellents. Thus the essential oils of eucalypts are toxic to most animals (in their raw state) – but koalas are equipped with a huge extension of the gut (the caecum) which is packed with symbiotic bacteria, and these detoxify all the noxious compounds in eucalypt leaves. Very few other creatures can cope with eucalypts, which is one reason why the trees are so successful. Because koalas can cope they have the entire run of Australia's host of eucalypts (600 species or so) almost (though not quite) to themselves. Indeed, koalas will eat very little else, and usually nothing else at all, *except* eucalypts – and different populations of koalas typically confine themselves to just one or a few eucalypt species, and reject others. Very few mammals are anything like so specialized. Pandas munch almost exclusively on bamboo but they will eat many other things besides if given half a chance, from omelettes to roast pork. But creatures that specialize in eating toxic tree leaves pay a price. Brains are particularly susceptible to toxins. Koalas, like the leaf-eating monkeys and the peculiar leaf-eating Amazonian birds known as hoatzins, have smaller brains than their more omnivorous relatives. Even the best friends of the koala have little praise for its intellect.

Legumes (Fabaceae) are among the most accomplished chemists of all. Many produce soap-like 'saponins' which interfere with animal digestion. Many produce 'flavones', some of which are strong insecticides – and again have been exploited commercially. Legumes often produce agents that limit the oestrogen hormones of mammals, so that sheep grazing on legume-rich pastures often become infertile. I know of no direct evidence of such activity in leguminous trees, but it would be very surprising if there were none. Many leguminous trees secrete similar flavones into the soil – not to kill insects, but to help establish good relations with the nitrogen-fixing bacteria that they seek to entice into their roots. Complex chemicals in general are versatile. Any one molecule, with or without further chemical adjustment, might kill one group of creatures and attract others.

Flavones impinge directly upon our senses, did we but know it. Among them is anthocyanin, the material which, in a myriad slightly different chemical forms, provides plants with a broad palate of pigments, all red, but ranging from gentle tints through the brightest scarlet to maroons and purples. Flowers and fruits are often red, of course – but so too, surprisingly often, are leaves. In particular, the young shoots of tropical trees are often red – as are, in the autumn, the dying leaves of many a temperate tree, perhaps most famously the maples that are the pièce de résistance in the glorious autumn colours of New England, one of the greatest natural shows on earth.

Anthocyanin is not produced gratuitously, it seems. It is a 'secondary metabolite' to be sure, but it is no mere accident. Chemically, it is expensive to produce. So why is it produced in young tree shoots (especially in the tropics) and in dying leaves (especially in temperate countries)? What do the two have in common?

New work based both in the US and in New Zealand shows beyond reasonable doubt that anthocyanin is protective. In particular it is an antioxidant. Creatures like us and trees need oxygen of course, in constant supply, to stay alive. We need to burn sugars to provide ourselves with energy. But oxygen does such a good job on sugars because it is so lively, and we admit it to our bodies at our peril. Left unchaperoned, oxygen gives rise to a whole range of 'free radicals' that destroy our flesh and corrupt our DNA – and do the same in trees. Any creature that aspires to live where there is oxygen (anywhere, in fact, that is not some murky swamp) must equip itself with anti-oxidants, to keep oxygen (or, rather, the radicals it gives rise to) under wraps. In people, these anti-oxidants include many a vitamin, including C and E. In plants, they include some enzymes, and the agents known as phenolics. And they include anthocyanin.

Trees, like all creatures, are most vulnerable to attack by oxygen radicals when under stress. Young leaves, before they have all their chemical and physical defences in place, are tender. High in tropical trees, exposed to the fiercest sun and sometimes short of water, young leaves are extremely vulnerable. Extra anthocyanin protects them.

But why should dying leaves produce such expensive stuff? Why should they be protected when their number is already up? Precisely because the state of dormancy that temperate trees enter in the autumn is *not* a simple shutting down. Before they pack up for the winter, deciduous trees withdraw as much nutrient as possible from their leaves. Chlorophyll, the principal protein, is broken down, the nitrogen it contains carefully drawn back into the body of the tree, to nourish next year's growth. But the breaking down is stressful. It opens the leaf to attack by oxygen radicals before the work is complete. So (modern theory has it) trees such as maple produce anthocyanins to keep the oxygen in check, and allow the withdrawal of chlorophyll to proceed in good order.

We might ask, of course, why *all* deciduous plants don't do as the maple does; why most of them are *not* bright red in autumn. The general evolutionary answer would be that nothing is for nothing. Anthocyanin helps the maple to sort out its problems – but it is expensive. Other trees may simply find that it is more trouble to produce anthocyanins in the autumn, than to endure a little damage. Or they may simply not have got around to producing anthocyanins in autumn, for not every species does everything that is theoretically possible.

The general notion that anthocyanin is actively protective fits neatly with a whole range of otherwise bizarre observations from other trees (and other plants) in many different circumstances. Thus it is that when insects bite the horopito tree *Pseudowintera colorata* of New Zealand the wound turns red: an outpouring of protective anthocyanin. It also fits the common observation that New England's autumn colours are most spectacular when the weather is cold, or has been so. It's then, after all, that the trees are under greatest stress. Horopito is a relative of winter's bark, in the Winteraceae.

Finally, though, we should note in passing that not *all* autumn colour is due to anthocyanin, or indeed to active production of extra pigments. The yellows and browns of autumn leaves seem simply to represent pigments, notably yellow carotene, that are left behind after the tree has stripped its leaves of chlorophyll. They really do reflect innate inefficiency. But we should be grateful for this too. The reds are beautiful; but are the more beautiful because there are browns and yellows and every other shade alongside.

There are many more twists as the game of chemistry unfolds. An array of plants has been shown to produce various terpenes only *after* insects have begun to feed on them: another economy. These terpenes discourage the invading pests from laying their eggs. But in addition they also *attract* the natural enemies of the pests, so that they fly in from far and wide to see the pests off: swarms of parasitic wasps or ladybirds summoned to feast on aphids. So far such effects have been shown in maize, cotton and wild tobacco. I know of no specific examples in trees – but again it would be very surprising indeed if they were not to be found. Most herbaceous plants have trees that are close relatives and, beyond doubt, it's just a question of finding the time and resources to look. Again there are commercial possibilities. Chemicals that summon help from insects that kill pests could surely be developed to protect crops.

Exactly how plants know that they are being attacked, and so tell their genes to produce more insect-repellent, is not known in detail. It is clear, though, that one essential ingredient in the chain of chemical communication is salicylic acid, which is widespread among plants (although willow bark is particularly rich in it: this is the source of aspirin). One modified form of salicylic acid is methyl salicylate, and this is volatile: produced in one plant it can float off in the atmosphere and so affect another. Thus it may serve as a 'pheromone', an airborne chemical signal. This means, though, that a tree that is being attacked can not only induce defences in other parts of the same tree – it can also warn other trees in the vicinity that trouble is afoot. So it is that when elephants in Africa feed from the mopane trees of Africa (*Colophospermum mopane*) they take just a few leaves before moving on to the next tree. Furthermore (so it is claimed) the elephants move *upwind* to a new tree. Evidently the mopane increases its output of tannins as the elephants browse, so the leaves become less and less palatable. Evidently, too, they emit organic materials (the identity of which I do not claim to know: perhaps the tannins themselves) that act as pheromones, and so warn other mopanes downwind that an attack is imminent and they too should produce more tannins. Some acacias are said to behave in the same way, in response to giraffes. We cannot hear the trees calling each to each, as T. S. Eliot claimed to hear the mermaids. But the air is abuzz with their conversations nonetheless, conducted in vaporous chemistry, and the ground too, via the bush telegraph of their roots.

But although it is clear that parasites are damaging, and although trees and their various tormentors go to such extraordinary lengths to overcome each other, we still must ask whether the attacks that we perceive to be so heinous are always as destructive as they seem. Thus, many trees – including those of the Rosaceae and perhaps most dramatically lime trees – are attacked every year by millions of aphids. The trees crawl with them. Linden leaves become sticky with 'honey-dew' – the surplus sugary sap that the aphids excrete. The stickiness attracts disfiguring sooty black fungi. Aphids also carry viruses. It seems downhill all the way. But perhaps, says Oxford ecology professor Martin Speight, the honeydew that falls on to the place below is good for the soil. Leguminous trees such as acacias, and non-legumes such as alder, harbour nitrogen-fixing bacteria in their roots. Many other plants, including many grasses, exude organic compounds from their roots that feed colonies of bacteria that remain free in the soil, but fix nitrogen nonetheless. Perhaps the honeydew dripping from the leaves of lime trees has just this effect: creating an environment in the soil beneath where nitrogen-fixing bacteria can thrive. Perhaps, taken all in all, the overall effect of the aphids is neutral – or even beneficial. Other insects too may help trees along by their nitrogenous excretions. Termites have nitrogen-fixing bacteria in their guts. Perhaps they bring extra nitrogen to tropical forest trees that are often starved of it, and so on the whole their attacks may be helpful.

Trees that seek to be pollinated by insects must sacrifice nectar, pollen, fats, or parts of the flower itself, all of which are metabolically expensive. Perhaps in the same spirit trees may sacrifice a proportion of their leaves – even half of them in a season – for the sake of soil fertility. Many thrive on marginal soils, after all, and perhaps they would not if it were not for pests, which perhaps are not so pestilential after all. Foresters hate the parasites that reduce growth rate. But does it matter to the tree? Trees, we may assume, are not proud. Does a mahogany *mind* if it develops a bushy top, instead of a fine straight bole? So long as it is still able to reproduce, why should it? Gardeners of a certain kind typically zap everything that moves with the vilest toxins they can find, and then spend their evenings cutting back the growth they perceive to be superfluous. Why not let the pests do the pruning? Organic gardeners who eschew all vileness often have the best gardens.

However – and it's a big 'however' – the whole equation changes if the conditions change. Pests or diseases that can be shrugged off as part of life's rich pattern in favourable circumstances, and may even bring net benefit, may be seriously bad news if the tree is under additional stress. For example, and notably, pests are kept at bay to some extent by the simple physical pressure of the sap. In drought, this pressure is reduced. Thus (says Professor Speight) bark beetles commonly make little headway into eucalypts – except in times of drought. Drought, and consequent stress, happen often enough to keep the beetles in business. Contrariwise, waterlogging is bad too.

Here is where our own restless and meddlesome species, *Homo sapiens*, plays many a part. We have carted entire battalions of pests around the world to fresh woods and pastures new, to where the native plants have had no time to adapt. 'Exotic' pests (and weeds) are often the worst. If it is true (as it seems to be) that the enormous heterogeneity of tropical forest is in part a response to disease – no tree can afford to be too near another of the same kind – then it follows as night follows day that monocultural plantations of tropical trees will be especially vulnerable (as is true of all monocultures). I do not know if mahoganies in plantations suffer more from shoot-borer moth than in the wild but it seems likely, and we know that many a mahogany grows straight and tall in the natural forest, out-topping its neighbours, without any help from pesticide.

More broadly, we are changing the climate. General warming is already enabling many an animal pest to move away from the equator into the realms of temperate trees that have had no time to adapt to them, and cannot of course run away. It is impossible to predict in detail how warming the whole globe will affect the climate of any particular spot, but we can be sure there will be plenty more drought, and plenty more waterlogging. Parasites in general adapt to changing conditions many times more quickly than trees can do. Some insects have a generation time measurable in days. In some bacteria it is minutes. Either may go through a thousand and more generations while a tree is still feeling its way. Trees are opportunists, but they are slow opportunists. It is the stock in trade of parasites to seize new opportunities very rapidly indeed. The armed but often amicable truce that has evolved over many a millennium between trees and their hordes of parasites could be horribly thrown off balance over the next few decades. There are a few things we can do – some specific controls, some replanting of more resistant species, and every possible attempt to minimize warming. But for the most part, we will just have to wait and see.

IV

Trees and Us

14

The Future With Trees

O if we but knew what we do
When we delve or hew –
Hack and rack the growing green!
Gerard Manley Hopkins,
'Binsey Poplars', 1879

Humanity is in a mess. The statistics are simple and stark: a billion chronically undernourished, a billion in slums (and growing), more than a billion who live on less than two dollars a day. Even worse, because potentially more final, is the collapse of the earth itself, the place where we live. Soil, air, sea, lakes, aquifers and rivers: all are under stress. Above all, potentially the *coup de grâce*, lies the reality of climate change. Global warming does not merely imply amelioration – Mediterranean beaches in hitherto chilly northern latitudes; vineyards in Scotland. It will bring a century and more of extremes, and of extreme uncertainty: flooding of reefs, mangroves and fertile coastal strips, of lowland countries and states from Bangladesh to Florida, and of cities from London to New York; stronger hurricanes, more often; the world's great wheat and corn fields, our 'breadbaskets', potentially reduced to dust bowls; absolute loss of ancient habitats, including the beautiful frozen lands of the polar bear, the northern forests of spruce and aspens, and – most destructively of all – tropical rainforest. The first signs are with us. I have two grandchildren and I fear for them.

Worst of all, though, is that the standard solutions to all our problems, bruited from on high by the world's most powerful governments, are in truth their prime cause. 'Progress' is the buzzword. If this meant improvement of human wellbeing and security then it would a fine concept, but in practice it does not. It has come to mean 'Westernization' – that all the world should be more like us: more industrialized, which means more global warming; and more urbanized, which means less care for the countryside and even bigger slums. By 2050, on current projections, there will be as many people in cities (6 billion) as now live on the whole earth. The cities cannot cope. Demonstrably, they are losing the race. At the same time the world's politics has become more abstract – geared not to physical realities, but to the abstraction of money. Governments measure their success in GDP – the total wealth created. Good endeavours can create wealth, of course, like building schools, and planting forest. But bad things create wealth at least as easily – like felling forest and flogging it off, or simply making war. The cash thus conjured up is all grist to the GDP and so, on paper, contributes to the great desideratum of 'economic growth'. Agrarian economies that have ticked along for thousands of years in perfect harmony with their surroundings are said to be 'stagnant', and 'ripe for development'. 'Development' in turn is equated with increase in cash. The fight against hunger, disease, oppression and injustice has been reduced (dumbed down, one might say) to 'the war on poverty'.

Yet very simple arithmetic shows that the Western goal of urbanized industrialized life and the promise of limitless wealth for everyone *cannot* be realized. It would need the resources of at least three earths to raise everyone in the world to the material standards of the present-day average Brit (or a less-than-average American). Furthermore, the means by which this dream is being imposed on the world (in the names of 'freedom', 'democracy' and 'justice', as well as 'progress') seem expressly designed to undermine the wellbeing of most of humanity. For most people in the world live and work in the country-side, at rural crafts of which the chief is farming. Industrialization simply puts them out of work. If India farmed the way the British do, then half a billion people would be unemployed – more than the total population of the newly expanded European Union; almost twice the population of the present United States. No other industry can employ so many people as farming can, and still does, and no other industry could conceivably be so sustained. Most of Africa is even more emphatically rural than India, and has even fewer realistic options. Yet Westernization – urbanization, industrialization, and a monetarized economy geared to GDP – is promoted by the world's most powerful governments as the universal algorithm for all of Africa, Asia and South America: the formula which, once applied, will solve all our problems. Even worse, the formula is applied with all possible haste, for the world's new economy is driven by the perceived need for 'competition'. But untempered competition is a crude concept indeed, as any worthwhile moral philosopher would attest, and any truly modern biologist could ratify.

The world needs a sea change. It isn't just a question of changing our leaders' minds, for that is too exhausting and we don't have enough time. It isn't a question of changing our leaders – for we are liable simply to land ourselves with more of the same. The world as a whole needs a different *kind* of governance; different kinds of people in charge. To ensure this, we (all of us, for it's no use relying on the current leaders, who are doing well according to the status quo) must find new ways of ensuring that new kinds of people are elected and given real power: people who respond to the real, physical needs of the world, and to the needs of humanity at large. Present-day leaders – politicians and captains of industry – are wont to suggest that any radical initiative, that takes account of the realities of soil, water and climate, is 'unrealistic', commonly because such initiatives may inhibit the plans of bullish industries and their governments, and hence inhibit 'growth'. But the word 'realistic' has been corrupted. It ought to apply to the realities that are inescapable – of physics, of biology – made manifest in the declining earth, and the creatures that live on it. It should apply to the realities of people's lives – whether they have enough to eat, and water, and shelter; whether they have control over their own lives, and worthwhile jobs, and can live in dignity. The 'reality' of which our current leaders speak is the reality of cash. But cash is not the reality. Cash is the abstraction.

Sea change, to mix the metaphors, may seem pie-in-the-sky. In truth, however, and encouragingly, most of the big and necessary ideas are already in place, or already being worked upon. Excellent ecological studies are in train. Truly 'appropriate' technologies, designed expressly to be sustainable – both low tech and high tech – are developing apace, and many (including many of the simplest) are quite wonderful. And, which is at least as necessary, many groups worldwide are working on new economic models which, though basically capitalist, are geared directly to human wellbeing, rather than to the gross abstraction of accumulating cash. Certainly, there can be no sure-fire formulae. We simply cannot tell, until we try, whether our ecological projections and economic plans will work out the way we want them to. But if we keep our eyes on the right targets – human wellbeing in a diverse world; humanity able to live well, effectively for ever – then we are in with a chance. If we proceed with reasonable caution, we can correct mistakes as we go. If we simply apply the present-day algorithms – high tech applied willy-nilly wherever it will make most money most quickly – then, surely, we and the world as a whole have had our chips.

Most importantly, the future endeavours of humanity must be geared to biological realities. The world's economies (and the endeavours of scientists and technologists) must serve those realities. Most obviously – once we start to think seriously about the fate of cities, and environmental stress in general, and human employment and dignity – we see that for the foreseeable future, and probably for ever, the economies and physical structure of the world must be primarily agrarian. In the current, crude, unexamined dogma, 'development' and 'progress' mean urbanization. The prime requirement, in absolute contrast, is to make agrarian living agreeable. It can be. It's just that at present, all the world's most powerful forces are against it.

Trees are right at the heart of all the necessary debates: ecological, social, economic, political, moral, religious. It isn't the case that trees are always a good thing. The wrong tree in the wrong place can do immense damage. It isn't true that everywhere in the world needs more trees. In some places, there could be too many – lowering water tables that are already too low. The wrong trees in the wrong places can poison rivers, undermine buildings and even cause soil erosion. Yet beyond any doubt, most landscapes and the world in general would benefit from many, many more trees than there are now, and the wholesale squandering of present-day forest seems like an act of suicide.

But it also true – marvellously and encouragingly so – that societies can build their entire economies around trees: economies that are much better for people at large, and infinitely more sustainable, than anything we have at present. Trees could indeed stand at the heart of all the world's economics and politics, just as they are at the centre of all terrestrial ecology. The more I have become involved with trees in writing this book, the more I have realized that this is so. In the future of humanity, and of all the world in all its aspects, trees are key players.

THE BIGGEST CHALLENGE OF ALL:
CLIMATE

The world's climate has fluctuated spectacularly this past few billion years, often to fantastical extremes. At times the whole world has been tropical, with palm trees in Canada, fringing the Arctic Ocean. At other times (long before the recent wave of ice ages) the entire surface may have frozen over. Beneath the streets of London – obligingly revealed by civil engineers this past few hundred years – are fossils of crocodilians and of nipa palms, denizens of tropical rivers; but also of woolly mammoths, creatures of the tundra. Those who suppose, as some of the world's leaders have chosen to do this past few decades, that the world is more or less bound to be the way it is now, should look at the evidence all around of past deviations, or at least take notice of those who have.

All kinds of evidence show that the fluctuations in temperature, from global tropical to global freezing and back again, are correlated with the concentration of carbon dioxide in the atmosphere. Some of the most convincing evidence comes from ancient ice in Greenland and Antarctica, which traps atmospheric gas from earliest times that can then be directly analysed. When the concentration of carbon dioxide was high, the fossils show the world was warm. When it was low, the world was obviously cool.

The physics is simple. The earth is constantly warmed by sunlight, and constantly radiates the warmth away again. The solar energy coming in contains light of all wavelengths. The radiant heat that the earth gives out is all in the infra-red spectrum. Carbon dioxide absorbs infra-red. So atmospheric carbon dioxide filters out only a fraction of the solar energy coming in – a proportion of the infra-red; but it traps a great deal of the energy going out, since most of what's going out is infra-red. So more energy is kept in than is kept out, and the net effect is to warm the atmosphere. The glass in a greenhouse operates in the same way, so this is called the 'greenhouse effect' and carbon dioxide is called a 'greenhouse gas'. Some other gases also have a greenhouse effect, including nitrous oxide, methane and water vapour. But carbon dioxide is the main variable.

It may seem odd that carbon dioxide can have such an effect. After all, its concentration in the atmosphere is low. The present concentration of carbon dioxide in the atmosphere is only around 370 parts per million. Yet is has been calculated that if the atmosphere contained no carbon dioxide at all, the average surface temperature would fall to $-18°$ C. At the time of the ice ages, atmospheric carbon dioxide was down to 190 to 200 parts per million; and it was indeed very cold. When plants first colonized land 400 million years ago, atmospheric carbon dioxide stood at around 7,000 parts per million – and the world was extremely warm. At the time when cyanobacteria first evolved photosynthesis, more than 2 billion years ago, the atmosphere contained no oxygen at all. Carbon dioxide was the chief gas of the atmosphere, apart from nitrogen. The world must then have been ridiculously hot. For creatures like us (or modern trees) the heat alone would have been lethal (let alone the noxiousness).

Modern meteorological records, and the notebooks of gardeners and naturalists, show that the world has been getting warmer this past 150 years or so. The Intergovernmental Panel on Climate Change (IPCC) reported in 2001 that the average temperature of the whole world increased by 0.6° C between 1900 and 2000. Such a rise may not seem great, but ostensibly small changes averaged over the whole world can imply huge changes locally and regionally – enough to have profound effects on environments and people. The trend seems inexorably upwards. 1998 was the warmest year since reliable records began; and the 1990s was the warmest decade.

Changes in temperature cause changes in rainfall. Precipitation rises overall (since more water evaporates from the surface of the oceans), but the distribution is uneven; so some places become much wetter, and others much drier. The IPCC noted an overall increase in precipitation (rain and/or snow) of 0.5 to 1 per cent *per decade* through the twentieth century: up to 10 per cent over the century. The increase was highest in the mid and high latitudes of northern continents. In the tropics (10° N to 10° S) the increase was only 0.2 to 0.3 per cent per decade – and in the northern subtropics (10° N to 30° N) rainfall actually *lessened*, by around 0.3 per cent per decade. The subtropics are hugely important to humanity for their agricultural crops, their forests, and as places where people live. Drought in that latitudinal band is seriously bad news. In the southern hemisphere, which lacks a vast continental land mass and has vast oceans instead, there were no comparable changes – or at least the changes were not so regular.

But overall averages are not all that matters. Extremes increased too – just as theory predicts must be the case, since the temperature of the world's surface changes unevenly. More and more places recorded their hottest days in history. The late twentieth century saw some of the worst-ever storms and hurricanes, floods, droughts and forest fires. The southern hemisphere is hugely affected – one might say plagued – by El Niño, the warm current that flows from west to east

across the Pacific every few years as a result of the 'southern oscilla-
tion'. The western South Pacific warms up more than the east, and
every so often the warm water in the west escapes, and flows to the
east. The El Niño current (so called because it tends to flow around
Christmas time – Niño means 'little boy' and refers particularly to the
Christ Child) causes floods in some places, droughts in others, and
also causes fish stocks to fail, with the consequent devastation of
seabirds, and potentially of local fisheries. El Niños seem to be getting
more frequent and more severe.

The changes in atmospheric carbon dioxide that caused huge climatic fluctuations in the deep past were brought about by natural
events. In general, carbon in the earth's surface takes the form of
carbon dioxide that floats free in the atmosphere, or is dissolved in
the oceans. Or it is present as 'organic' carbon, locked up in the bodies
(including the dead bodies) of living creatures – flesh, leaf litter, and
timber; and carbonates, in rocks. Rainfall, volcanic action and the
movements of tectonic plates ensure that carbon flows between the
four sources, changing chemically as it does so – from carbonate to
carbon dioxide to flesh (plus dead leaves and timber) to carbon dioxide
and back to carbonate.

The general cooling over the past 40 million years or so has apparently been caused by the (fairly) steady loss of carbon dioxide from
the atmosphere; and this has been ascribed to the rise of the Himalayas
and the Tibetan Plateau, caused by the slow tectonic crunch of India
into the south of Asia. The Tibetan Plateau and the Himalayas are a
huge mass of rock, which interrupts the saturated winds that blow
from the Pacific, and thus causes the rains that manifest as the monsoons. This rain is steeped in atmospheric carbon dioxide, which
reacts with the rocks of the Himalayas, and then washes into the sea.
Thus, carbon dioxide has been steadily leached from the atmosphere,
and the world has cooled accordingly.

But carbon dioxide has been rising again this past 200 years or so
– not from natural causes but mostly through the restless activity of
humanity. Atmospheric carbon dioxide now stands at around 370
parts per million, but in 1750, at the start of the Western world's
Industrial Revolution, it was a mere 280 parts per million. The annual
rise averages around 0.4 per cent. Methane in the same period has

risen by 151 per cent, and nitrous oxide by 17 per cent, and both continue to rise. By 2080, at the current rate of increase, carbon dioxide levels will be double present levels, around 750 parts per million. This will make the world between 1.4 and 5.8°C warmer than it was in 1990, which will melt the polar ice caps (as is already happening) and increase sea levels by anywhere between 9 and 88 centimetres. The higher figure, the best part of a metre, would be enough to wipe out many a small island (and not a few major cities). The calculations are rough for all kinds of reasons – not least because more warmth means more cloud, and more cloud would be cooling, and so would have a negative feedback effect. But future changes in cloud cover are difficult to calculate.

In recent years, too – in fact since 2001 – American meteorologists in particular have realized that over the past few decades the full effects of global warming have been masked by 'global dimming'. It transpires that atmospheric pollution by particles – variations on a theme of soot – has been reducing the energy input from the sun by an astonishing 30 per cent. The world, currently, is cleaning up the soot – which, technically, is fairly easy to do. Catalytic converters on cars do much of what is needed, and these are encouraged because they are big business, and their development and sale increases GDP. But the world is not commensurately reducing its output carbon dioxide. The Kyoto Protocol of 1997 was and is intended to reduce carbon dioxide output, or at least to ensure that whatever extra carbon dioxide is produced, is mopped up. The Protocol was not ratified until the beginning of 2005. The United States, by far the world's biggest producer of industrial carbon dioxide, has not signed up to it. But the Protocol would not go far enough, even if everyone did sign up, and acted upon it. Meanwhile, most of the rest of the world has yet to industrialize, and evidently feels that industrialization is necessary. China has already become the world's second-greatest carbon dioxide emitter.

To reduce carbon dioxide on the necessary scale would mean severe curtailment of the use of fossil fuels, and although this can be made profitable (house insulation could and should become a significant industry), on the whole it is cheaper and easier to continue business as usual. The carbon dioxide output of the United States is virtually unabated, and, although we have yet to see which way China will go at present, it seems hell-bent on further industrialization. As the dimming soot is removed and carbon dioxide continues to rise, both the magnitude and the speed of global warming seem liable to exceed even the most extreme predictions of the late twentieth century.

Trees, as ever, are or should be at the heart of all discussion on climate change. The changes in carbon dioxide, in temperature, and in patterns of rainfall will each affect them in many ways – and each paramater interacts with all the others, so between them these three main variables present a bewildering range of possibilities.

Rising carbon dioxide alone – even if temperature and rainfall stayed the same – should accelerate photosynthesis. This in theory must increase the amount of carbon in each tree, and reduce the amount in the atmosphere – and if enough trees photosynthesized more quickly, then they could in theory reduce atmospheric carbon dioxide enough to lower global temperature. That would produce a negative feedback loop: an excellent outcome. We know that rising carbon dioxide can indeed stimulate photosynthesis. Commercial growers sometimes ply their crops with extra carbon dioxide to boost their growth. In large-scale experiments in forests and plantations, groups of trees that have been semi-enclosed and then given extra carbon dioxide also grow faster, as they fix more carbon.

But all is not so simple. Key players in photosynthesis are the stomata, the holes in the leaf surface that allow the carbon dioxide to enter. Rising levels of carbon dioxide, and increasing photosynthesis, stimulates the stomata to close. This, of course, will slow the increase in photosynthesis. So whatever effect the increase in photosynthesis may have in reducing atmospheric carbon dioxide will be limited. At some point, metaphorically speaking, the trees will cry, 'Enough!' They could simply be overwhelmed.

And as carbon dioxide increases so too will temperature – and this complicates the picture again, in many ways. Heat in general increases the rate of all chemical reactions, and since metabolism is the sum total of the body's chemistry, rising temperature means faster metabolism. Rising temperature should stimulate photosynthesis along with every-thing else, which is fine: the warmer it gets, the faster the trees should absorb CO_2, and so prevent further warming.

But – there is always a but! – rising temperature will also stimulate respiration – the burning of sugars; and respiration may speed up more quickly than photosynthesis does, and if it gets too warm there could be net *loss* of carbon from any one plant. Worse: the many creatures that live in the soil – bacteria, fungi, invertebrates – will also respire faster as temperature rises. Since they feed primarily on leaf litter, they will break it down faster – and so release the carbon it contains more rapidly. Thus rising temperature could easily cause a net loss of carbon from the forest as a whole – partly from the trees, and partly from the ground. Finally, if temperatures rise too high, then essential enzymes within the plant are damaged, and the plant starts to die. As this happens, photosynthesis stops and the living parts decay. Then, the net release of carbon into the atmosphere becomes massive – and the temperature rises even more. Again, there is some direct experimental evidence to support this scenario.

Of course, tropical trees are adapted to heat, and should not be killed by the kind of temperatures that are envisaged; and all trees can adapt to change to some extent, given time. But what's truly serious, not to say terrifying, is the *rate* of climate change. We may well see significant temperature rises in the next half-century, or even sooner if the global dimming hypothesis turns out to be correct. Few trees indeed could adapt in such a time, and none could make the necessary, generation-by-generation genetic changes. Only the creatures with short lifetimes – bacteria, flies, perhaps mice – can effect significant change in such a period. So we could see a massive die-off, pushing even more carbon dioxide into the atmosphere, resulting in higher and higher temperatures, leading to more deaths, and so on.

Then there's the matter of water. If plants have a design fault, it lies with their stomata. The same neatly guarded apertures that allow the essential carbon dioxide to enter also, inevitably, allow evaporating water to leave. This of course is exacerbated by rising temperatures. So plants that are warmed more than they are used to, in regions where water is the limiting factor (which is often the case even in rainforest – for many rainforests have dry seasons) will wilt and eventually die: the precise opposite of what is desired. Again, there is direct evidence from the field. Scientists in Brazil, including Dr Yadvinder Malhi of Oxford University, have deprived whole hectares of trees of water by covering the ground in polythene panels, so that most of the rain runs off into drains around the edges. As might be predicted (but these things always need to be demonstrated), the trees quickly show signs of suffering. Growth stops – meaning that respiration starts to exceed photosynthesis, meaning that carbon dioxide, in net, is being lost.

As a final botanical complication, it seems likely that when conditions favour more photosynthesis – more carbon dioxide, more warmth (within limits), and more water – then the epiphytes, the mass of ferns, bromeliads, orchids and others that grow on trees, will grow quicker than the trees themselves. This may not matter for the wellbeing of the planet as a whole – so long as *some* plants absorb more carbon, it does not matter which they are. But overgrowth of epiphytes would be bad for trees, and would profoundly affect overall forest ecology (it could in theory increase biodiversity, though all change is more likely to be destructive in the short term). This last complication proves another general point: that prediction is immensely difficult because we just don't know enough about the physiology of wild creatures in general, let alone the result of all their interactions; and if we did know enough, the computers that help to make sense of masses of data would be very hard pressed indeed to simulate reality reliably, on the local as well as on the grandest scale.

Then there is fire. More and more fires these days are started by human beings, deliberately or inadvertently, although even when deliberate and for legitimate agricultural or conservational purposes, they sometimes get out of hand. But fires, commonly though not exclusively triggered by lightning, are part of nature. In the places where they naturally occur – virtually everywhere that has anything to burn and is not permanently wet – the local plants (and animals, to a greater or lesser extent) tend to be adapted to them. Grasses need to have their tops burned off if grazing animals do not do the job for them, or the tops become senescent and stifle the fresh growth beneath. As we have already seen, many trees are highly fireproof, like redwoods and eucalypts, and the seeds of many pines and other species will not germinate unless first effectively cooked, whereupon they 'know' they can sprout in the nutrient-rich ash provided by their immediate predecessors.

But as with every input – including water, general warmth, light, carbon dioxide, and many minerals that are in small doses essential – there can be too much of a good thing. Fire is lethal to the trees that are not adapted to it, of course – and even for those that need it, timing and intensity are all. If, for whatever reason, the fires come too frequently, or too rarely, or burn too intensely, then the best adapted trees are overwhelmed. Human beings are altering the world in ways that are very much against the interests of wild trees, upsetting even those that seem well adapted.

Human beings have been using fire to influence vegetation probably for about 500,000 years: that at least is the age of the oldest fires that are thought to have been started by humans. Human beings in those earliest days were not modern, even anatomically. They had smaller brains than we have. But they knew about fire. Sometimes our ancestors set fire to the local plants, as Australian Aborigines may still do, to drive wild animals into traps, or simply to freshen the vegetation (grasses) to attract more grazers. Sometimes they set fire to trees just to get rid of them, and then sow crops or plant grass for cattle in the ashes. Sometimes people just seem to want a better view. As the world becomes more and more crowded and complicated, there are many conflicts of interest, each requiring a different attitude to fire. Conservationists in general want to keep trees, but recognize the need for occasional fires, for example to stimulate germination. Tour operators and people who live in the country commonly want to keep trees, but hate the idea of fires, which reduce some of the forest to ashes and threaten their houses. Farmers commonly want to get rid of trees – but prefer it if the ones that do remain do not catch fire and burn their farms. People in general tend to feel that fires are a bad thing, and politicians have an eye to their votes.

Even in the absence of any specific policy, however, human beings have an immense influence on the likelihood and frequency of fires. Thus fires in forests tend to begin with the leaf litter; and in the savannah, they commonly begin with the grasses. Over the past few decades the Brazilians (and North Americans too) have introduced several grasses from Africa, which they feel make better fodder for cattle. Some of these grasses have crept in to the Cerrado, the dry forest, primarily along the disturbed ground along the verges of the roads that now criss-cross the country. From there they spread into the surrounding land. These particular grasses, as it happens, burn more slowly than the native grasses: and this prolonged burning is far more damaging to other vegetation than the quick, albeit hotter flames generated by the native grasses.

That is the first problem. Then, through the 1980s, the fire brigade of Brasilia, which is in the Cerrado, decided to show that it could suppress fires altogether. For fourteen years it succeeded. Then came the general election – won by the present president, Luiz Inacio Lula da Silva, known as Lula. Everyone was given the day off to vote. Including the fire brigade.

While the firefighters were away, the Cerrado caught fire. Since the Cerrado has been catching fire for as long as it has existed (at least since the last ice age, 10,000 years ago, and probably far longer), the trees are adapted to it. But two things had changed. First, the grasses were now slow-burning. Secondly, thanks to the heroic efforts of the fire brigade, there was a huge backlog of litter and dead grass. So the fire raged as never before. The trees of the gallery forest, which runs along the rivers, are not fire-adapted, but on the whole they don't need to be: their innate wetness keeps normal fires at bay. But this was a super-fire, and it swept through the gallery forest as well. The result was, of course, devastating. Eucalypts, imported some decades ago from Australia, are now spreading happily. (Though thankfully not exclusively. The natives are fighting back too.)

For this kind of reason – accidents, changes in the vegetation, and policy that perhaps is misguided – fires these days are often bigger than ever before. North America, Australia and southern Europe have seen some horrendous blazes in recent decades. The fire in Indonesia in 1987 in tropical rainforest that should be free of fires, left a haze that hung around for months, as if in the wake of a volcano or a nuclear explosion. In very big fires, as with nuclear bombs, special physics comes into play: the rising heat creates an updraft that drags in air from all around, a veritable wind, and so produces yet another positive feedback loop: heat of a degree that makes nonsense of all adaptations, reducing everything remotely flammable to ashes.

Global warming, brought about by the greenhouse effect, will make fires worse in several ways. Firstly, general warming, sometimes accompanied by drying, obviously increases the risk; and more frequent tropical storms, with lightning, will provide the trigger. Secondly, more photosynthesis means more leaf litter, which means more tinder to set the forest ablaze.

One solution might be to remove the leaf litter, but this would be immensely difficult on all but the smallest scale and could in the end be damaging in other ways. The leaf litter provides the soil organisms with their supply of carbon; and these organisms include the nitrogen-fixing bacteria, which are such a vital source of fertility. On the small scale, I have seen the effects of such tidiness in Brazil, where some coffee farmers, obsessed with hygiene, remove the dead leaves from the ground – and so lose their fertility. Cocoa is particularly damaged. The flies that pollinate its flowers breed in the leaf litter. Remove it, and there will be no fruit.

All in all, then, fire is a perennial problem: and global warming is making it worse. If and when trees burn, the carbon trapped within them is of course released: all the good work of their growth is undone. The ground that's left behind them is bare. The organisms in the soil are constantly releasing carbon dioxide as they respire: they are releasing carbon from the organic material in the forest floor. When the trees are gone there is nothing to absorb this carbon dioxide, and the organisms of the soil add yet more to the climate's woes. In any case, sometimes fire enters the soil itself and burns the organic material directly, sometimes persisting underground for months. It seems most unlikely, but it happens nonetheless.

Yet fire is not the only threat from global warming. There will also be more storms – as Hurricane Ivan, most unseasonably, proved in the Caribbean and the southern United States in the late summer of 2004. Again, forests in general are adapted to occasional, partial wipe-outs: the pioneer trees in particular (like *Cecropia* and mahogany) depend on them to provide occasional glimpses of the sky. But again, frequency and intensity are all. Huge storms – in northern climates as well as in the tropics – lead to total wipe-out, with masses of trees left to rot and hundreds of square kilometres of ground respiring away their stored carbon.

Global warming, too, cannot come about smoothly. From the outset, scientists have predicted that the general increase in global temperature will, for a time, perhaps a very long time, lead to sudden outbursts of weather that in the historical records at least is quite unprecedented – extremely hot or extremely cold, or wet or dry, or simply out of season. Again the Cerrado around Brasilia provides a small but cogent example. Thus, on 3 September 2004, at the start of what should have been the rainy season, it did indeed rain more or less on schedule for an hour and a half. The plants sent out their green shoots. Flowers appeared out of nowhere, as desert flowers so miraculously do. But there was no follow-up. That was all the rain there was. The new shoots and flowers were burnt up in the sun. Such scenes must be occurring a thousand times around the globe – but this particular occasion was recorded by Dr Stephen Harris, again from Oxford. Here is the tropical equivalent of the false spring followed by a late frost, so often experienced in notoriously fickle climates like that of Britain and which all gardeners fear.

Plants will be fooled, too, on the global scale, just as we have seen that city trees may be deceived when the streetlights come on. Trees in high latitudes are geared to alternate patterns of long days and short days, accompanied by fairly predictable swings of temperature: warm when the days are long, cold when they are short. Global warming changes the rules. Soon, the short days of northern winters and spring will be warm too. In general, the effect should be less damaging than it would be the other way around: if trees prepare to produce their buds in answer to the lengthening days of spring, only to encounter sharp late frosts. Even so, it does not augur well. Trees in all latitudes are finely adapted to the kind of climate that prevailed this past few thousand years. A sudden change – and the threatened changes are very sudden by biological standards – will take the rug from beneath their feet.

Finally, it seems very likely that the insect and other pests that now find life difficult in northern climes will find it progressively easier. Animals that can move quickly and easily do not need to adapt genetically. They simply up sticks and move on. Again, there are plenty of examples, both in agriculture and forestry, of an apparent migration northwards. We will just have to wait and see what damage is wrought by this.

In short, global warming needs to be taken very seriously; and although the matter is not open and shut, the sum of evidence, plus common sense and basic biological theory, suggest that the more forest we retain, and the more new forest we plant, the better. Europe in general is planting more trees, after hugely reducing its cover in the 'enlightened' eighteenth century and the zealously industrializing nineteenth century. But other countries, anxious to join the party of neoliberal economics, are still reducing their forests in the interests of what they see, and are encouraged to see, as modernity. Though Brazil now has an enlightened president in Lula, it too has mooted a scheme to reduce the forest of Amazonia, the biggest and most important tropical forest on earth, by about 50 per cent over the next few decades. Brazil of course has to get its own economy straight. But the world as a whole has to help the Brazilians to get straight without felling their trees. Brazil is a long way from where most of the rest of the world lives, and only a minority are lucky enough to visit it. But we all need its forest.

WATER AND SOIL: THE PARTICULARITIES OF RAIN AND FLOOD

Trees shift a prodigious quantity of water – from the soil up into the leaves, out through the stomata, and away into the air. The water, drawn as we have seen in long thin threads up the xylem, generally flows at less than 6 metres per hour but can sometimes reach 40 metres an hour: enough to reach the top of the tallest tree in two hours. A big tree can transpire 500 litres in a day. A hectare of wood or of plantation with 100 well-grown trees (planted 10 metres apart – a modest stocking rate) pushes out 50,000 litres or 50 cubic metres: enough to fill, say, a hotel swimming pool. One square kilometre of such woodland (100 hectares) would send out 5,000 cubic metres – enough to fill two Olympic-sized swimming pools (for Olympic pools must measure 50 metres × 25 metres × 2 metres). This is *per day*. The catchment area of a river that feeds into a village may cover scores or hundreds of square kilometres. Vast amounts of water are thus sent up into the atmosphere that otherwise would add to the ground-water and run away into the rivers. Thus the danger that the rivers will overflow their banks is reduced. The water that is sent up into the sky forms clouds and will fall again some other day, or in some other place; but so long as the downfall is spread out over time and space, the ground should not be overwhelmed.

Trees also re-route the rain as it falls. Many epiphytes, perched high in tropical trees with no roots to the ground, go to great lengths to trap whatever water they can: bromeliads in particular trap the falling rain in their pineapple-like whorls of leaves (and mosquitoes and tree frogs may breed in the pools that they create: aquatic ecosystems in miniature, high above the forest floor). But the tropical trees themselves commonly contrive to jettison surplus water, their leaves fitted with drip-tips. Yet on all trees, a fair proportion of any one shower is caught in the leaves, and since the leaves are hung out high above the ground like washing on a balcony, the water evaporates again before it reaches the forest floor. It is returned to the atmosphere whence it came – to fall again as rain somewhere else, or in the same place on some other day. Several smaller showers, spaced out, are easier to cope with than one downpour. Thus the weather is ameliorated. Forest floor tends to be permeable, too: less trampled than open grassland, and penetrated by many a root. So the water that does reach the ground is more likely to sink in, and less likely to run away, than on pastureland. Once the water is in the soil then, as we have seen, it is summarily sucked up again and shot back into the atmosphere.

All in all, then, mature forest is wondrously drying. Where dryness is a problem, you can have too many trees, as many a householder has discovered, as the desiccated clay under their home contracts and the foundations crack. Eucalypts are famously desiccating, with long taproots reaching down to the groundwater, transpiring long after other trees have given up. They may create drought around them and kill the surrounding trees if planted in the wrong places, as I have seen them doing in the dry Cerrado forest around Brasilia. But if you live at the foot of a mountain, on the rainy side, then the more trees there are on the slopes above, the better. Of course, if the rain is prodigious and persistent, as it may be in some monsoons, then the forest and the ground it stands in can be saturated like anything else, and the surplus will run off just it would run from bare ground. But trees prevent many a smaller flood; and even when the forest is over-whelmed and sheds its surplus water it should hang on to its soil.

In August 2004 London suffered a flash-flood that discharged 3 million litres of sewage into the Thames. This of course is small beer compared to many a flood of late, but what's significant are the suggested remedies: water butts to slow the flow of water from roofs into the streets and drains; porous pavements to allow the ground beneath to take up some of the surplus. These measures do not of course reduce the total mass of water. But they slow it up. They dole it out more slowly into the streets, drains and rivers, so they are not so likely to be overwhelmed. All this would involve extensive civil engineering, simple in principle but extremely expensive. Yet this is the service that trees provide in every river catchment throughout the world *for free*. In principle, all we have to do is leave them alone. Of course, too, the water that finally does flow from a forest is generally clean. When bare hillside is flooded it's mud that flows off, and that is more damaging and dangerous by far.

On the other hand, nothing is ever quite straightforward. In general, trees ameliorate the ill effects of too much rain: of course they do. In general, too, their roots help to hold the soil in place. How could it not be so? But this does not mean that any old tree or group of trees, planted any old how, will always be helpful. Teak, for example, typically has huge leaves, as big as dinner plates. Each one can catch a prodigious quantity of water – which then flows from the tips in commensurately enormous drops. These fall from on high like glass marbles, and on impervious soils may batter the surface into a glassy crust – which increases run-off and eventually, if there is any slope at all, may simply slide away. If the soil is more permeable the huge drops penetrate and eventually carve deep gulleys.

Then again, people seeking to establish new forests too quickly have sometimes approached the task too eagerly. Thus in China, zealous peasants in recent years have sometimes begun by ploughing the land they wish to plant – and if this is on hillside, which it often is, then by the time the trees have established themselves the newly bared soil has already washed away. Forests in general work best when they are diverse and when they are given time to do their own thing. Attempts to push ahead too quickly, or to oversimplify the planting, can cause the kinds of disasters that give proactive forestry a bad name. A surprising variety of vested interests are ranged against trees and forests. It is as well not to give them further ammunition.

Finally, trees can help to ameliorate global warming if grown as a prime source of energy (and help us to solve our energy problems as fossil fuels run out). Of course when wood is burnt it is oxidized to carbon dioxide – the very stuff that is causing global warming. But the carbon it releases is only what the tree itself has stored through a life of photosynthesis, and so, overall, trees grown for fuel are carbon neutral. Fossil fuels, by contrast, were created by plants (and other creatures in the case of oil) that lived hundreds of millions of years in the past; and within a few decades, as we burn them, we release all the carbon that those ancient organisms took many millions of years to fix. Exactly how much energy we could and should derive from trees and other biomass (as opposed to other renewables – wind power, solar panels, tidal power) has yet to be determined, but it must be a great deal more than now.

A WORLD BUILT ON TREES

Great architecture (and great ships) began with timber. Yet, in our frenetic search for new wonders, we have spent the past 200 years developing alternatives, like steel and plastics. They have their place, of course, but an economy rooted in such high tech uses far too much energy. More broadly, economies rooted in industrial chemistry (the kind that produces steel and plastic) can now be seen to be old-fashioned. Future economies must be rooted in biology. In construction, especially but not exclusively of buildings, we must reverse the trend away from timber.

Timber cities would lock up a great deal of carbon. Even more to the point: although it requires energy to turn a tree-trunk into a finished beam (sawing, planing, transport), it takes roughly twelve times as much to make a steel girder that is functionally equivalent. So it surely would pay us to use timber as much as possible instead of steel. By no means would timber necessarily be inferior. It is possible, for example, that if the joists in New York's Twin Towers had been of teak, suitably protected, they would have withstood the inferno of 9/11 for longer than the steel did, for steel buckles when heated, while thick wood takes a very long time to burn through. With more time, more people could have got out. In short, even the most prestigious buildings of the future might with advantage incorporate as much timber as possible.

Of course the world already has many fine timber buildings: the meeting halls of the Maoris, with their carved gables and pillars of totara; the lovely colonial churches of New Zealand and the United States; their stunning modern counterparts in Scandinavia; many beautiful houses all over the world – and commercial buildings too: I recall a wondrous winery in California with a roof as broad as an aircraft hangar, built like a barrel. Britain is now acquiring some serious timber buildings, but on the whole we seem to retain folk memories of September 1666, when, in the Great Fire, street after street of London's timber-framed houses were reduced to ash within five days. Other people – notably in New Zealand, so I'm told – harbour morbid fears of dry rot. But again, life doesn't have to be like that. Even in Britain many an ancient beam has held up many a cathedral roof for the better part of a thousand years, and with good technique and modern technology fire and rot can largely be avoided. Many a concrete building, by contrast, has run its course in thirty years, and iron buildings burn too (as London's Crystal Palace so spectacularly did in November 1936).

Timber in its modern forms provides architects with an aesthetic challenge as great as that posed in the early twentieth century by concrete and steel. There surely should be major prizes for buildings made entirely from wood and glass – buildings comparable in stature with, say, the Sydney Opera House, or the Guggenheim Museum at Bilbao. Indeed entire cities can be sylvan. Modern Beijing, for example, is largely open to the sun (I have been there only in summer) and terribly harsh – but the diplomatic quarter, viewed from above, looks like an orchard: all the houses beautifully shaded. At ground level this favoured area is a delight. Not all cities can be quite like this (water is an obvious limitation), but many more could be. The problems that builders often raise – roots undermine foundations – are all there to be solved (for example by growing the trees in containers, sunk in the ground). Sylvan cities and great timber buildings are an exercise in Greenness. The Green movement is often perceived to be too earthy. Green activists have nick-names like 'Swampy'. Yet for all their rags and anoraks true Greens are aesthetes – like the Romantic poets, in love with nature. Clearly, too, Green architecture can ascend the heights of urban refinement.

Building and civil engineering are among the world's biggest industries. Biggest of all, though, and the most important, is farming. Farmers and foresters have all too often been at loggerheads. But when farming and forestry are judiciously combined, they can complement each other beautifully.

FORESTS AND FARMING IN TANDEM: THE PROMISE OF AGROFORESTRY

Forests provide us with a great deal more than timber – and all the rest are collectively known as 'non-timber forest products'. The total inventory of resins, fibres and chemically-potent agents requires several fat books of its own, and indeed is the subject of many a library and not a few research centres. Suffice to say that a high percentage of all modern drugs are derived from plants, a large proportion of which are trees. Crops grown for spices, perfumes and medicines provide high value from a small area and are easy to transport: a tremendous bonus in principle for small farmers – including agroforesters.

However, the world has yet to sort out the practicalities of such production – technical, legal, ethical. Many valuable, recondite materials from plants need serious pharmacological development and this, in practice, is carried out by specialist university departments and commercial companies.

Some of the necessary high tech is to be found in the countries where the valued trees grow, but much of it is not. Not even the richest countries harbour every kind of expertise that's needed. So partnerships are needed, typically between tropical (usually impoverished) countries that grow the relevant plants and rich countries with the necessary high tech – with as much work as possible carried out in the countries where the crops originate. Some such partnerships exist, that are beneficial to all parties. But greed and opportunism have too often reared their horrible heads. Henry Wickham's expropriation of rubber tree seeds from Brazil in 1876 might charitably be seen as a somewhat equivocal case, but many others have been just plain theft, of the kind known as 'bio-piracy'. Battalions of lawyers are now employed to give bio-piracy the veneer of legality and this surely is to plumb the venal depths, for when the law itself is on the side of palpable injustice there is nowhere else for humanity to run. The fear of bio-piracy is such that many countries refuse to consider even legitimate partnerships that could do them good; and non-government organizations, which in general rank among the world's most valuable institutions, have sometimes prevented very good deals from going through. Thus because of frank (if sometimes legalized) banditry on the one side and suspicion on the other (sometimes justified and sometimes not), the riches of wild plants in general are far less exploited than they might be. This is all very sad. There is little trust, and little basis for trust, and legal nicety (lovely jubbly for lawyers) prevails where simple respect and honesty should be enough.

Food from trees is significant already – but again, we ain't seen nothing yet. It's obvious that human beings could not have become anything like so numerous as we are unless we had learned to grow our own food. The archaeological record suggests that large-scale, settled farming began around 10,000 years ago in the Middle East. By then, our species was already ancient and had spread out of Africa and Eurasia to Australia and the Americas – and yet the world population at that time is estimated at only around 10 million. By the time of Christ, after 8,000 years of settled farming, there were somewhere between 100 and 300 million of us. Now we number around 6,000 million. Historians typically suggest that the kind of farming that could generate such numbers depended on cereals and pulses: that is, the farming that really counted was arable. In the Middle East, wheat and barley prevailed; in the Far East, it was rice; in North America, maize. Grains have obvious advantages and have long dominated world agriculture – indeed, rice, wheat and maize currently provide humanity with half our total energy and two-thirds of our protein. All other crops (even soya, beef and potatoes) are also-rans by comparison. The seeds and fruits of trees are reduced to footnotes.

But it's a mistake to read history by extrapolating backwards from the status quo. Even in the modern Mediterranean the olive is a significant source of calories, as well as of delectation. I find it entirely plausible that people of 10,000 years ago – and indeed well before – would have regarded olives as a staple. Dead goat and herbs have a lot to commend them nutritionally – protein, vitamins, minerals – but they are low in calories. Baste them in olive oil and they become substantial. Today, the coconut of India and the Pacific, the macadamia of Australia, and the mongongo nuts of the Kalahari are serious staples for local people. People of the Mediterranean and eastwards into Asia would between them have leaned heavily on pistachio, walnut, cashew and almond – all of which still feature strongly in Middle Eastern cooking. People further north had hazel and chestnuts: hazel flour is a significant presence in traditional German cooking, and the chestnut stuffings that now eke out the traditional goose might once have been the centrepiece, with the goose as the garnish. North Americans had walnuts and hickories, including pecans. The seeds of many pines are good too. Nuts in general are rich both in fat (calories) and in protein. As a bonus, maple and birch and others give us syrups. The flesh of fruit, too, is not just for delectation and vitamins. Many are significant sources of fat, the richest of all sources of energy, including the coconut, olive and avocado. Some fruits contain significant protein.

In short, it seems too cavalier by half to give all the credit to grains for the rise of farming, and hence for the expansion of the human species. Obviously grains played a huge part. They are also very convenient, lending themselves to simple technologies for mass production and processing – notably the plough and the millstone. They have short generation times too, giving plenty of scope for genetic improvement – which traditional farmers achieve simply by selecting the best, without formal knowledge of Mendelian genetics. But, I suggest, if there had been no grains at all – no wheat, no barley, no oats, no maize, no rice, no rye, and indeed no sorghum, millet, teff, quinoa or amaranth – then the human species might well have flourished just the same, with an 'agriculture' built on trees. After all, if the same effort had been put into the walnut as has been put into wheat, then walnut trees by now would be taking hundreds of forms

– some mighty trees, some dwarfed, some grown like grapes as vines, some like peaches in espaliers, clinging to the wall: and there would be walnuts of all shapes, sizes, and flavours, and a range of liquors (walnut beer and walnut whisky) fermented and distilled from them.

I am floating these ideas partly through whimsy but mainly to emphasize that although food from trees at present plays only a small part in human affairs (at least if you judge from global statistics), this is largely an historical and economic accident. Grains clearly do have advantages, but they have become as dominant as they have largely through their own momentum. In particular, once the plough was developed (at least 5,000 years ago) arable farming became the norm, and everything else became secondary. But if trees had only been taken more seriously, they could have become an enormous food resource – and might be now, if only the coin of history had flipped differently. Indeed, trees have many advantages over grains – they keep the soil in place, they help to keep the climate equable – and we should be growing as many as possible. So it is important to change our mindset – move away from the *idée fixe* which says that grains must be central and everything else is marginal. At present, much of Brazil, both rainforest and Cerrado, is being cleared to make way for vast estates of soya, which is now Brazil's biggest agricultural export (and is destined not to feed people, but to fatten European cattle – and not because we need the cattle or because the cattle need feeding, but because Brazilian soya is, for the present, cheap). But modern research is showing that the Cerrado could yield far more, and do its society much more good, if its people were encouraged and helped to develop and exploit its native plants. In *Frutas do Cerrado* (2001) EMBRAPA lists fifty-seven native species of fruit, with recipes. The trees among them range from four kinds of araticum (various species of *Annona*, custard-apple relatives) through banha-de-galinha (*Swartzia langsdorfii*) to pitomba-do-cerrado (*Talisia esculenta*) and puca (*Mouriri pusa*). But *Frutas do Cerrado* is only a brief brochure: there are scores of other species. Professor Carolyn Proenca of the University of Brasilia lists 120 – but stresses that this, too, is just a sample.

Trees provide a great deal of food by indirect routes, too – and again could give us much more. In traditional agrarian economies the leaves, twigs, branches and seeds – including many that human beings find fairly unpalatable, such as acorns – support significant herds of livestock. In India you commonly see long files of women and girls carrying huge bundles of branches from the forest to feed their cattle. The technology could surely be improved – the work seems mighty hard – but the general idea, that cattle (and sheep, and indeed goats) can be raised on trees is at least salutary to westerners who think that only grass will do – and is hugely important, since woodland is often preferable to grassland. Pigs and poultry will eat seeds of the kind that humans generally prefer to leave alone, such as acorns. At present, half the world's cereals (and 90 per cent of the soya) are used to feed livestock. Since people can live perfectly well on cereals (plus a few extra vitamins), the animals thus become our rivals: and indeed the United Nations calculates that by 2050, when the human population seems likely to reach around 9 billion, livestock will be consuming an amount equivalent to another 4 billion – increasing the world's food burden by nearly 50 per cent. But if animals are fed on grass – or on bits of trees – then they *add* to our food supply, since we cannot usefully eat grass or trees. Research is in progress to provide better trees for fodder (more nutritious, less toxic), but nothing like the amount that is being done to add another fraction of a per cent to the yield of soya (which makes more money, though only for a few people).

In various ways production of trees for all purposes – timber, food, resins, whatever – can be combined with production of livestock *or* of conventional food crops in systems known as 'agroforestry'. Agroforesty is in all ways tremendously intriguing. It is ancient in principle – much of the economy of medieval Europe was based on 'forestry', and marauding herds of pigs (in particular) were very much a part of it. But it is also one of the great hopes for the future – and is now at least beginning to attract the kind of research funding that it deserves.

There is a spectrum of agroforestry. In England, traditionally, farmers often left some trees (notably elms) to grow tall in hedgerows, to provide a useful source of timber in decades to come. Copses in field corners served the same purpose. In northern France, the rows of poplars along the field edges define the character of the entire landscape. They have cash value and in the short term serve as wind-breaks – and, odd though it may seem, they form better windbreaks when there are gaps between them than they would if they formed a solid barrier, for solid fences create turbulence. In southern Europe you commonly see broad beans and other crops grown among olive trees. In Andalusia and Portugal cork oaks, valuable in their own right, become even more valuable as black pigs wax fat on their acorns. At the Food Animals Initiative in Oxfordshire, chickens are being raised under young trees – birches, beeches, hazels. Chickens prefer woodland: they are descended from Indian jungle fowl. Allegedly free-range chickens are often reluctant to take to the great outdoors precisely because they feel threatened if there is no shelter – and rightly, because even in Britain the main threat comes not from foxes on the ground but from above, notably from crows but also from herring gulls (and to a far lesser extent, from birds of prey). But the aerial invaders prefer a clear run. A cover of trees deters them.

Yet the tropics surely have most to gain from agroforestry. The best coffee and tea is grown under shade. Some spices and medicinal plants grow in woods, including cardamoms, an important local industry in Kerala. Leguminous trees are commonly grown for shade: as nitrogen-fixers they also help to fertilize the crops around them, and their nitrogen-rich leaves make particularly fine fodder. Enormous herds of cattle, pigs and poultry could be raised in plantations to the benefit both of the trees (which would thereby be manured) and of the animals, which would find food (largely cut for them from the trees) and much-needed shade. The value of shade for livestock can hardly be overestimated. Of the common domestic livestock, all except sheep are descended from forest animals. Of all wild cattle, only the yak and the North American bison take naturally to the great open spaces – and America's bison is descended, and only in relatively recent years, from the European bison, which is a forest animal and still roams in the forests of Poland. These broad biological observations translate into hard-nosed commerce. Research in Costa Rica has shown that the milk yield of tropical dairy cattle can increase by 30 per cent if they are shaded. Contrast this with the parched and desperate herds that traditionally run on the unprotected prairies, pampas and savannahs. We all like cowboys, driving their dogies across the plains of Texas and Wyoming. But as a way of raising cattle, this is both wasteful and cruel.

Indeed, agroforestry benefits everybody, in all ways. Traditional foresters must typically wait around thirty years before seeing any return on their investment. But if they use the space between the trees to raise other crops (including valuable spice and medicinal crops), they can gain a short-term income too – while all the time the trees grow steadily, to provide a bonanza in the future. Again, I have come across this in Kerala. Again, one wonders why it is not the norm. In most of the world, in most circumstances, agroforestry makes obvious sense. Contrariwise, the present separation of forest and farming that is customary, with the farms more and more monocultural, often makes very little sense at all, except in the short term to a few entrepreneurs. But get-rich-quick is not what farming is supposed to be about. Yet there is still a case for growing trees in dedicated plantations – and for harvesting them judiciously from wild forest.

HOW TO GROW TREES

We can grow and manage trees specifically for our own use, and/or we can help them to grow just because they have a right to live, and are good for other species. Of course, too, by managing the climate and soil, trees are likely to bring us benefit even if we are not specifically growing them just for ourselves.

Trees for our own use can be raised in plantations, and it makes sense to identify the kinds that produce the timber we need and grow quickest in the land available. This was the thrust of the colonial approach to tropical forestry, of the nineteenth and most of the twentieth century. Often the favoured trees were exotics, planted in lands not their own: and thus the world at large now has many thousands of hectares of eucalypts, originally transplanted from Australia, while Australia has estates of Indian sandalwood; South and Central America have acquired teak from India and Burma; huge estates of America's Monterey pine can be found everywhere; there are plenty of European poplars in India; there's a great forest of Caribbean pine on the outskirts of Brasilia; Britain's Forestry Commission planted vast areas of uplands (and sometimes lowlands) with American Sitka spruce after the First World War; and so on.

Such exotic planting has many advantages. Eucalypts judiciously planted in Africa and Asia commonly produce ten times as much timber per year as native species – and this in theory should take an enormous burden from the native forests. Commonly, too, species brought from abroad leave their usual pests behind them, so that teak grown in Amazonia, for instance, seems free of the defoliator moths that cause such havoc in India. But there are many drawbacks too. Britons have objected on aesthetic grounds to the military ranks of Sitka spruce. Exotic trees do provide shelter for local creatures and often supply some food, but in general they are far less hospitable to local wildlife than native species. Sometimes they are positively hostile: conifers on Scottish hillsides increased the acidity of the soil and compromised local flora, and eucalypts on dry soils can rob their neighbouring plants of water. Exotic timber trees have often escaped to become weeds – like many a eucalypt and acacia. Parasitic sandalwoods from India grown in Australia have sometimes escaped to attack the local eucalypts, while native sandalwoods in India have sometimes attacked eucalypt plantations. Human beings may choose to manipulate nature this way and that, but the natural battles of ecology continue.

Finally, although exotics often grow supremely well in ideal land, they often fare no better than the natives when planted on poor land: but forest is commonly relegated to poor land, since the best is reserved for agriculture. Then, native species would often be preferable because of their general friendliness to local people and wildlife – although they have often in practice been ousted by forestry plantations effectively as a matter of routine. In Harare, Zimbabwe, Gus Le Breton runs Phytotrade Africa, devoted to the development of native species, not least of the wonderfully versatile local baobab, *Adansonia digitata*. In 2004, at the Forestry Research Institute at Dehra Dun, the then director Dr Padam Bhojvaid was seeking to reinstate as many as possible of the 400 or so native Indian species that have been used commercially in the past, many of which have been sidelined through the emphasis on teak (native to India to be sure, but still grown in colonial style in huge monocultural plantations). In truth, there is and always will be a place for plantations of exotics, and the world has good cause to be grateful to the traditional 'colonial' foresters who laid the groundwork in science and technique and nowadays tend often to be under-appreciated. But it's a mistake to apply even the best ideas slavishly and under all conditions – as the old-style foresters would certainly have agreed.

Wild forest ideally should surely be kept pristine because pristine forest is, as serious actors say of live theatre, what it's all about. Even so, in a crowded world we are obliged to make use of the natural forest as much as possible without destroying it. Besides, many people worldwide, particularly in Africa, tropical America and Asia, but also in much of Europe, make their living in the forest. Some traditional forest people are true foresters – producers of charcoal, for instance. Others specialize in wild foods or medicines. Nowadays tourism can be a huge source of income – it is the biggest earner by far in Kenya, for example, although it is easier to take tourists around the savannah than through, say, the forests of South-East Asia (although a ride I took in a cable car at canopy height in the subtropical forest of Yunnan in central China is one of the great memories of my life, the bamboos stretching endlessly below and endlessly above, the air awash with dragonflies, smoky pink and iridescent blue). But it is important that the cash that tourism generates should benefit local communities, and the habitats themselves, and this is often far from the case.

Above all, wild forest can be and always has been a prime source of timber – but whereas in the past foresters all too often just took what they wanted (and clear-felled vast areas of North America, for instance, often with gratuitous profligacy) the trend now is to log selectively and sustainably. Tropical forest, with its huge range of species, all of different ages, poses the greatest challenge – but it is one that Brazil's researchers at EMBRAPA is rising to. Thus the foresters under EMBRAPA's jurisdiction divide the Amazon forest around Belém into sections. Each section is harvested only at thirty-year intervals – thirty years being a rule of thumb, but a sensible one. They do not simply fell whatever looks biggest and likeliest, or seems roughly to conform to the species they want. They go to great lengths to ensure they know the particular species of each tree. The reason is as described earlier; that what seems to be one kind of tree may in reality be several, and if the logging is indiscriminate then the rarer types might be driven to extinction in passing. In any one round of harvesting, the foresters are careful to leave a good representation of each kind of tree – and preserve especially the most fruitful 'mother trees', to give rise to the next generation.

I spent a day with Brazilian foresters, round Santarém, guided by EMBRAPA's Ian Thompson; and indeed their work is most impressive. The trees to be felled have all been previously identified, marked, and their position plotted on a computerized map. Because there are so many different species, and only a few of each can be harvested at any one time, the target trees may be widely scattered. Each one has to be located, felled, and then dragged out of the forest to the dispersal point – a task performed by dedicated tractors known as 'skidders', with caterpillar tracks and grappling hooks. In the bad old days (and still, where the harvesting is less controlled) likely-looking trees were simply felled, and the skidders crashed through the undergrowth to look for them – creating mayhem, and often missing some. Now, the skidders follow the most economical route to the trees that have been felled and drag them out decorously, with immense skill, touching no other trees on either side. In the bad old days, logs were often half stripped of bark by the time they were dragged out – and so, it might be inferred, were the trees they had crashed against en route. Now they emerge unblemished.

But even with the best of intentions, it is hard to work always to the highest standards. In Brazil, I went out with a pukka team linked to EMBRAPA, and truly dedicated to the cause of sustainable harvesting. Even they were obliged to cut corners. The men work eleven days on and two days off. They sleep in camps. The food is wholesome – beef, chicken, beans, rice – but it's more or less the same every day. Brazilians are very friendly, but these lumberjacks were too tired even to look up. Under the rules the skidders should not work if the ground is wet, for then they make deep furrows where water gathers and mosquitoes breed, and they crush the soil and generally screw up the natural drainage. (For the same reason, northern lumberjacks traditionally work only in winter, when the ground is iron hard.) But on the day I was there it had been raining. The skidders dug deep. What option did they have? The men and machines were in the field and there were quotas to fill. But the team I was with was among the best, with a benign and enlightened foreman (a local man). I was told of a team nearby that was contracted to cut sixty trees a day – one every few minutes: far too many to allow serious reflection, or best practice. Recently a man on that team had been killed, hardly surprisingly, and his mates were not allowed a day off to attend his funeral. Thus we see in microcosm the tragedy of the modern world: how good ideas, and life itself, are sacrificed to the all-powerful gods of profit and competition. We can't afford to run the world like that: it's too vicious, too dangerous.

It is also tragically and abundantly the case that the neat, clever, well-planned mode of harvesting I witnessed in Santarém is not the norm in Brazil, or in the tropics as a whole. In Brazil, 60 per cent of felling in that vast and difficult country is carried out illegally. Greenpeace estimates that in the state of Pará, it is 90 per cent. President Lula acknowledges that most people in Brazil, as in the Third World as a whole (which means most of the world), are agrarian. He wants to build an agrarian economy based on small farms and forest-farms, not unlike that of the United States in the early to mid twentieth century. But the big-time ranchers and loggers, including or especially the illegal loggers, have others plans, and pursue them ruthlessly. Those who speak for small settlers are liable to be murdered, like the rubber tapper Chico Mendez in 1988 or, most recently, Sister Dorothy Stang, originally from Ohio, aged seventy-four, who had campaigned for thirty years for the poor people of Amazonia. She was gunned down on 12 February 2005. Assassination is a regular trade in rural Brazil. The hitmen are called *pistoleiros*. I met one who retired to drive a taxi.

Loggers and ranchers, legal and illegal, have already removed 20 per cent of the 4 million square kilometres (400 million hectares) of Amazonia's rainforest. The International Monetary Fund, which lent billions of dollars to Brazil after its recession of 2002, is urging yet more clearance in the cause of economic growth – as if economic growth, defined in crude cash, was necessarily reflected in human wellbeing. As in Brazil, so in Indonesia: the police are estimated to intercept only about 3 per cent of the shiploads of wild timber that stream out of the Indonesian province of Papua (the western half of New Guinea), each one returning a profit of around $100,000. Overall, at this moment, the fight to maintain the world's wild forests, and to leave a world that can support our grandchildren, is being lost.

But at least now there are rules, which weren't there a few years ago, and they are catching on – for reasons of sound business as well as a more general sense of enlightenment. More and more importers now insist that timber from all sources is certified by the Forestry Stewardship Council, launched in 1989, to guarantee that properly identified species have been harvested in a sustainable manner. Still, there are plenty of drawbacks, not least that the smallest providers cannot always afford to comply with the protocols: and the problem may sometimes lie with the paperwork, rather than the practice. Nonetheless, FSC guidelines are now applied to more than 16.5 million hectares in more than thirty countries. The industry is at least trying to clean up its act, and credit where it's due. With luck, the race can be won. Perhaps what's needed now above all are better-informed consumers: people worldwide who recognize that Brazil's angelim, say, is a very special timber and are prepared to pay handsomely for it – provided it is certified as the right species, properly harvested. If, at the same time, producer countries ensure that the cash flows back to the communities on whose lands the trees were growing, then we truly have the basis for a benign industry that benefits everybody. The same is true, of course, in food production. If consumers pay well for properly raised chickens (chickens raised in woods are ideal) and for fair-trade organic coffee, and if the farmers get the money, then the world can truly improve. If the producers are paid too little, and production is cut-price and careless to keep costs down and maximize profit margins, and/or the money is siphoned off by middlemen, then the world will go to pot and all of us will go with it.

In temperate and extreme northern forests the ecology is in principle much simpler. Latvia is a fine example. It is neatly poised between temperate and boreal; forestry is the biggest industry. I went walking with Latvian foresters, late in 2004. All the vast woods are dominated by just a few species: silver birch, Norway spruce, Scots pine, and a couple of species of alder. Plus red and roe deer, moose, wolves, lynxes, and a host of beavers which really do dam the rivers in a most spectacular fashion. Amazing creatures. Sometimes the foresters just harvest individual trees. Sometimes they clear-fell entire areas, usually not too much at a time, and then replant. They replant only native species, barring the odd larch from Russia. Indeed, although the country is small, the foresters have notionally divided it into four regions and do not transfer trees from one region to another, since they may have different adaptations in different places.

The Latvian forests are replenished from vast nurseries – I visited one whose properly proud owner claimed it was 'only small' although she produced 180,000 birch per year, about a quarter of a million spruce and half a million pines (plus a pleasing array of ornamentals – cypress, juniper, rowans, and so on). However, she raises her trees from elite parent stock – trees grown in the wild, but nonetheless more robust than usual. Thus the forest is not exactly replaced in pristine form. There is some genetic improvement along the way (meaning in effect that the wild trees are turned into landraces). On the other hand it seems perverse to plant seed from trees that are known to be feeble, even if there is some loss of genetic diversity when the feeble ones are left out, so this seems a reasonable compromise. There is huge contrast with the tropics, too, in the rate of growth. The Brazilians are hoping to produce worthwhile crops of teak in eighteen years; eucalypts commonly reach harvestable size in less than a decade. The common forest trees of northern-temperate Latvia are typically expected to take around a century. Northerners plant for the next generation but one.

But not all pristine forest should be exploited, either by tourists or by loggers. We need heartlands left entirely to whatever forest people are indigenous to them, and to the wild creatures (albeit with rights of entry for dedicated scholars, for it is always important to improve understanding). The wild creatures have rights of their own and besides, without those heartlands, the slightly less wild places that we do exploit will surely lose their diversity, however dutifully we strive to keep them intact.

Great forestry cannot be a matter simple of aesthetics, however, and cannot be left simply to common sense. Both must be abetted by excellent science.

THE RIGHT KIND OF SCIENCE

Modern forest science can be breathtaking. It operates both on the very largest scale, and on the most minute. Satellites now fly far overhead, measuring the height of individual trees to within a few centimetres – and so are able to monitor growth over vast areas, which is especially useful in times of climate change; and are able, too, from the reflected light, to some extent to identify individual species. The canopy is being opened up by towers, cranes, ingenious systems of ropes borrowed and adapted from rock-climbers, and by gas-filled balloons that hover overhead and lower each scientist into the branches like a worm on a fishing line. The excitement and the promise is the same as it was half a century ago when scuba diving first opened up the coral reefs – except of course that the canopies are even richer than the reefs. Permanent gauges ticking twenty-four hours a day monitor the flow of gases of all kinds, including the volatile organic materials produced by the trees and by the ground litter, providing continuous data on growth and general health, year after year. Bigger and bigger computers extract more and more from the data. All in all, the instruments and the ingenuity are providing a continuous overview that even a couple of decades ago would have been beyond imagining. Without these data, we would have very little insight at all into the effects of global warming. As it is we can see the changes unfolding before our eyes, though it is hard to grasp the complexity.

Science operates on the smallest scale too, as demonstrated at EMBRAPA. It's all very well to identify specific species of trees, and then remove a few – but what effect does this have on the genetic diversity of the ones that are left? After all, if trees of any one species are widely scattered, the total population in any one area is unlikely to be large, and each individual may be making a significant contribution to the overall gene pool. At EMBRAPA, Dr Milton Kanashiro coordinates a programme known as Dendrogene (adapted from a comparable strategy developed in Europe). The idea is to analyse the DNA in the cambium of the trees, and see whether selective logging leads to any change in the total genetic variation in the population as a whole. If the results show that diversity is being lost, then the harvesting can be adjusted. Thus can science improve on common-sense rules of thumb.

So the developing science of forestry is wondrous. The future of biology surely lies at least as much in these broad arenas as in the minutiae that at any one time are fashionable (biotech is the present-day flavour of the month). Yet we should not get carried away by forest science, or by science in general. Science does not, as is so often supposed, provide an undeviating, flawless, royal road to truth. At the deepest level, modern philosophers of science point out that all its theories are uncertain – all provisional, waiting to be upset by new insights. John Stuart Mill pointed out that however much we know, we can never be sure that we haven't missed something vital. Always there are known unknowns – and unknown unknowns, and even unknowable unknowns. When it comes to dealing with living systems – and particularly with systems as complex as tropical forest – the unknowns and the unknowables multiply. Even to acquire the most basic data is extremely difficult and time-consuming: note from earlier discussions how hard it is even to judge how many species of trees there are in the American tropics. Yet the basic inventory of species is only the beginning. The tales related in the last chapter show how complicated the relationships between different creatures can be. After half a century of close study the subtleties of figs and their dialogue with wasps are still being unravelled. But there are millions of species out there, each directly and indirectly interacting with millions of others – and among them, for good measure, bacteria and viruses can often be crucial players, and of them we have virtually no inkling at all except when particularly obvious types attack particular species that we happen to take an interest in.

Yet there is worse. The modern theory of chaos shows that even a few simple forces, when left to interact, may produce endlessly complex and diverse outcomes, and the complexities and diversity are *innately* unpredictable. In forests there aren't just a few simple forces. There are interactions of countless species, each subject to its own pressures. We can sometimes guess within broad limits the outcome of any one exigency – climate change, or particular strategies of logging – but in detail we certainly cannot. One casual introduction can make all the difference – like the European wasps that have been taken to New Zealand, and feed (among other things) on the resin from the totara and other conifers, and seem to be wiping out entire food chains of specialist insects that used to feed on resin, and in turn were preyed upon by birds. Such outcomes cannot be predicted.

Science, in short, for all its wondrousness, is innately limited: the picture it can give of the universe, or of life, or of trees and forests, is always biased and incomplete, and we can never tell how incomplete it is. Yet to conserve forest, and to take from it what we need, we are obliged to manage it. Clearly, even the best forest managers can never achieve the precision of the engineer. They are, at best, like physicians, who are obliged to act if their patients are in trouble, but must always do so with imperfect information. They just have to use their judgement.

So we can adjust the world's economic structure in ways that are sensible – building it largely around trees. There is plenty of good, traditional husbandry out there. It will never be possible to control wild forest absolutely, even if it was aesthetically desirable to do so – but ecology is coming on apace. Science in general, of the right kind, can abet all human endeavour.

What matters in the end, though, is politics – politics in the broad sense: the creation of societies that actually work, and have fruitful relations with each other. It matters who leads those societies, and what the leaders do with their power. Above all we must never stop asking the question that seems to have gone missing. What do we actually *want*? What are we trying to achieve?

I don't believe the world can get significantly better if we leave politics to career politicians. That is not what 'democracy' means. I also nurse the conceit (for which there is abundant evidence) that human beings are basically good (a belief that I have been intrigued to find of late is fundamental to Hindus). It seems to follow that if only democracy can be made to work – if the will of humanity as a whole can prevail – then the world could be a far better place: that it could, after all, come through these next few difficult decades; that our grandchildren can indeed live as they will want to do, and as people should.

It follows that the most important initiatives are those that are called 'grass roots'. Indeed they always have been, when you take a cool view of history: the suffragette movement; the trade unions; organic farming. Things start to work well when people at large take matters into their own hands. All these principles are exemplified by the Green Belt Movement in Kenya. Appropriately, it was begun by a woman from rural Kenya, Wangari Maathai; and it is built around trees.

WANGARI MAATHAI AND THE GREEN BELT MOVEMENT

Wangari Maathai was awarded the Nobel Peace Prize in 2004. She is not of course the first African to win it – others in recent years include Nelson Mandela, Archbishop Desmond Tutu and Kofi Annan – but she is the first African woman. She began the Green Belt Movement in 1977 – partly, she said in her Nobel acceptance speech, 'responding to needs identified by rural women, namely lack of firewood, clean drinking water, balanced diets, shelter and income'. The people in this world whose opinions really count are those who are closest to the action and in Africa, as Professor Maathai pointed out, 'Women are the primary caretakers' and so 'they are often the first to become aware of environmental damage as resources become scarce'.

From the outset the movement focused on planting trees, to supply the things that once were taken for granted but within the past half century, in Wangari Maathai's lifetime, have gone missing. But the psychology is vital too, as all managers everywhere recognize, for 'tree planting is simple and guarantees successful results within a reasonable amount of time. This sustains interest and commitment.'

Since 1977 the women (primarily) of Kenya's Green Belt Movement have planted no fewer than 30 million trees. The trees do indeed 'provide fuel, food, shelter and income to support their children's education and household needs': everything that was hoped of them. More that that, though, as Professor Maathai told a meeting in London early in 2005, they have made the whole environment more agreeable. Kenyan people, women in particular, must still walk many miles carrying water and provisions. Whether you walk in burning sun or in shade makes all the difference – not only to the individual's comfort, but to social life. The women in treeless places had to some extent lost the habit of standing and talking. It was just too hard. Now they do it again. The temper of the whole society is improved. In the same way, Plato and Aristotle both taught their pupils in groves of trees around Athens. Mood is everything.

The political implications are momentous – and for those who care about humanity, as opposed to those concerned only with personal power, they are all to the good. The women, by palpably improving the whole environment, have also vastly improved 'their social and economic position and relevance in the family'. More broadly, when the Green Belt Movement first began, the Kenyan people 'were conditioned to believe that solutions to their problems must come from "outside"', and they were 'unaware of the injustices of international economic arrangements'. But 'through the Green Belt Movement, thousands of ordinary citizens were mobilized and empowered to take action and effect change. They learned to overcome fear and a sense of helplessness and moved to defend their democratic rights.' In short, the Green Belt Movement has relaid the foundations of autonomy and promoted democracy. If 'development' means anything worthwhile at all then surely this is it.

Indeed, 'The tree became a symbol for the democratic struggle. In Nairobi's Uhuru Park at Freedom Corner, and in many parts of the country, trees of peace were planted to demand the release of prisoners and a peaceful transition to democracy. In time, too, the tree became a symbol for peace and conflict resolution, especially during ethnic conflicts in Kenya. The Green Belt Movement used peace trees to reconcile disputing communities. The elders of the Kikuyu carried a staff from a thigi tree that, when placed between two disputing sides, caused them to stop fighting and seek reconciliation.' This tradition is widespread in Africa. Other societies worldwide, not least in North America, mediate discussions through 'talking sticks'. Only those with the stick in hand may talk, and when they do, everyone else must listen.

Eventually, in 2002, Kenya elected a new government, more deliberately committed to the ideals of democracy and autonomy. Wangari Maathai serves in the Ministry of Environment and Natural Resources. In 2004 she established the Wangari Maathai Foundation to continue the work on a global scale. It has an office in London, at the Gaia Foundation, as well as in Nairobi.

Kenya's Green Belt Movement is not alone. There have been and are comparable initiatives elsewhere in the world, not least in India. But it encapsulates everything that seems to matter most. It is a people's movement. It deals with realities – 'real' realities: those of day-to-day life, of human beings and of other living creatures. All the abstractions – the need to create agrarian economies, rooted in biological reality and a true concern for human wellbeing, are put into practice. The abstractions can then be put where they belong, far in the background. Reality is far more interesting. What is happening in Kenya could be re-enacted, in a thousand different forms, all over the world: people themselves creating a world that is good to live in. The contrast with the grand schemes that are now imposed *de haute en bas*, in the name of 'progress', is absolute.

Trees are of course at the heart of things. How could it be otherwise? The human lineage began in trees. We have left our first ancestors far behind but we are creatures of the forest still.

Notes and Further Reading

First, a shortlist of books that I refer to constantly (I have noted titles specific to chapters below):

Jeffery Burley, Julian Evans and John A Youngquist (eds), *Encyclopedia of Forest Sciences* (4 vols; Elsevier, Oxford, 2004). As comprehensive as can reasonably be imagined, with many fine essays. There is promise of constant updating.

D. V. Cowen, *Flowering Trees and Shrubs in India* (Thacker and Co. Ltd., Bombay, 1950; 6th edn, 1984). A lovely piece of publishing.

V. H. Heywood (ed.), *Flowering Plants of the World* (Oxford University Press, Oxford, 1978). A classic, to be found on the shelves of a high proportion of the botanists I have visited. I refer to it as 'Heywood'.

Hugh Johnson, *The International Book of Trees* (Mitchell Beazley, London, 1978).

Walter S. Judd, Christopher S. Campbell, Elizabeth A. Kellogg, Peter F. Stevens and Michael J. Donoghue (eds), *Plant Systematics* (Sinauer Associates Inc., Sunderland, Massachussetts; 2nd edn, 2002). This book has become an instant classic. This is the book I call 'Judd'.

William T. Stearne, *Botanical Latin* (David and Charles, Newton Abbot, 1966; and Timber Press, Portland, Oregon, 2004). A luxury, but a pleasure for those who like words.

I TREES IN MIND

1. Biologists contrive to estimate the number of trees in the tropics by applying a few statistical tricks from the few areas of certainty. One approach is to look at the number of species discovered in a single family since records began, and see whether the curve is levelling out. Thus within the Sapotaceae (the big tropical family that includes chicle, the chewing gum tree), effectively

no species at all were known to European science by 1700. But as naturalists and then modern scientists got stuck in, the numbers of known types rose logarithmically, which in effect means ever more rapidly, so that by 1990 300 species of Sapotaceae had been recorded in the neotropics. If in the last decade of that period the number of new discoveries had levelled off, we could conclude that nearly all the neotropical Sapotaceae must by then have been known. But there was no levelling off. In 1990 the number of newly recorded types was increasing more rapidly than ever. So there is no obvious top end on the number of Sapotaceae species that might be out there. Judd suggests that the present inventory stands at 1,100 (in 53 genera). But there could be many thousands.

Another approach is to look in the world's herbaria. For obvious reasons, common species tend to appear in many different herbaria, while rarer ones turn up only in a few. The very rare species are represented in no herbaria at all – meaning that they are as yet unknown. Again, it's reasonable to suppose that a great many are unknown – precisely because they are rare (and/or extremely inaccessible). In fact there is good reason to suppose that *most* of the species in any one genus are rare. Since the rarest are the least likely to be found this in turn implies that within any one genus, only a minority of species (the commonest ones) have so far been identified (even after 300 years of exploration which in some areas has been fairly intense).

Dr Mike Hopkins and his colleagues at the Brazilian EMBRAPA research centre applied comparable chains of reasoning to the known distribution of various species of trees from several families and genera. As with all plants, a few species of Amazonian trees are fairly common over quite a wide area; some are common in any one place but rare elsewhere; some are rare all over, but widely scattered; and some are rare, and seem to occur in only one or a few places. Any one area is liable to contain various species of any one genus, some common and some rare, some belonging to widespread species and some very local.

From the limited data available – good studies of a few places such as the Ducke Reserve, and scattered observations from elsewhere – and by making a few assumptions that are largely commonsensical, the botanists are able at least roughly to work out what the distribution of each species is liable to be. Thus if specimens of one particular species are known only from areas 1,000 km apart, it is obvious that there must be others in between, even though the others are not known: either that, or the two specimens that are known are the last of their species, and simply waiting to go extinct. By such cogitations (reinforced by computer models) the botanists are able at least roughly to estimate the distribution of the trees they have identified, and how densely they are liable to occur even in places that no botanist has yet visited.

By putting all such data and estimates together, and applying a series of mathematical projections whose details I won't try to convey, the botanists can then work out at least to a first approximation how many species of plants (or trees) there are likely to be in any one place, and how greatly the list of species is liable to differ between any two places. They are able to do this (approximately) even though they have not visited most of the places they are making guesses about, let alone studied them. They are also able to estimate how many species in any one place are still unknown. Finally, and importantly, they are able to guess where the 'hot spots' are liable to be: the places where there are liable to be most species.

All this may sound too rarefied for words – guesswork running miles ahead of data, pulling itself along by its own bootlaces. Indeed, such ways of thinking are known generically as 'bootstrapping'. But in fact the method is more robust than it may seem from this necessarily rough description. More importantly, the estimate arrived at is not just a guess, left hovering in space. It offers testable hypotheses, which are the stuff of real science. That is, if today's botanists guess that such-and-such an area ought to contain a high number of species, with a high proportion of a particular genus, then, in the fullness of time, when more grant money is available, future botanists will be able to go out and see if the projections are true. The more the predictions do prove to be true, the more the calculations on which they are based are vindicated; and if they prove untrue, then that is instructive too. In the short term, such estimates could have great significance for conservation – not least because it's the hot spots, so far identified only on theoretical grounds, that seem most worthy of protection. In short: even if the estimates are wrong, they are definitely better than nothing.

2 KEEPING TRACK

Jose Eduardo L. da S. Ribeiro (ed.), *Flora da Reserva Ducke* (INPA-DFID, 1999).

1. At least a dozen different species of tree are marketed as 'angelim'. All are from the family Fabaceae – but they do come from two different subfamilies. Thus from the subfamily Papilionoideae come at least half a dozen different species of *Hymenolobium*; at least another three from the genus *Vatairea*; plus a *Vataireopsis* and an *Andira*. From the subfamily Mimosoideae come *Zygia racemosa*, *Dinizia excelsa* and the magnificent *Parkia pendula*. Similarly, the valued taurai tree commonly includes at least five species (and probably many more) from the Brazil nut family, Lecythidaceae. At least two

of the alleged 'taurais' are from the genus *Cariniana*, and another three (probably more) from *Couratari*.

3 HOW TREES BECAME

1. Martin Ingrouille, *Diversity and Evolution of Land Plants* (Chapman & Hall, London, 1992). An excellent general outline of plant evolution.

4 WOOD

Aiden Walker (ed.), *The Encyclopedia of Wood* (Quantum Publishing, London, 2001).

5 TREES WITHOUT FLOWERS

Aljos Farjon, *World Checklist and Bibliography of Conifers* (Royal Botanic Gardens, Kew, 2nd edn, 2001).

6 TREES WITH FLOWERS: MAGNOLIAS AND OTHER PRIMITIVES

1. In particular I have in mind Heywood's *Flowering Plants*.

7 FROM PALMS AND SCREW PINES TO YUCCAS AND BAMBOOS

E. J. H. Corner, *The Natural History of Palms* (Weidenfeld & Nicolson, London, 1966). A classic by one of the twentieth century's most original botanic thinkers. In the text I call it 'Corner'.

9 FROM OAKS TO MANGOES: THE GLORIOUS INVENTORY OF ROSE-LIKE EUDICOTS

Thomas Pakenham, *The Remarkable Baobab* (Weidenfeld & Nicolson, London, 2004).

1. Thus while Heywood in the 1970s placed both the currants and the hydrangeas within the Saxifragaceae family, Judd (writing in 2002) separates the currants into the Grossulariaceae family, and gives the hydrangeas their

own family, the Hydrangaceae – which for good measure he transfers to a quite different rosid order, the Cornales. Then again, while Heywood groups the American sweet gum (*Liquidambar*) in with the witch hazels (*Hamamelis*) within the Hamameliaceae family, Judd puts the sweet gums together with the ramara tree (*Altingia*), within the Altingiaceae family. At this stage of taxonomic history, with molecular studies and computer-assisted cladistic discipline rapidly coming on board, life can be very confusing.

2. *Cerciphyllum* is often traditionally classified alongside *Trochodendron* in the Trochodendraceae.

3. Rosaceae is such a big and various family that it has often been divided into tribes: Judd recognizes the Rosoideae (the group with roses) and Maloideae (the group with apples) and several smaller groupings that don't seem to fit in either. The Maloideae in particular have a remarkable tendency to hybridize not only within species but even between genera. *Crataegus* demonstrates this in spades: Judd speaks of 265 species worldwide, but others estimate nearer 400 – in any case it is extremely hard to see which is a true species and which a hybrid, and which true species have arisen as hybrids of others. The same is true to a lesser extent of *Amelanchier* (with about thirty-three recognized species); and for good measure, *Crataegus* and *Amelanchier* seem at times to have hybridized with each other.

4. Botanists have struggled ever since the time of Linnaeus to bring some order to the prodigious variety of oaks. There are more than twenty different classifications in the literature. But there are big problems. Some individual species of *Quercus* are enormously variable. *Quercus*, too, is among those many genera of trees that are prone to hybridize, so it can be hard to see where one species ends and the next begins. As we saw in Chapter 1, the concept of 'species' seems to be far more flexible among trees (and indeed plants in general) than among animals – although even animals hybridize far more than was traditionally supposed. To cap it all, taxonomists cannot agree on which features reveal true evolutionary relationships, and which are incidental. As things stand, many taxonomists at present split the genus into three 'series': the 'red oaks', restricted to North America; the 'white oaks'; and a mixed bag of intermediates. Some kinds, however, including the holm oak, *Q. ilex*, do not seem to fit comfortably into any of these groupings. It seems best to treat this genus-splitting as work in progress, and hope that in time the new molecular studies will throw more light.

5. From their origins in South-East Asia oaks spread all ways and by the Eocene, around 55 million years ago, the fossils show they were common in China, Europe and North America. This was a warm period: there were cycads, primarily trees of the tropics, as far north as Alaska. The world has been cooling fairly steadily ever since (although there have been a few warm

spells). The pending greenhouse effect is returning us (roughly) to the climate of 40 to 50 million years ago. Oaks diversified rapidly between 35 and 5 million years ago, a cooler and drier period which gave rise to the vast expanse of modern grasslands. Most of the several hundred modern species of oak were probably extant by 14 million years ago.

11 HOW TREES LIVE

M. R. Macnair, 'The Hyperaccumulation of Metals by Plants' in *Advances in Botanical Research*, vol. 40 (Elsevier, Amsterdam, 2003), vi–105. On plants that tolerate nickel and other metals.

Peter H. Raven, Ray F. Evert and Susan E. Eichhorn, *Biology of Plants* (Worth Publishers, New York, 5th edn, 1992). A good general introduction to plant physiology.

1. The evolution of photosynthesis, somewhere around 2 billion years ago, changed the course of life on earth. The evolution of all creatures changed direction. The cyanobacteria (or their ancestors) that first evolved photosynthesis clearly lived at first in the absence of free oxygen – since before they developed photosynthesis, there wasn't any. As those first photosynthesizers put more and more oxygen into the atmosphere, so they and all other creatures had to adapt to it (or stay out of its way, as some microbes still do, living in the depths of airless marshes). The adaptation of ancient creatures to the constant presence of oxygen gas was a huge physiological leap, of enormous evolutionary importance. Oxygen is very lively stuff, very reactive, and for creatures that can make use of it, it is extremely useful. In particular, creatures like us use it to break down sugars to provide energy by the method known as 'aerobic respiration'; and aerobic respiration is fast and efficient. But for creatures that are not adapted to it, oxygen is highly toxic; one of the quickest and surest killers there is. (Creatures like us, who do make use of oxygen, still need to pack our bodies with 'anti-oxidants', such as vitamin C, to protect our flesh against its corrosiveness.)

As described in Chapter 3, the chloroplasts that contain the chlorophyll within green leaves, have evolved from cyanobacteria which, in the deep past, lodged in the host cell.

2. The two forms are known as Pr and Pfr. Pr absorbs red light and Pfr absorbs 'far-red' (which effectively means infra-red) light. Pfr is biologically active, and causes things to happen. Pr is inactive, its presence leaving the plant unmoved. Light flips the pigment between its two forms, and thus provides an on – off switch. Red light shone on to (inactive) Pr converts it to

(active) Pfr; and far-red light shone on to (active) Pfr converts it to (inactive) Pr.

Sunlight contains both red and far-red, so Pr and Pfr are normally in equilibrium. At noon, with the sun at its brightest, about 60 per cent of the phytochrome is in the form of Pfr. But in the dark of night, as the hours pass, Pfr steadily spontaneously degrades to become Pr: the active form decays into inactivity. Yet one brief flash will reconvert the accumulating Pr back into the active Pfr. In a long-day (short-night) plant, the burst of Pfr induces flowering. In a short-day (long-night) plant, it suppresses flowering.

12 WHICH TREES LIVE WHERE, AND WHY

Robin L. Chazdon and T. C. Whitmore (eds), *Foundations of Tropical Forest Biology* (University of Chicago Press, Chicago, 2002).

Verna R. Johnston, *California Forests and Woodlands* (University of California Press, 1994).

Plant Phylogeny and the Origin of Major Biomes. Discussion meeting at the Royal Society, London, March 2004, organized by Toby Pennington, Quentin Cronk and James Richardson. Proceedings published by the Royal Society in October 2004.

1. Recorded in J David Henry, *Canada's Boreal Forest* (Smithsonian Institution Press, Washington, 2002).

13 THE SOCIAL LIFE OF TREES

Egbert Giles Leigh, Jr., *Tropical Forest Ecology: A view from Barro Colorado Island* (Oxford University Press, Oxford, 1999).

Ghillean T. Prance, 'The Pollination of Amazonian Plants' in G. T. Prance and T. E. Lovejoy (eds), *Key Environments, Amazonia* (Pergamon Press, Oxford, 1985).

1. S. Patino, E. A. Herre and M. T. Tyree, 'Physiological determinants of *Ficus* fruit temperature and implications for survival of pollinator wasp species', *Oecologia* 100, 13–20. Other key sources for their researches on figs and fig-wasps are:

Edward Allen Herre and Stuart A. West, 'Conflict of interest in a Mutualism: Documenting the elusive fig-wasp–seed trade-off', *Proceedings of the Royal Society*, Series B, 267 (1997), 1,501–7.

Edward Allen Herre, 'Population Structure and the Evolution of Virulence in Nematode Parasites of Fig Wasps', *Science* 259 (1993), 1,442–5.

Carlos A. Machado, Emmanuelle Jousselin, Finn Kjellberg, Stephen G. Compton and Edward Allen Herre, 'Phylogenetic Relationships, Historical Biogeography and Character Evolution of Fig Pollinating Wasps', *Proceedings of the Royal Society*, Series B, 268 (2001), 685–94.

E. A. Herre, N. Knowlton, U. G. Mueller and S. A. Rehner, 'The evolution of Mutualisms: Exploring the paths between conflict and cooperation', *Trends in Ecology and Evolution* 14 (1999), 49–53.

2. Stanley A. Temple, 'Plant–Animal Mutualism: Coevolution with dodo leads to near extinction of plant', *Science* 197 (1977), 885–6,. See, too, the protest in *Science* 203, 1,364 from A. W. Owadally of the Forestry Service, Mauritius, and Dr Temple's reply in the same issue.

14 THE FUTURE WITH TREES

Global Environment Outlook 3 (Earthscan/UNEP, 2002).

M. Ibrajim and J. Beer (eds), *Agroforestry Prototypes for Belize* (CATIE, Turrialba, Costa Rica, 1998).

Wangari Maathai's Nobel Acceptance Speech and other articles about her work can be found on the website of the Green Belt Movement of North America: www.gbmna.org

Andrew W. Mitchell, Katherine Secoy and Tobias Jackson (eds), *The Global Canopy Handbook* (Global Canopy Programme, Oxford, 2002).

L. Szott, M. Ibrahim and J. Beer, *The Hamburger Connection Hangover* (CATIE, Turialba, Costa Rica, 2000).

Glossary

A

allele Many genes are 'polymorphic', meaning they may take more than one form. An allele is any one of the possible variants.

alternation of generations All land plants practise alternation of generations, in which a diploid generation (the sporophyte) gives rise to a haploid generation (the gametophyte) which in turn gives rise to another sporophyte generation, and so on. (Some animals also exhibit alternation of generations including the Cnidarians, which include jellyfish and anemones. But the basis of this is quite different.)

analogous Applied to structures that are similar in function, but originate in different ways. Thus the wings of a fly are merely analogous to the wings of a bird. Many plants, including some acacias and conifers, have phyllodes instead of leaves: they do the same job, but originate differently.

angiosperm Technically, the term refers to plants whose seeds are completely enclosed within an ovary. More casually (but accurately) angiosperm simply means 'flowering plant'.

aril A covering around the seed that is often formed by outgrowth from the base of the ovule. Arils are often brightly coloured, and lure animals that disperse the seeds. Yew 'berries' are arils, and so is mace, the aromatic lacy covering of the cinnamon seed.

B

broadleaf The term colloquially applied to a dicot tree.

bryophyte A primitive land plant that lacks specialized conducting tissue ('tracheary elements') for internal transport of water and nutrients. The most conspicuous generation is the gametophyte. The living examples are hornworts, liverworts and mosses. In earlier classifications these three were grouped together in the formal taxon 'Bryophyta', spelt with a capital 'B'. But the three do not necessarily share a specific common ancestor, and so do not form a true clade, and so should not be presented as a formal group. But they are all of the same 'grade', which can be denoted by the informal 'bryophyte', spelt with a small 'b'.

C

cambium A meristem that gives rise to parallel rows of cells. The cambium of coniferous or angiospermous trees is responsible for 'secondary thickening', producing xylem tissue on the inside and phloem tissue on the outside.

carbon fixation The process by which hydrogen is combined with carbon dioxide from the atmosphere to produce organic molecules.

catkin An inflorescence of single-sexed flowers arranged as on a spike. Catkins are found primarily in woody plants, including trees such as willows and oaks.

chlorophyll The green pigment that mediates photosynthesis.

chloroplast The organelle that contains the chlorophyll

chromosome Chromosomes are long, thin structures that carry the genes. During most of the life of the cell, each chromosome is spread throughout the nucleus, and in this relaxed form they cannot be seen under the light microscope. But during cell division (mitosis and meiosis) the chromosomes contract to form short rods which, when suitably stained, can be seen clearly. In this visible, contracted form, each chromosome has its own characteristic size and shape; and each organism has its own characteristic number of chromosomes, each with its own characteristic size and shape. In truth, each chromosome consists of one enormously long molecule (or rather 'macromolecule') of DNA.

circadian rhythm Regular rhythms of growth and activity over a roughly 24-hour period.

clade 'Clade' derives from the Greek *clados*, meaning 'branch'. In biology a clade is a taxon of all the creatures that are descended from a common ancestor, plus the common ancestor itself. In modern taxonomy, no group is admitted to be a 'true' taxon unless it is a true clade, as defined here. Small clades nest within bigger clades: so species are contained within genera which are contained within families and so on all the way up to domain; and each rank of taxon is itself a clade (provided the taxonomists have done their work properly).

cladistics The set of techniques that are intended to help taxonomists decide whether the creatures they are attempting to classify do or do not form a true clade; and also to show how different clades relate to each other.

class A class is a large taxon (clade) between phylum (animals) or division (plants), and order (*see* Linnean classification).

clone As a noun: a group of genetically identical cells or individuals. The term is also applied to each of the individuals. It may also be used as a verb (so that taking cuttings is an exercise in cloning).

community All the organisms that share a particular environment, and interact with one another. Different members of the community may be of many different species.

conifer A cone-bearing tree.

convergence (convergent evolution) Frequently, species from different lineages of creatures adapt to their surroundings in very similar ways, and so come to resemble each other, wholly or in part. This is convergence.

cork Tissue with polygonal cells that are infused with suberin, which is a waxy material. When they are mature, the cork cells are dead – but in plants, and especially in trees, dead cells often contribute a great deal. Cork in general is protective. The waxiness repels and excludes water, but the gaps between the cells allow the passage of air.

cotyledon The seed leaf; the first leaves of the embryo. Typically though not invariably, dicots have two cotyledons, and monocots have one.

cultivar A domestic variety of plant: a variety produced in cultivation, and generally maintained only in cultivation. Many garden trees are cultivars.

D

day-neutral plants Plants that flower without regard to day-length.

deciduous Trees (and other plants) that shed their leaves periodically are said to be deciduous. Many temperate and boreal trees shed their leaves in autumn, and some tropical trees shed their leaves before the dry season.

dicotyledon (dicot) Basically, dicots are flowering plants whose embryos have two cotyledons. Traditionally, flowering plants were divided into two classes, dicots and monocots. But the dicotyledonous condition is now known to be primitive, so the 'dicots' as originally defined do not form a true clade. (This is explained at length in the text.)

differentiation The process by which embryonic ('stem') cells or tissues become specialized for particular functions. Differentiation is generally associated with loss of totipotency.

dioecious Unisexual. In dioecious trees and other plants, individuals contain either male flowers, or female, but not both. Holly is an example of a dioecious tree.

diploid, diploidy A cell with two sets of chromosomes is said to be diploid. The adjective also applies to an organism that has diploid cells. Diploidy is the abstract noun, referring to the state of being diploid.

division In botany, 'division' is a large taxon, smaller than a kingdom but bigger than a class. Thus it is equivalent to the 'phylum' of zoology (*see* Linnean classification).

DNA (deoxyribonucleic acid) The stuff of which genes are made. DNA provides the code for proteins.

dormancy A special condition in which seeds or buds (or, in principle, any organ) enter a state of arrested development. Growth is actively suppressed (typically by hormones) and will not resume until the plant has been subjected to particular environmental signals. For instance, the seeds of many temperate trees and other plants need to be subjected to bursts of cold (sometimes extreme cold) before they will germinate. Such seeds may not germinate at all if the winters are too warm – which may happen because of global warming.

double fertilization A strange characteristic of angiosperms. A male sex cell from the pollen fuses with a female sex cell from the ovule to form an embryo, as is normal in sexual reproduction. But at the same time, in angiosperms, a subsidiary cell in the pollen fuses with another cell from the ovule. The subsidiary male cell is haploid, while the subsidiary ovule cell is diploid, so the result is a triploid cell. This then divides to form the endosperm of the seed, which serves as a food store for the developing embryo. Extraordinary.

E

ecology The study of all the interactions between different creatures, often of many different species, that share the same environment; and between those creatures and the physical environment as a whole. The word derives from the Greek *oikos*, meaning household, which is also the root of the word 'economy'.

ecosystem The sum total of an environment and all the creatures within it.

enzyme A protein that serves as a catalyst, regulating the individual reactions that constitute the metabolism.

epiphyte A plant that grows on another plant, but is not necessarily a parasite. Trees all over the world are commonly festooned with epiphytes of all kinds – mosses, ferns, orchids, bromeliads. There are even some epiphytic cacti, such as *Zygocactus*.

ethylene A simple organic gas that is a major hormone in trees and other plants, and may also serve as a pheromone.

eukaryote, eukaryotic Literally, 'good cell'. In a eukaryotic cell the DNA is contained within a specialized region called the nucleus, surrounded by a protective, discriminating membrane. An organism that possesses eukaryotic cells is a eukaryote. Plants, animals, fungi, seaweeds, protozoans and so on, are eukaryotes. Bacteria and Archaea are 'prokaryotes'.

evolution The process by which organisms change over time, from generation to generation: Darwin spoke of 'descent with modification'. He proposed that evolutionary change is brought about largely or mainly by natural selection, which leads to adaptation. Other mechanisms of a non-adaptive kind also play a large part, however, including 'genetic drift'.

F

family A taxon of middling size, smaller than an order but bigger than a genus (*see* Linnean classification).

fertilization (1) In reproduction, fertilization is fusion of two gametes to form a diploid zygote. (2) In plant nutrition, fertilization means increasing the nutrient content of the soil (and the term is also sometimes applied to improvement of soil texture).

flower The reproductive structure of angiosperms. The whole 'complete' structure consists of four whorls: the outer calyx, made up of sepals; the corolla, with petals; the male stamens; and the female carpels. Many flowers are 'incomplete', however, and lack one or more of the whorls.

freeloader In ecology: a creature that cashes in on other creatures' mutualistic relationships – taking what's on offer but giving nothing back.

fruit 'Fruit' is a term that should belong exclusively to angiosperms. Fruits may be fleshy or hard or papery, but in any case they are formed from the ovary, plus any other surrounding structures that may become incorporated. Reproductive structures of other plants – or even other non-plants such as fungi – are sometimes called 'fruiting bodies'.

G

game theory A body of mathematical analyses that attempts to quantify the outcome of any encounter between two or more different games players – or two or more wild creatures. Through game theory military strategists and ecologists attempt to define the strategies that are most likely to succeed in any one circumstance.

gamete A haploid sex cell which fuses with another haploid sex cell to form a diploid zygote. In some primitive organisms all individuals produce gametes of the same size. But in organisms that traditionally were said to be 'higher', males produce very small, motile (mobile) gametes known as spermatozoa (or 'sperm'), and females produce large gametes, sometimes enlarged even further with considerable quantities of nutritious yolk, known as eggs. This is true of animals and of plants including cycads and ginkgoes. In conifers and angiosperms, however, the male sex cell is contained within a multi-celled structure known as 'pollen', and the female sex cell is contained within the multi-celled ovule. The male sex cell is then conveyed to the female sex cell via a 'pollen tube'.

gametophyte The generation of plants that produces gametes. In mosses, the predominant generation is the gametophyte. In ferns, the gametophyte is generally small. In angiosperms and conifers, the gametophyte is subsumed within the pollen and ovule.

gene The unit of heredity. Genes are constructed of DNA.

gene pool The total catalogue of all the alleles (genetic variants) within a population of creatures that are interbreeding sexually.

genetic drift The processes by which some alleles are lost from the gene pool by means other than those of natural (or artificial) selection. Most notably: any one individual passes on only half of its genes to each of its offspring. It is possible therefore that some genes are not passed on at all. Particularly in small populations, and particularly in K-strategists (which have only a few offspring) it becomes quite likely that some of the rare genes (alleles) will be lost from the population entirely. Loss of genetic variation by genetic drift leads to evolutionary change, often significant evolutionary change, that is not primarily adaptive – and may indeed lead to the decline of the population and the extinction of the species.

genome The total apportionment of genes within any one organism.

genotype Related organisms with roughly similar genes are said to be of the same genotype.

genus A small taxon, smaller than a family but bigger than a species. The adjective from 'genus' is 'generic' (*see* Linnean classification).

grade Taxonomists speak of 'clades' (defined above) and 'grades'. 'Grade' is a descriptive term that refers to the general level of organization of a creature: how complex it is, structurally and physiologically. Thus mosses, liverworts and hornworts have much in common, both in what they possess and in what they lack: they are all small, green plants that practise very clear alternation of generations, with the main generation being the gametophyte; and they lack specialized conducting tissues (phloem and xylem). They may not be closely related to each other, and so do not seem to belong to the same clade. But in general form and way of life they are much of a muchness, and so can be said to be of the same 'grade' – the grade that is commonly called 'bryophyte'. Similarly, in zoology, the many various creatures commonly referred to as 'reptiles' do not form a single, coherent clade. Tortoises have very different origins from snakes. But again they have much in common – leathery skin, relatively simple brains – and it makes sense to think of them together, and give them the common name of 'reptile'. But 'reptile', like 'bryophyte', is the name of a grade.

gymnosperm A seed plant whose seeds are not fully enclosed within an ovary. The living gymnosperms are the cycads, ginkgoes and conifers.

H

habitat The place and environment where creatures live.

haploid A cell with only one set of chromosomes is said to be haploid. Gametes are haploid (at least when produced by diploid organisms). So are the body cells of gametophytes, as in mosses.

hardwood The forester's term for the timber of broadleaved (dicotyledonous) trees.

heartwood The central core of the trunk of a mature tree, consisting of dead xylem tissue and ray tissue, often impregnated with tannins or other materials. Heartwood forms the greater part of timber and generally by far the most valuable part.

herbarium A central repository where plant material is stored (most typically dried), and can be clearly identified, studied and referred to.

hexaploid A cell that contains six sets of chromosomes (or an individual composed of such cells).

homologous, homology Organs of different creatures that have the same evolutionary and embryonic origin are said to be homologous, whether or not they have the same function. Thus, the wing of a bird is homologous with the arm of a human being (but not with the wing of a fly). The state of being homologous is homology.

homologous chromosomes In a diploid cell (or organism), one of the two sets of chromosomes is derived from the mother, and the other set is from the father. Being of the same species, the two haploid sets are very similar; and each chromosome in each set has a corresponding partner in the other set. The sets of partners are said to be 'homologous'.

hormone A chemical agent produced in one cell or tissue that affects the physiology or behaviour of another cell or tissue (or indeed affects the physiology or behaviour of the whole organism).

host An organism on which a parasite or epiphyte lives.

hybrid Offspring of two genetically distinct parents. A hybrid between individuals from different genera is said to be 'intergeneric'; a hybrid between individuals from different species is 'interspecific'; a hybrid between different varieties from the same species is 'intraspecific'. Many but by no means all hybrids are sexually sterile. Many otherwise sterile hybrids become sexually fertile by becoming polyploid.

I, K

inbreeding Breeding between two closely related organisms, such as siblings, or parents and offspring. Plants sometimes inbreed by self-pollination.

inflorescence A flower cluster. The form of the inflorescence is characteristic of each species. In the Asteraceae (Compositae) the arrangement of individual flowers is so tight that the whole inflorescence (as in a daisy) resembles a single flower (and each individual flower within the inflorescence is then called a 'floret').

kingdom The largest taxon recognized by Linnaeus (who proposed only two kingdoms – Plantae and Animalia). Nowadays, however, kingdoms are grouped within even larger 'domains'; and are divided into divisions (plants) or phyla (animals) (*see* Linnean classification).

L

legume The name colloquially applied to members of the family formerly known as the Leguminosae, but now properly called Fabaceae.

lenticel Holes in the surface tissues of stems or roots, loosely packed with cork cells, that allow the free exchange of gases between the inside of the plant and the outside. Common in many plants but of special significance in the roots of mangroves.

liana A large woody vine that climbs on other plants (and sometimes weighs them down).

lignin A polymer containing nitrogen that binds cellulose fibres together and so provides enormous strength. Wood is basically cellulose toughened with lignin.

Linnean classification The hierarchical system of taxonomy devised by the Swedish biologist Carolus Linnaeus in the middle decades of the eighteenth century. First, he made formal the 'binomial' system of naming living creatures, which had been unfolding over the previous few centuries. In this system, each creature is given two names: the first is 'generic' – that is, the name of the genus or 'kind'; and the second name is 'specific' – the name of the particular species. Thus the common oak of Britain is *Quercus robur* – *Quercus* being generic, and referring to all oaks (all 450 species of them); and *robur* denoting the particular English kind in question. (Humans on the same system are *Homo sapiens*.)

Secondly, Linnaeus proposed a hierarchy of 'taxa' (groups), in which smaller ones nest within larger ones, and so on. Linnaeus proposed five 'ranks'. The biggest in his system was the kingdom, which was divided into classes, which were further subdivided into orders, which then divided into genera and finally into species.

More ranks have been added since Linnaeus's day, and the modern 'Linnean' classification should really be called 'neolinnean' (though so far as I am aware, I am the only person so far to adopt this term). The complete modern sequence runs: domain; kingdom; phylum (animals) or division (plants); class; order; family; genus; species. Species may be further subdivided into subspecies or, less formally, into races. Races of plants are also sometimes called 'varieties'. But varieties of plants produced by artificial breeding in captivity are called 'cultivars'. Varieties of animals produced by artificial breeding are called 'breeds'. Varieties of animals or plants that are produced by informal selection on traditional farms are called 'landraces'.

long-day plant A plant that will not flower unless first exposed to a minimum number of hours of daylight (although in fact long-day plants respond to short nights rather than to long days; see text).

M

macronutrient An inorganic nutrient that a plant requires in large amounts, such as nitrogen, phosphorus, sulphur and potassium.

meiosis The form of cell division in which a diploid sex cell divides to form two haploid gametes.

meristem Undifferentiated plant tissue from which new cells arise. The 'apical meristem' is the growing tip.

metabolism The sum of all chemical processes occurring in a living cell or organism.

micronutrient An inorganic chemical element that is essential to the growth of the organism but is required only in very small amounts. Also known as 'trace element'. Among the essential micronutrients for trees are chlorine, iron, copper, manganese, zinc, molybdenum and boron.

mineral A general term for any element or naturally occurring non-organic compound.

mitochondrion Mitochondria are organelles within a eukaryotic cell where most of reactions of respiration are carried out. Sometimes colloquially called 'the power houses of the cell'.

mitosis The process by which a diploid (or polyploid) cell divides to form two 'daughter' cells containing exact copies of all its chromosomes.

monocotyledon (monocot) Basically, an angiosperm with only one cotyledon in the seed. The monocots all descend from a common ancestor and so form a true clade within the Angiospermae.

monoecious Refers to a plant with single-sex flowers, but in which both sexes occur on the same tree. Pines and oaks are among the many monoecious trees.

mutualism The form of symbiosis in which both partners in the relationship gain net benefit.

mycelium The total of all the hyphae in a fungus. The mycelium of a single fungus may extend over many hectares, and form mycorrhizal relationships with hundreds of trees.

mycorrhiza (*pl.* **mycorrhizae**) A symbiotic relationship between fungi and the roots of plants. Many trees, from pines to oaks to acacias (and a great many others) rely heavily or absolutely upon their fungal associates for optimal growth or even for survival.

N

natural selection The process that Charles Darwin proposed is the chief adaptive force in evolution. The basic idea is that all creatures have the potential to produce more offspring than the environment can support, and so there is *competition*, which he also called a 'struggle for existence'; that among these offspring there is *variation*: that inevitably, some of the variants will be more closely *adapted* (or 'fitted' as the Victorians tended to say) to the prevailing conditions than others, and so are more likely to survive and leave offspring of their own; and so as the generations pass the lineage becomes better and better adapted to the prevailing conditions (until and unless the conditions change).

neolinnean classification *see* Linnean classification.

niche A particular conceptual space within a habitat, offering opportunity for exploitation by specialist organisms.

nitrogen fixation The process by which certain bacteria convert atmospheric nitrogen into soluble ions, notably ammonium, which is further converted to nitrate in the soil and can be used by plants as a macronutrient. Many plants including many trees harbour nitrogen-fixing bacteria within special nodules in their roots.

nucleic acid *see* DNA and RNA.

nucleus The special region of the cell, surrounded by a specialized double membrane, within which the chromosomes (DNA) reside.

nut A fruit that is dry, hard and indehiscent (meaning it does not split naturally to release the seed inside, but must be actively prised open).

O

order A taxon smaller than a class and bigger than a family; *see* Linnean classification.

organelle A discrete, specialized structure within a cell, such as a nucleus, chloroplast or mitochondrion.

organic Chemists use the term 'organic' to mean any compound containing carbon (or at least, containing carbon, hydrogen and oxygen, with carbon as the principal component). More generally the term applies to any living material (or at least to material that was formerly alive).

osmosis The net diffusion of water from a region where the concentration of dissolved materials is low, to where the concentration is high.

outcrossing Fertilization effected between different individuals (as opposed to inbreeding).

ovary The enlarged base of the carpel (or the fused bases of adjacent carpels) which forms a chamber that contains the ovule or ovules.

ovule The structure within the carpel of a seed plant that contains the female gamete (egg cell); and which matures after fertilization to become the seed.

P

palaeobotany The study of ancient plants, generally conducted through the study of plant fossils (including fossil pollen).

parallel evolution Sometimes two separate lineages of creatures that live in similar habitats evolve over time in similar ways, so that at any one time in their history each resembles the other. This is parallel evolution.

parallel venation The condition characteristic of monocot plants in which the principal veins in the leaf run roughly in parallel from the leaf base to the tip.

parasite An organism that lives on or within another organism, usually of a different species, and derives nutrient from it. Parasitic relationships always benefit the parasite and do varying degrees of harm to the host. When the host also benefits from the presence of the parasite, the relationship is said to be 'mutualistic'. Thus mycorrhizal fungi might be said to be parasitic on the roots of plants, but they also bring great benefit to the plant.

pathogen Any organism that causes disease in another.

phenotype The overall form of an organism. Two or more organisms of similar genotype may nonetheless look or behave differently – which means that although they are genetically similar, they have different phenotypes.

pheromone A chemical agent that passes from one organism to another and influences the physiology or the behaviour of the recipient. Essentially, an airborne (or water-borne) hormone.

phloem The specialized tissue outside the cambium which contains elongated cells that transfer foods, particularly organic foods such as sugars, around the plant. The phloem forms the inner (living) part of the bark.

photoperiodism The mechanism by which plants respond to day-length and so adjust their life cycle to the seasons.

photosynthesis The process, mediated by chlorophyll, whereby plants harness the energy of the sun to split molecules of water into hydrogen and oxygen, and then attach the hydrogen to carbon dioxide (from the atmosphere) to form organic materials.

phototropism The tendency of plants to grow towards or away from light. In general, stems grow towards the light (positive phototropism), and roots grow away from it (negative phototropism).

phyllode A flat, expanded area of petiole (leaf stacks) or stem that carries out photosynthesis. In some trees, including the coniferous celery pines, leaves have been abandoned in favour of phyllodes.

phylogeny The term derives from the Greek *phylos* meaning tribe; so 'phylogeny' literally refers to the origin of different taxa. In practice phylogeny has come to refer to the true evolutionary relationships between different taxa. The entire methodology of cladistics in particular is designed to identify these true relationships. Modern, neolinnean classification, guided by cladistic principles, is intended to produce a taxonomy firmly rooted in true phylogeny.

phylum A large taxon of animals, positioned between kingdom and class, and equivalent to 'division' in botany; *see* Linnean classification.

physiology The study of all the functions and metabolic processes of living organisms.

phytochrome A plant pigment (or collection of pigments) that absorbs red and infra-red light, and is heavily involved in the mechanism that controls photoperiodism.

pigment Any chemical that absorbs light of particular wavelengths, and reflects what it does not absorb, so that it appears coloured. In general, all the many responses of plants to light are mediated by pigments of various kinds.

pollen A small structure produced by conifers and angiosperms, containing the male sex cell.

pollen tube When an individual pollen grain lands on the stigma of a flower it germinates to produce a pollen tube that grows down through the tissues of the style to reach the ovule beneath.

pollination The process by which pollen is transferred from the anther to the stigma.

polyploid A cell that contains more than two sets of chromosomes; or an organism that is compounded from such cells.

population The term is used in various ways. It is best used to describe a group of individuals from the same species, living in the same place at the same time, who interbreed (or may be considered capable of doing so when circumstances permit).

predator A creature that preys on another creature. The term is commonly taken to refer to carnivores, which prey on other animals. But herbivores such as giraffes, koalas and many a caterpillar may be seen as predators of trees.

prokaryote An organism in which the DNA is not sequestered within a cell nucleus. There are two domains of prokaryotes: the Bacteria, and the Archaea. Plants, including trees of course, belong to the third domain, the Eucaryota, which have eukaryotic cells.

R

radicle The root of the plant embryo.

rank, ranking In Linnean (or neolinnean) classification, 'genus' is a higher rank than 'species', 'family' is higher than 'genus', and so on all the way up to 'domain', the highest rank (or ranking) of all.

reaction wood Wood with an abnormal structure that develops in response to particular stresses and strains. It is manifest in trees that lean, and as 'compression wood' beneath the boughs of conifers, and as 'tension wood' above the boughs of angiospermous trees.

rhizome A more or less horizontal underground stem, frequently swollen and serving as a storage organ.

RNA (ribonucleic acid) DNA, ensconced within the nucleus in a eukaryotic cell, provides the code for the construction of proteins. RNA in various forms ferries the information out from the nucleus and into the cytoplasm where the proteins are put together.

S

sap The fluid within the xylem, which is released when the stem is cut. 'Cell sap' is the fluid within individual cells.

sapwood Outer part of the stem or trunk in which the xylem conducting tissue is still alive, and flowing with sap. Sapwood is commonly lighter in colour than heartwood, sometimes strikingly so as in the yew tree and some species of ebony.

savannah Grassland with scattered trees.

secondary growth Growth arising from division of cells of the cambium, which increases the girth of the trunk.

short-day plant A plant that will not flower unless first exposed to days that are shorter than some critical day-length (or, more accurately, is exposed to nights that are longer than some critical length).

shrub A perennial woody plant with several stems arising at or near the ground. The distinction between large shrubs and trees is arbitrary.

softwood A forester's name for the timber of conifers.

species The basic ranking of living creatures; *see* Linnean classification, and discussion in text.

spore A cell (usually just a single cell) that is typically diploid, is released from the parent plant, and can then grow directly into a new plant. Thus spores serve as agents of asexual reproduction.

sporophyte The generation of the plant that produces spores. In fact, although seed plants are properly called 'sporophytes', they do not produce free-living spores. They produce seeds instead. See text.

strobilus A reproductive structure constructed from a number of leaves or scales, commonly arranged in a spiral, as in a cone.

subspecies A subdivision of a species: generally synonymous with 'race'; *see* Linnean classification.

substrate Whatever the plant is growing on.

succession In ecology, the sequence of species over time as new land is colonized.

succulent Refers to plants that have fleshy leaves or stems, swollen with water.

sucker A sprout arising from the roots that can give rise to a new plant – as in aspens and many willows.

syconium Broadly speaking, the fruit of the fig tree. This is not quite accurate, however, since a fruit only becomes a fruit after the ovules within are fertilized. But the term 'syconium' also refers to the fleshy inflorescence, even before fertilization.

symbiosis Literally means 'together life'. Refers to the often close association of different organisms generally of different species. Although the term is commonly taken to imply mutual benefit, technically it can include parasitism, in which the host organism suffers. Symbiotic relationships from which both (or all) partners benefit are properly called 'mutualistic'.

systematics Essentially synonymous with 'taxonomy'.

T

taxon, taxonomy Taxon literally simply means 'group'; and taxonomy is the craft and science of placing creatures into groups – that is, of classification. A taxon (*pl.* taxa) may be of any 'grade': so 'species' is a taxon; 'family' is a taxon; 'order' is a taxon, and so on all the way up to domain. Ever since Darwin, however, mainstream taxonomists have insisted that a taxon cannot be formally acknowledged as such unless all the creatures within it are literally related to each other – meaning that they all share a common ancestor. Cladists have refined the idea so that no group can nowadays be admitted as a 'true' taxon unless it includes *all* the descendants of a common ancestor, *and* the ancestor itself, *but* includes no other creatures that are *not* part of the lineage.

tension wood Reaction wood that develops in response to tension, which develops on the upper side of big branches of dicot trees, and on the upper surface of leaning trunks, and in buttress roots (which in truth act more like guy-ropes than buttresses).

tetraploid A cell with four sets of chromosomes; or an organism compounded of tetraploid cells.

tissue, tissue culture A tissue is any group of cells of similar type and function. (Organs commonly consist of several different kinds of tissue working in concert.) Tissue culture is the craft and science of maintaining tissues in cultures, used both to study cells and, increasingly, as a means of asexual reproduction: whole new trees (of teak, coconuts, and many more) may be grown from cultured cells taken from outstanding ('elite') trees.

totipotent Cells of embryos, and other specialized cells known as 'stem' cells, are able to differentiate to form any of the tissues of the organism. Such cells are said to be 'totipotent'.

trace element, trace mineral A chemical element that is essential for nutrition but is required only in very small amounts. For plants, molybdenum and manganese are examples.

tracheid The xylem of conifers is composed of elongated cells known as tracheids (pronounced with three syllables: 'track-ay-ids').

translocation The transfer of water or foodstuffs within a plant. (Used more commonly to refer to the transfer of foodstuffs within the phloem.)

transpiration The loss of water from the aerial parts of the plant. Most of the loss occurs through the stomata.

triploid A cell with three sets of chromosomes, or an organism with triploid cells. Cultivated bananas are triploid (and sexually sterile). So are the cells of the endosperm of angiosperm seeds.

trophic Refers to feeding. Thus plants, which synthesize their own food, are 'autrotrophs'. The different layers of creatures in a food chain – carnivores eating smaller carnivores eating herbivores eating plants and so on – are referred to as 'trophic levels'.

tropism Movement, or more accurately growth, towards or away from an external stimulus.

turgid, turgor pressure 'Turgid' means swollen, as in a cell that has taken up a lot of water so that it presses hard against the cell wall. Herbaceous plants stay upright primarily because of water pressure in their cells, known as 'turgor pressure'. The leaves and young shoots of trees are also held in shape by turgor pressure. Herbs and tree leaves wilt when too much water is lost.

U, V

unicellular An organism with only one body cell.

variety A subdivision of a species; *see* Linnean classification.

vascular, vascular plant 'Vascular' refers to any part of a plant that contains conducting tissue, either xylem or phloem. Living vascular plants include all land plants except the bryophytes – mosses, liverworts and hornworts.

vector Anything that carries anything else. Thus bees are vectors of pollen and fruit bats are vectors of figs (and aphids are vectors of viruses that cause many diseases in plants).

vegetative A general term that applies to all parts of the plant apart from those involved in sexual reproduction. 'Vegetative reproduction' is reproduction via suckers or rhizomes, or (in cultivation) via cuttings or tissue culture.

vein, venation A vein is the conducting and supporting tissue within a leaf, or in other flat structures. 'Venation' refers to the pattern of veins within a leaf.

vessel An open-ended tubular cell of the kind that forms the main conducting tissue within the xylem of an angiosperm. (Also used loosely in this book, and strictly speaking inaccurately, to refer to the corresponding structures in conifers, which should be called 'tracheids'.)

W, X, Z

weed Any plant growing in places where, in the opinion of the botanist, gardener or forester, it should not be growing. Any plant that the botanist, gardener or forester regards as a nuisance.

whorl A circle of leaves or flower parts. The basic flower contains four whorls: of sepals, petals, stamens and carpels.

wood Botanists apply the term 'wood' to secondary xylem.

xylem Conducting tissue that carries water (and minerals dissolved in the water).

zygote A cell formed by the fusion of male and female gametes. In effect, a single-celled embryo.

Index